# State Space Systems With Time-Delays Analysis, Identification, and Applications

Emerging Methodologies and Applications in Modelling, Identification and Control

# State Space Systems With Time-Delays Analysis, Identification, and Applications

**YA GU**
College of Information, Mechanical and Electrical Engineering, Shanghai Normal University, Shanghai, China

**CHUANJIANG LI**
Shanghai Normal University, Shanghai, China

Series Editor

**QUAN MIN ZHU**

ELSEVIER

ACADEMIC PRESS
An imprint of Elsevier

Academic Press is an imprint of Elsevier
125 London Wall, London EC2Y 5AS, United Kingdom
525 B Street, Suite 1650, San Diego, CA 92101, United States
50 Hampshire Street, 5th Floor, Cambridge, MA 02139, United States
The Boulevard, Langford Lane, Kidlington, Oxford OX5 1GB, United Kingdom



ISBN: 978-0-323-91768-1

For Information on all Academic Press publications
visit our website at https://www.elsevier.com/books-and-journals

*Publisher:* Mara E. Conner
*Acquisitions Editor:* Sophie Harrison
*Editorial Project Manager:* Moises Carlo P. Catain
*Production Project Manager:* Nirmala Arumugam
*Cover Designer:* Matthew Limbert

Typeset by MPS Limited, Chennai, India

# Contents

# About the authors

**Ya Gu** is an associate professor at the College of Information, Mechanical and Electrical Engineering, Shanghai Normal University, Shanghai, China. She received her PhD degree in automatic control from Jiangnan University, Wuxi, China, in 2015. She was a visiting PhD student from 2014 to 2015 at the University of Alberta, Edmonton, AB, Canada. She is the reviewer of *Signal Processing*, *IET Control Theory & Applications*, and *Journal of the Franklin Institute*. Her current research interests include model identification, parameter estimation, and adaptive control. E-mail: guya@shnu.edu.cn.

**Chuanjiang Li** is a professor at the College of Information, Mechanical and Electrical Engineering, Shanghai Normal University, Shanghai, China. He obtained his BS degree from Henan University of Science and Technology, China, in 2000, MS degree from Henan University of Science and Technology, China, in 2003, and PhD degree from Shanghai University, China, in 2014. His current research interests are robot control, intelligent control theory, and application. E-mail: licj@shnu.edu.cn.

# Acknowledgment

This work was supported by the National Natural Science Foundation of China (No. 61903050), the Natural Science Foundation of Shanghai (No. 22ZR1445300).

I wish to express my sincere appreciation to the following professors who have encouraged and guided me throughout this book: Prof. Feng Ding, Jiangnan University, and Prof. Quanmin Zhu, University of the West of England.

# Introduction

## 1 An overview of each chapter

System identification and parameter estimation are mostly applied to the mathematical model of the system with input and output variables. However, in the industrial system, many controlled objects need to be abstracted into state space models of the system, and in general, a physical phenomenon with time delay accompanies the controlled object, time delay is considered to be the most difficult to control, and the identification of delay system is also very difficult. In recent years, delay system has been an active research area, the application is very extensive especially in the industrial process, and it is still a research focus in the field of international process control. Therefore delay system is also the research hotspot in system identification. It is a subject worthy of further study to use system identification methods consciously and reasonably to identify time delay. This book focuses on "the identification of state space system with time delay," and the proposed algorithms have theoretical value and application prospects. This book gives the detailed discussion and research, and the following results are obtained. A brief introduction of the major research contents and achievements is outlined in the following.

In Chapter 1, for the state space model with a unit time delay, by extending the state equation, using the properties of the shift operators, and eliminating some state variables, the state space system with time delay is converted into the form of the input–output expression, which is the identification model of the system. When the states are unmeasurable, they are replaced with their estimates, by using the parameter estimation and known input–output data to estimate states. This chapter proposes the auxiliary model–based least squares identification method, the stochastic gradient identification algorithm, and the least squares–based iterative parameter estimation algorithm for the state space model with a unit time delay. The simulation examples verify the effectiveness of the proposed algorithms.

In Chapter 2, for the state space model with d-step time delay, by using the least squares identification algorithm to estimate the parameters and states, first, the identification model is derived, the unknown noise and state variables in the information vectors are replaced with their estimates, and the parameters are estimated. Then, the parameter estimates are used to compute the states. For the multivariable state space model with d-step time delay, the number of variables is large, and the structure of the model is complex; while some single variable system identification methods can be applied to some of the multivariable systems, these identification algorithms for multivariable systems are not enough. To guarantee the identification accuracy of the algorithm, we also need to do some improvements to the original algorithm. This chapter mainly uses hierarchical identification theory to improve the computational efficiency of identification algorithm. Finally, simulation examples show the effectiveness of the proposed algorithms.

In Chapter 3, for the state space model with multistate delays, the number of delays is large, and the theoretical derivation of multistate delays is more complex than that of a unit time delay. This chapter uses hierarchical gradient–based iterative identification algorithm and hierarchical least squares–based iterative identification algorithm to estimate the parameters of the system. Decomposing the system into two subsystems reduces the calculation. For the dual-rate state space model with multistate delays, we use the data filtering technique; compared with the auxiliary model-based least squares algorithm, the filtering-based least squares algorithm can produce more accurate parameter estimates.

Chapter 4 considers the identification problem of the state space model with d-step delay for multivariable systems and presents a state estimation-based recursive least squares parameter identification algorithm by using the hierarchical identification principle. Combining the linear transformation and the property of the shift operator, a state space system is transformed into an equivalent canonical state space model, and its identification model is derived. Finally, an example is provided to validate the proposed theorems.

Chapter 5 researches parameter estimation problems for a Hammerstein input nonlinear system with state time delay. Combining the linear transformation and the property of the shift operator, the Hammerstein state space system is equivalent to a bilinear parameter identification model. The gradient-based and least squares–based iterative parameter estimation algorithms are used for identifying the state time-delay system, and the proposed iterative algorithms make full use of all data at each iteration, which can produce highly accurate parameter estimation. Finally, the example is provided to validate the proposed algorithms.

In Chapter 6, for the uncertain state space model with time delay, we mainly adopt two kinds of methods, one is the moving horizon estimation algorithm, and the basic idea is to derive the cost function and optimize the objective function; compared with the general Kalman filter algorithm, the proposed algorithm can simultaneously estimate the continuous state and discrete time delay. Another method is the expectation maximization algorithm, and it has E step and M step. Step E is used to calculate the expectation of the complete data which is often referred to as Q function, and step M is used to maximize the Q function. These two steps iterate until it converges. Finally, simulation examples are given.

CHAPTER 1

# One-unit state-delay identification

## 1.1 Auxiliary model identification method for a unit time-delay system

State-space models have wide applications in many areas, for example, system modeling, system identification, signal processing, adaptive filtering, and control (Devakar and Lyengar, 2011). There exist many estimation methods for state-space models, such as the least squares methods, the auxiliary model identification methods, and the stochastic gradient methods (Ding et al., 2012).

It considers identification problems of time-delay control systems based on the auxiliary model identification idea. The auxiliary model method is a new-type parameter estimation one and can deal with identification problems with the information vector including unknown internal variables. It presents an auxiliary model-based identification algorithm of the input–output representations corresponding to state-space systems with time delays. The basic idea is, by means of the property of the shift operator, to transform the state-space model with a time delay into an input–output representation and then to identify the parameters of the input–output representation based on the auxiliary model identification technique. The proposed method has the advantage of handling the unmeasured variables in the information vector.

### 1.1.1 The input–output representation

Let us define some notations. $\hat{\theta}(t)$ represents the estimate of $\theta$ at time $t$; $A =: X$ or $X := A$ stands for $A$ is defined as $X$; the symbol $I(I_n)$ stands for an identity matrix of appropriate sizes ($n \times n$); $z$ represents a unit forward shift operator: $zx(t) = x(t+1)$ and $z^{-1}x(t) = x(t-1)$; the superscript T denotes the matrix/vector transpose; $1_n$ represents an n-dimensional column vector whose elements are all unity; adj[$X$] denotes the adjoint matrix of the square matrix $X$: $\text{adj}[X] = \det[X]X^{-1}$; and det[$X$] denotes the determinant of the square matrix $X$.

*State Space Systems With Time-Delays Analysis, Identification, and Applications*
DOI: https://doi.org/10.1016/B978-0-323-91768-1.00005-8

Consider the following state-space system with a time delay,

$$\boldsymbol{x}(t+1) = \boldsymbol{A}\boldsymbol{x}(t) + \boldsymbol{B}\boldsymbol{x}(t-1) + \boldsymbol{f}u(t), \tag{1.1}$$

$$y(t) = \boldsymbol{c}\boldsymbol{x}(t) + du(t) + u(t), \tag{1.2}$$

where $x(t) \in R^n$ is the state vector, $u(t) \in R$ is the system input, $y(t) \in R$ is the system output, $v(t) \in R$ is a random noise with zero mean, and $A \in R^{n \times n}$, $B \in R^{n \times n}$, $f \in R^n$, $c \in R^{1 \times n}$, and $d \in R$ are the system parameter matrices/vectors.

The following transforms the time-delay state-space model in (1.1) and (1.2) into an input–output representation and gives its identification model.

**Lemma 1.1**: For the state-space model in (1.1) and (1.2), the transfer function from the input $u(t)$ to the output $y(t)$ is given by

$$\begin{aligned} G(z) &:= \boldsymbol{c}\left(z^2\boldsymbol{I}-\boldsymbol{A}z-\boldsymbol{B}\right)^{-1}\boldsymbol{f}z + d \\ &= \frac{\boldsymbol{c}\,\mathrm{adj}\left[z^2\boldsymbol{I}-\boldsymbol{A}z-\boldsymbol{B}\right]\boldsymbol{f}z + \det\left[z^2\boldsymbol{I}-\boldsymbol{A}z-\boldsymbol{B}\right]d}{\det[z^2\boldsymbol{I}-\boldsymbol{A}z-\boldsymbol{B}]} \end{aligned}$$

**Proof** Using the properties of the shift operator $z$, Eq. (1.1) can be rewritten as

$$z\boldsymbol{x}(t) = \boldsymbol{A}\boldsymbol{x}(t) + z^{-1}\boldsymbol{B}\boldsymbol{x}(t) + \boldsymbol{f}u(t).$$

Multiplying both sides by $z$ gives

$$z^2\boldsymbol{x}(t) = \boldsymbol{A}z\boldsymbol{x}(t) + \boldsymbol{B}\boldsymbol{x}(t) + \boldsymbol{f}zu(t),$$

or

$$\boldsymbol{x}(t) = \left(z^2\boldsymbol{I}-\boldsymbol{A}z-\boldsymbol{B}\right)^{-1}\boldsymbol{f}zu(t).$$

Substituting the above $x(t)$ into (1.2) gives the output equation:

$$\begin{aligned} y(t) &= \boldsymbol{c}\left(z^2\boldsymbol{I}-\boldsymbol{A}z-\boldsymbol{B}\right)^{-1}\boldsymbol{f}zu(t) + du(t) + v(t) \\ &= \left[\boldsymbol{c}\left(z^2\boldsymbol{I}-\boldsymbol{A}z-\boldsymbol{B}\right)^{-1}\boldsymbol{f}z + d\right]v(t) + v(t) \\ &= G(z)v(t) + v(t). \end{aligned} \tag{1.3}$$

Then, we have the transfer function of the system from the input $u(t)$ to the output $y(t)$:

$$\begin{aligned} G(z) &= \boldsymbol{c}\left(z^2\boldsymbol{I}-\boldsymbol{A}z-\boldsymbol{B}\right)^{-1}\boldsymbol{f}z + d \\ &=: \frac{\beta(z)}{\alpha(z)}, \end{aligned} \tag{1.4}$$

where $\alpha(z)$ is the denominator of the transfer function, i.e., the characteristic polynomial of the system, and $\beta(z)$ is the numerator of the transfer function, and they are defined by

$$\begin{aligned}\alpha(z) &:= z^{-2n}\det\left[z^2\boldsymbol{I} - \boldsymbol{A}z - \boldsymbol{B}\right] \\ &= z^{-2n}\left(z^{2n} + \alpha_1 z^{2n-1} + a_2 z^{2n-2} + \cdots + \alpha_{2n}\right) \\ &= 1 + a_1 z^{-1} + a_2 z^{-2} + \cdots + a_{2n} z^{-2n},\end{aligned} \tag{1.5}$$

$$\begin{aligned}\beta(z) &:= z^{-2n}\boldsymbol{c}\ \mathrm{adj}\left[z^2\boldsymbol{I} - \boldsymbol{A}z - \boldsymbol{B}\right]\boldsymbol{f}z + a(z)d \\ &= \beta_0 + \beta_1 z^{-1} + \beta_2 z^{-2} + \cdots + \beta_{2n} z^{-2n}.\end{aligned} \tag{1.6}$$

Substituting (1.4) into (1.3) gives the input—output representation of the time-delay state-space model in (1.1) and (1.2), that is, an output error model:

$$y(t) = \frac{\beta(z)}{\alpha(z)}u(t) + v(t) \tag{1.7}$$

The objective of this paper is to study new identification methods to estimate the parameters/coefficients $(\alpha_i, \beta_i)$ in the input—output representation in (1.7).

**Example 1.1**: For a second-order state-space system with a time delay

$$\begin{aligned}\boldsymbol{x}(t+1) &= \boldsymbol{Ax}(t) + \boldsymbol{Bx}(t-1) + \boldsymbol{f}u(t), \\ y(t) &= \boldsymbol{cx}(t) + u(t),\end{aligned}$$

where $A = \begin{bmatrix} a_1 & a_2 \\ a_3 & a_4 \end{bmatrix} \in R^{2\times 2}$, $B = \begin{bmatrix} b_1 & b_2 \\ b_3 & b_4 \end{bmatrix} \in R^{2\times 2}$, $f = \begin{bmatrix} f_1 \\ f_2 \end{bmatrix} \in R^{2\times 1}$, $c = [c_1, c_2] \in R^{1\times 2}$.

From Eq. (1.3), we have

$$\begin{aligned}y(t) &= \boldsymbol{c}\left(z^2\boldsymbol{I} - \boldsymbol{A}z - \boldsymbol{B}\right)^{-1}\boldsymbol{f}zu(t) + u(t) \\ &= \frac{\boldsymbol{c}\ \mathrm{adj}\left[z^2\boldsymbol{I} - \boldsymbol{A}z - \boldsymbol{B}\right]\boldsymbol{f}z}{\det[z^2\boldsymbol{I} - \boldsymbol{A}z - \boldsymbol{B}]}u(t) + u(t),\end{aligned}$$ where

$$
\begin{aligned}
&\det\left[z^2\boldsymbol{I}-\boldsymbol{A}z-\boldsymbol{B}\right]=\det\left\{\begin{bmatrix} z^2 & 0 \\ 0 & z^2 \end{bmatrix}-\begin{bmatrix} a_1 z & a_2 z \\ a_3 z & a_4 z \end{bmatrix}-\begin{bmatrix} b_1 & b_2 \\ b_3 & b_4 \end{bmatrix}\right\}\\
&=\det\begin{bmatrix} z^2-a_1 z-b_1 & -a_2 z-b_2 \\ -a_3 z-b_3 & z^2-a_4 z-b_4 \end{bmatrix}\\
&=\left(z^2-a_1 z-b_1\right)\left(z^2-a_4 z-b_4\right)-\left(-a_2 z-b_2\right)\left(-a_3 z-b_3\right)\\
&=z^4-(a_1+a_4)z^3+(a_1a_4-a_2a_3-b_1-b_4)z^2\\
&+(a_1b_4-a_2b_3-a_3b_2+a_4b_1)z+b_1b_4-b_2b_3,\\
&\boldsymbol{c}\,\mathrm{adj}\left[z^2\boldsymbol{I}-\boldsymbol{A}z-\boldsymbol{B}\right]\boldsymbol{f}z=[c_1,c_2]\begin{bmatrix} z^2-a_4 z-b_4 & a_2 z+b_2 \\ a_3 z+b_3 & z^2-a_1 z-b_1 \end{bmatrix}\begin{bmatrix} f_1 \\ f_2 \end{bmatrix}z\\
&=\left[c_1z^2-c_1a_4z+c_2a_3z-c_1b_4+c_2b_3,\ \ c_2z^2+c_1a_2z-c_2a_1z+c_1b_2-c_2b_1\right]\begin{bmatrix} f_1 \\ f_2 \end{bmatrix}z\\
&=(f_1c_1+f_2c_2)z^3-(f_2c_2a_1-f_2c_2a_2-f_1c_2a_3+f_1c_1a_4)z^2\\
&-(f_2c_2b_1-f_2c_1b_2-f_1c_2b_3+f_1c_1b_4)z.
\end{aligned}
$$

Its corresponding input—output representation is given by

$$
\begin{aligned}
y(t)&=\frac{(f_1c_1+f_2c_2)z^3-(f_2c_2a_1-f_2c_1a_2-f_1c_2a_3+f_1c_1a_4)z^2-(f_2c_2b_1-f_2c_1b_2-f_1c_2b_3+f_1c_1b_4)z}{z^4-(a_1+a_4)z^3+(a_1a_4-a_2a_3-b_1-b_4)z^2+(a_1b_4-a_2b_3-a_3b_2+a_4b_1)z+b_1b_4-b_2b_3}u(t)+v(t)\\
&=\frac{(f_1c_1+f_2c_2)z^{-1}-(f_2c_2a_1-f_2c_1a_2-f_1c_2a_3+f_1c_1a_4)z^{-2}-(f_2c_2b_1-f_2c_1b_2-f_1c_2b_3+f_1c_1b_4)z^{-3}}{1-(a_1+a_4)z^{-1}+(a_1a_4-a_2a_3-b_1-b_4)z^{-2}+(a_1b_4-a_2b_3-a_3b_2+a_4b_1)z^{-3}+(b_1b_4-b_2b_3)z^{-4}}u(t)+u(t).
\end{aligned}
$$

### 1.1.2 The auxiliary model-based squares algorithm

This section derives an auxiliary model-based identification algorithm to estimate the parameters of the input—output representation. The basic idea is to replace the unknown intermediate variables in the information vector with the outputs of the auxiliary model. The details are as follows.

Define an intermediate variable:

$$
s(t):=\frac{\beta(z)}{\alpha(z)}u(t), \tag{1.8}
$$

or

$$
\alpha(z)s(t)=\beta(z)u(t) \tag{1.9}
$$

Define the parameter vector $\theta$ and the information vector $\phi(t)$ as

$$
\boldsymbol{\theta}:=\left[\alpha_1,\ \alpha_2,\cdots,\alpha_{2n},\beta_0,\ \beta_1,\ \beta_2,\cdots,\beta_{2n}\right]^{\mathrm{T}}\in\mathbb{R}^{4n+1},
$$

$$
\varphi(t):=[-s(t-1),-s(t-2),\cdots,-s(t-2n),u(t),u(t-1),u(t-2),\cdots,u(t-2n)]^{\mathrm{T}}\in\mathbb{R}^{4n+1}.
$$

Substituting (1.5) and (1.6) into (1.9) gives

$$\begin{aligned} s(t) &= [1 - \alpha(z)]s(t) + \beta(z)u(t) \\ &= -\alpha_1 s(t-1) - \alpha_2 s(t-2) - \cdots - \alpha_{2n} s(t-2n) \\ &\quad + \beta_0 u(t) + \beta_1 u(t-1) + \beta_2 u(t-2) + \cdots + \beta_{2n} u(t-2n) \\ &= \varphi^{\mathrm{T}}(t)\boldsymbol{\theta}. \end{aligned} \tag{1.10}$$

From (1.7) and (1.8) and the above equation, we have

$$y(t) = \varphi^{T}(t)\boldsymbol{\theta} + u(t).$$

Minimizing the criterion function,

$$J(\theta) = \sum_{j=1}^{t} \left[y(j) - \varphi^{\mathrm{T}}(j)\boldsymbol{\theta}\right]^2.$$

We can obtain the following recursive least square (RLS) algorithm for estimating the parameter vector $\theta$

$$\hat{\boldsymbol{\theta}}(t) = \hat{\boldsymbol{\theta}}(t-1) + \boldsymbol{L}(t)[y(t) - \varphi^{\mathrm{T}}(t)\hat{\boldsymbol{\theta}}(t-1)], \tag{1.11}$$

$$\boldsymbol{L}(t) = \boldsymbol{P}(t)\varphi(t) = \frac{\boldsymbol{P}(t-1)\varphi(t)}{1 + \varphi^{\mathrm{T}}(t)\boldsymbol{P}(t-1)\varphi(t)}, \tag{1.12}$$

$$\boldsymbol{P}(t) = \boldsymbol{P}(t-1) - \frac{\boldsymbol{P}(t-1)\varphi(t)\varphi^{\mathrm{T}}(t)\boldsymbol{P}(t-1)}{1 + \varphi^{\mathrm{T}}(t)\boldsymbol{P}(t-1)\varphi(t)}. \tag{1.13}$$

Because the information vector $\phi(t)$ contains the unknown intermediate variables $s(t-i)(i = 1, 2, \ldots, 2n)$, the algorithm in (1.11)–(1.13) cannot be implemented. The solution is to replace the unknown $s(t-i)$ in $\phi(t)$ with the outputs $\hat{s}(t-i)$ of the following auxiliary model,

$$\hat{s}(t) = \hat{\varphi}^{T}(t)\hat{\boldsymbol{\theta}}(t),$$

where

$$\hat{\varphi}(t) = [-\hat{s}(t-1), -\hat{s}(t-2), \cdots, -\hat{s}(t-2n), u(t), u(t-1), u(t-2), \cdots, u(t-2n)]^{\mathrm{T}} \in \mathbb{R}^{4n+1}.$$

Replacing $\phi(t)$ in (1.11)–(1.13) with $\hat{\phi}(t)$, we can obtain the auxiliary model-based recursive least squares (AM-RLS) algorithm for estimating the parameter vector $\theta$ as follows:

$$\hat{\boldsymbol{\theta}}(t) = \hat{\boldsymbol{\theta}}(t-1) + \boldsymbol{L}(t)\left[y(t) - \hat{\varphi}^{\mathrm{T}}(t)\hat{\boldsymbol{\theta}}(t-1)\right], \tag{1.14}$$

$$\boldsymbol{L}(t) = \boldsymbol{P}(t)\hat{\varphi}(t) = \frac{\boldsymbol{P}(t-1)\hat{\varphi}(t)}{1 + \hat{\varphi}^{\mathrm{T}}(t)\boldsymbol{P}(t-1)\hat{\varphi}(t)}, \tag{1.15}$$

$$\begin{aligned}\boldsymbol{P}(t) &= \boldsymbol{P}(t-1) - \frac{\boldsymbol{P}(t-1)\hat{\boldsymbol{\varphi}}(t)\hat{\boldsymbol{\varphi}}^{\mathrm{T}}(t)\boldsymbol{P}(t-1)}{1+\hat{\boldsymbol{\varphi}}^{\mathrm{T}}(t)\boldsymbol{P}(t-1)\hat{\boldsymbol{\varphi}}(t)} \\ &= \left[\boldsymbol{I} - \boldsymbol{L}(t)\hat{\boldsymbol{\varphi}}^{\mathrm{T}}(t)\right]\boldsymbol{P}(t-1), \quad \boldsymbol{P}(0) = p_0\boldsymbol{I},\end{aligned} \tag{1.16}$$

$$\hat{\boldsymbol{\varphi}}(t) = [-\hat{s}(t-1), -\hat{s}(t-2), \cdots, -\hat{s}(t-2n), u(t), u(t-1), u(t-2), \cdots, u(t-2n)]^{\mathrm{T}}, \tag{1.17}$$

$$\hat{s}(t) = \hat{\boldsymbol{\varphi}}^{\mathrm{T}}(t)\hat{\boldsymbol{\theta}}(t) \tag{1.18}$$

The steps of computing the estimate $\hat{\theta}(t)$ in the AM-RLS algorithm in (1.14)–(1.18) are listed in the following.

1. Let $t=1$, set the initial values $\hat{\theta}(0) = \mathbf{1}_{4n+1/p_0}$, $P(0) = p_0 I$, $\hat{s}(t-i) = 0$, $u(t-i) = 0$, and $y(t-i) = 0$ for $i = 1, 2, \ldots, 2n$.
2. Collect the input–output data $u(t)$ and $y(t)$, and form $\hat{\phi}(t)$ by (1.17).
3. Compute the gain vector $L(t)$ by (1.15) and the covariance matrix $P(t)$ by (1.16).
4. Update the parameter estimation vector $\hat{\theta}(t)$ by (1.14).
5. Compute $\hat{s}(t)$ by (1.18).
6. Increase $t$ by 1 and go to step 2; continue the recursive calculation.

The flowchart of computing the parameter estimation vector $\hat{\theta}(t)$ is shown in Fig. 1.1.

### 1.1.3 Example

Consider the following second-order simulation system:

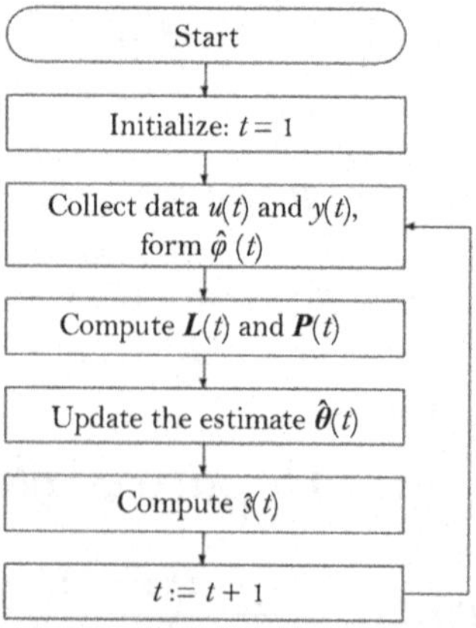


**Figure 1.1** The flowchart of computing the AM-RLS parameter estimate $\hat{\theta}(t)$.

$$\boldsymbol{x}(t+1) = \begin{bmatrix} 0 & 1.00 \\ -1.20 & -0.60 \end{bmatrix} \boldsymbol{x}(t) + \begin{bmatrix} 0.20 & 0.30 \\ -0.30 & 0.40 \end{bmatrix} \boldsymbol{x}(t-1) + \begin{bmatrix} 1.00 \\ 0.20 \end{bmatrix} u(t),$$

$$y(t) = [1.20, \quad 1.75]\boldsymbol{x}(t) + v(t).$$

Thus, we have

$$\begin{aligned} y(t) &= \frac{\beta(z)}{\alpha(z)} u(t) + v(t) \\ &= \frac{\beta_1 z^{-1} + \beta_2 z^{-2} + \beta_3 z^{-3}}{1 + \alpha_1 z^{-1} + \alpha_2 z^{-2} + \alpha_3 z^{-3} + \alpha_4 z^{-4}} u(t) + v(t) \\ &= \frac{1.550 z^{-1} - 1.140 z^{-2} - 1.003 z^{-3}}{1 + 0.600 z^{-1} + 0.600 z^{-2} + 0.540 z^{-3} + 0.170 z^{-4}} u(t) + v(t), \end{aligned}$$

$$\begin{aligned} \boldsymbol{\theta} &= \left[\alpha_1, \ \alpha_2, \ \alpha_3, \ \alpha_4 \ \beta_1, \ \beta_2, \ \beta_3\right]^{\mathrm{T}} \\ &= [0.600, \ 0.600, \ 0.540, \ 0.170, \ 1.550, -1.140, -1.003]^{\mathrm{T}}. \end{aligned}$$

In simulation, the input $\{u(t)\}$ is taken as an uncorrelated persistent excitation signal sequence with zero mean and unit variance and $\{v(t)\}$ as a white noise sequence with zero mean and variance $\sigma^2 = 0.50^2$. Apply the AM-RLS algorithm to estimate the parameters of this example system. The parameter estimates and their estimation errors are shown in Table 1.1, and the parameter estimation errors $\delta$ versus $t$ are shown in Fig. 1.2. When the noise variance is $\sigma^2 = 0.50^2$, the noise-to-signal ratio of the system is $\delta_{ns} = 16.45\%$.

From the simulation results in Table 1.1 and Fig. 1.2, we can draw the following conclusions:

1. The parameter estimation accuracy becomes higher as the data length $t$ increases—see Table 1.1.
2. The parameter estimation errors $\delta$ become smaller (in general) with the increasing of the data length—see Fig. 1.2.
3. The fluctuations of the parameter estimation errors become small, and the accuracy of the estimated parameters becomes high with the data length $t$ increasing—see Table 1.1 and Fig. 1.2.

### 1.1.4 Conclusions

This chapter discusses the identification problems for linear systems based on the transfer function models with unknown parameters. The auxiliary model-based least squares algorithm is presented from input−output data for estimating the parameters of the systems. The simulation results show that the proposed algorithms are effective.

**Table 1.1** The parameter estimates and errors ($\sigma^2 = 0.50^2$).

| $t$ | $\alpha_1$ | $\alpha_2$ | $\alpha_3$ | $\alpha_4$ | $\beta_1$ | $\beta_2$ | $\beta_3$ | $\delta(\%)$ |
|---|---|---|---|---|---|---|---|---|
| 100 | −0.12368 | 0.56497 | 0.14037 | 0.02588 | 1.50941 | −2.25515 | 0.44408 | 83.88026 |
| 200 | −0.00061 | 0.54214 | 0.19993 | 0.01994 | 1.56587 | −2.04930 | 0.12773 | 67.35368 |
| 500 | 0.29254 | 0.56771 | 0.37452 | 0.09922 | 1.57782 | −1.61222 | −0.41326 | 34.88358 |
| 1000 | 0.52579 | 0.58913 | 0.49117 | 0.14623 | 1.56509 | −1.24475 | −0.85703 | 8.45354 |
| 2000 | 0.59258 | 0.60214 | 0.53519 | 0.16497 | 1.55886 | −1.13800 | −0.96938 | 1.51590 |
| 3000 | 0.59732 | 0.59878 | 0.53803 | 0.16388 | 1.54956 | −1.13659 | −0.98639 | 0.76631 |
| True values | 0.60000 | 0.60000 | 0.54000 | 0.17000 | 1.55000 | −1.14000 | −1.00300 | |

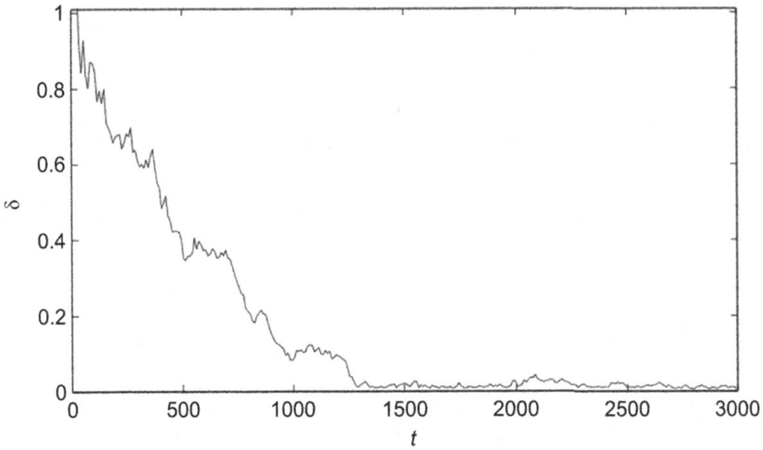


**Figure 1.2** The parameter estimation errors $\delta$ versus ($\sigma^2 = 0.50^2$).

## 1.2 Parameter and state estimation algorithm for one-unit state-delay system

Recursive algorithms or iterative algorithms are often adopted in solving a matrix equation such as the Jacobi iteration, modeling a system, filtering and control, and parameter estimation (Shi and Fang, 2010). The Kalman filtering is a typical recursive algorithm for state estimation and uses the available input—output data to estimate the states of dynamic systems. State estimation of the state-space model is of great significance in designing controllers based on the state feedback.

There exist a variety of parameter estimation methods for system modeling, for example, the recursive least squares methods for identifying equation-error type models, the auxiliary model identification methods for estimating the parameters of output-error type models, and the iterative identification methods for identifying systems with unknown noise terms in the information vectors (Ashiba et al., 2011).

Time-delay systems exist in industrial processes. The identification of the time-delay systems is important for system control and system analysis and has received much research attention in the area of control for decades (Ding, 2013).

It combines parameter and state estimation technique to study identification problems of state-delay systems. The basic idea is to transform a state-space system with a state delay into its observability canonical form and then to derive a parameter and state estimation algorithm by means of the least squares principle and the Kalman filtering theory.

### 1.2.1 The canonical state-space model for state-delay systems

Let us introduce some notation. $A =: X$ or $X := A$ stands for $A$ is defined as $X$; the symbol $I(I_n)$ stands for an identity matrix of appropriate size $(n \times n)$; $z$ represents a unit forward shift operator: $zx(t) = x(t+1)$ and $z^{-1}x(t) = x(t-1)$; the superscript $\tau$ denotes the matrix/vector transpose; $1_n$ represents an $n \times 1$ vector whose elements are all unity; and $\hat{\vartheta}(t)$ denotes the estimate of $\vartheta$ at time $t$.

Consider the following state-space system with one-unit state time delay,

$$\overline{\boldsymbol{x}}(t+1) = \overline{\boldsymbol{A}}\overline{\boldsymbol{x}}(t) + \overline{\boldsymbol{B}}\overline{\boldsymbol{x}}(t-1) + \overline{\boldsymbol{f}}u(t), \tag{1.19}$$

$$y(t) = \overline{\boldsymbol{c}}\overline{\boldsymbol{x}}(t) + v(t), \tag{1.20}$$

where $\overline{x}(t) \in R^n$ is the state vector, $u(t) \in R$ is the system input, $y(t) \in R$ is the system output, $v(t) \in R$ is a random noise with zero mean and variance $\sigma^2$, and $\overline{A} \in R^{n \times n}$, $\overline{B} \in R^{n \times n}$, $\overline{f} \in R^n$, and $\overline{c} \in R^{1 \times n}$ are the system parameter matrices/vectors.

Since Eq. (1.19) contains the term $\overline{x}(t-1)$, we say that this state equation has one-unit time delay. If we change $\overline{x}(t-1)$ into $\overline{x}(t-d)$, it is a d-step state-delay state-space system (d is an integer).

Assume that $(\overline{c}, \overline{A})$ is observable and there exists a nonsingular matrix $\tau \in R^{n \times n}$ such that under the linear transformation $\overline{x} = \tau x(t)$, the system in (1.19) and (1.20) can be transformed into an observability canonical form:

$$\begin{aligned} \boldsymbol{x}(t+1) &= \boldsymbol{T}^{-1}\overline{\boldsymbol{x}}(t+1) = \boldsymbol{T}^{-1}\left[\overline{\boldsymbol{A}}\overline{\boldsymbol{x}}(t) + \overline{\boldsymbol{B}}\overline{\boldsymbol{x}}(t-1) + \overline{\boldsymbol{f}}u(t)\right] \\ &= \boldsymbol{T}^{-1}\overline{\boldsymbol{A}}\mathrm{T}\mathrm{x}(t) + \boldsymbol{T}^{-1}\overline{\boldsymbol{B}}\mathrm{T}\mathrm{x}(t-1) + \boldsymbol{T}^{-1}\overline{\boldsymbol{f}}u(t) \\ &=: \boldsymbol{A}\boldsymbol{x}(t) + \boldsymbol{B}\boldsymbol{x}(t-1) + \boldsymbol{f}u(t), \end{aligned} \tag{1.21}$$

$$\begin{aligned} y(t) &= \overline{\boldsymbol{c}}\overline{\boldsymbol{x}}(t) + v(t) = \overline{\boldsymbol{c}}\mathrm{T}\mathrm{x}(t) + v(t) \\ &=: \boldsymbol{c}\boldsymbol{x}(t) + v(t) \end{aligned} \tag{1.22}$$

where

$$A := T^{-1}\overline{A}T = \begin{bmatrix} 0 & 1 & 0 & \cdots & 0 \\ 0 & 0 & 1 & \ddots & \vdots \\ \vdots & \vdots & \ddots & \ddots & 0 \\ 0 & 0 & \cdots & 0 & 1 \\ a_n & a_{n-1} & a_{n-2} & \cdots & a_1 \end{bmatrix} \in \mathbb{R}^{n\times n},$$

$$B := T^{-1}\overline{B}T = \begin{bmatrix} b_1 \\ b_2 \\ \vdots \\ b_n \end{bmatrix} \in \mathbb{R}^{n\times n}, \quad b_i \in \mathbb{R}^{1\times n}, \qquad f := T^{-1}\overline{f} = \begin{bmatrix} f_1 \\ f_2 \\ \vdots \\ f_n \end{bmatrix} \in \mathbb{R}^{n},$$

$$c := \overline{c}T = [1, 0, 0, \cdots, 0] \in \mathbb{R}^{1\times n}.$$

The transformation matrix is given by

$$T := \begin{bmatrix} \overline{c} \\ \overline{c}\overline{A} \\ \vdots \\ \overline{c}\overline{A}^{n-1} \end{bmatrix}^{-1} \in \mathbb{R}^{n\times n}.$$

Eqs. (1.21) and (1.22) can be written as

$$\begin{bmatrix} x_1(t+1) \\ x_2(t+1) \\ x_3(t+1) \\ \vdots \\ x_n(t+1) \end{bmatrix} = \begin{bmatrix} 0 & 1 & 0 & \cdots & 0 \\ 0 & 0 & 1 & \ddots & \vdots \\ \vdots & \vdots & \ddots & \ddots & 0 \\ 0 & 0 & \cdots & 0 & 1 \\ a_n & a_{n-1} & a_{n-2} & \cdots & a_1 \end{bmatrix} \begin{bmatrix} x_1(t) \\ x_2(t) \\ x_3(t) \\ \vdots \\ x_n(t) \end{bmatrix} + \begin{bmatrix} b_1 \\ b_2 \\ b_3 \\ \vdots \\ b_n \end{bmatrix} x(t-1) + \begin{bmatrix} f_1 \\ f_2 \\ f_3 \\ \vdots \\ f_n \end{bmatrix} u(t), \tag{1.23}$$

$$y(t) = [1, 0, 0, \cdots, 0]x(t) + v(t). \tag{1.24}$$

The model in (1.19) and (1.20) contains $2n^2 + 2n$ parameters, but the canonical model in (1.23) and (1.24) contains only $n^2 + 3n$. When $n = 10$, they are 220 and 130 parameters, respectively.

## 1.2.2 The identification model

This section derives the identification model of the canonical state-space model in (1.23) and (1.24) for the state-delay system. The details are as follows.

From (1.23), we have

$$\begin{cases} x_1(t+1) & = x_2(t) + \boldsymbol{b}_1\boldsymbol{x}(t-1) + f_1 u(t), \\ x_2(t+1) & = x_3(t) + \boldsymbol{b}_2\boldsymbol{x}(t-1) + f_2 u(t), \\ x_3(t+1) & = x_4(t) + \boldsymbol{b}_3\boldsymbol{x}(t-1) + f_3 u(t), \\ \vdots & \\ x_{n-1}(t+1) & = x_n(t) + \boldsymbol{b}_{n-1}\boldsymbol{x}(t-1) + f_{n-1} u(t), \\ x_n(t+1) & = a_n x_1(t) + a_{n-1}x_2(t) + \cdots + a_1 x_n(t) + \boldsymbol{b}_n\boldsymbol{x}(t-1) + f_n u(t). \end{cases} \tag{1.25}$$

Using the properties of the shift operator $z$, multiplying the $i$th equation of (1.25) by $z^{-i}$ gives

$$\begin{cases} x_1(t) & = x_2(t-1) + \boldsymbol{b}_1\boldsymbol{x}(t-2) + f_1 u(t-1), \\ x_2(t-1) & = x_3(t-2) + \boldsymbol{b}_2\boldsymbol{x}(t-3) + f_2 u(t-2), \\ x_3(t-2) & = x_4(t-3) + \boldsymbol{b}_3\boldsymbol{x}(t-4) + f_3 u(t-3), \\ \vdots & \\ x_{n-1}(t-n+2) & = x_n(t-n+1) + \boldsymbol{b}_{n-1}\boldsymbol{x}(t-n) + f_{n-1} u(t-n+1), \\ x_n(t-n+1) & = a_n x_1(t-n) + a_{n-1}x_2(t-n) + \cdots + a_1 x_n(t-n) + \boldsymbol{b}_n\boldsymbol{x}(t-n-1) + f_n u(t-n) \\ & = \boldsymbol{a}\boldsymbol{x}(t-n) + \boldsymbol{b}_n\boldsymbol{x}(t-n-1) + f_n u(t-n). \end{cases}$$

Define the parameter vector $\vartheta$ and the information vector $\phi(t)$ as

$$\begin{aligned} \vartheta &:= \left[\boldsymbol{b}_1,\ \boldsymbol{b}_2,\ \cdots,\ \boldsymbol{b}_{n-2},\ \boldsymbol{a}+\boldsymbol{b}_{n-1},\ \boldsymbol{b}_n,\ \boldsymbol{f}^{\mathrm{T}}\right]^{\mathrm{T}} \in \mathbb{R}^{n^2+n}, \\ \varphi(t) &:= \left[\boldsymbol{x}^{\mathrm{T}}(t-2),\ \boldsymbol{x}^{\mathrm{T}}(t-3),\ \cdots,\ \boldsymbol{x}^{\mathrm{T}}(t-n+1),\ \boldsymbol{x}^{\mathrm{T}}(t-n),\ \boldsymbol{x}^{\mathrm{T}}(t-n-1)\right], \\ &\quad u(t-1), u(t-2), \cdots, u(t-n)^{\mathrm{T}} \in \mathbb{R}^{n^2+n}. \end{aligned}$$

Adding all expressions of (1.25) gives

$$\begin{aligned} x_1(t) &= \boldsymbol{a}\boldsymbol{x}(t-n) + \boldsymbol{b}_1\boldsymbol{x}(t-2) + \boldsymbol{b}_2\boldsymbol{x}(t-3) + \cdots + \boldsymbol{b}_{n-1}\boldsymbol{x}(t-n) + \boldsymbol{b}_n\boldsymbol{x}(t-n-1) \\ &\quad + f_1 u(t-1) + f_2 u(t-2) + \cdots + f_n u(t-n) \\ &= \varphi^{\mathrm{T}}(t)\vartheta. \end{aligned}$$

From (1.24), we have

$$y(t) = x_1(t) + v(t) = \varphi^{\mathrm{T}}(t)\vartheta + v(t). \tag{1.26}$$

This is the identification model of the state-space model with one-unit time delay.

### 1.2.3 The parameter and state estimation algorithm

For the case with unmeasurable states, the basic idea is to estimate the system parameters by using the estimated states and available input–output data and to estimate the system states by using the parameter estimates and available input–output data. The details are as follows.

For the case with the unmeasurable state vector $x(t)$, the information vector $\phi(t)$ contains the unknown state vectors $x(t-i)$, applying the hierarchical identification principle to replace the unknown state vector $x(t-i)$ in $\phi(t)$ with its estimate $\hat{x}(t-i)$. Define the estimated information vector $\hat{\phi}(t)$ and the parameter vector $\hat{\vartheta}(t)$ as

$$\hat{\varphi}(t) := \left[\hat{\boldsymbol{x}}^{\mathrm{T}}(t-2), \hat{\boldsymbol{x}}^{\mathrm{T}}(t-3), \cdots, \hat{\boldsymbol{x}}^{\mathrm{T}}(t-n-1), u(t-1), u(t-2), \cdots, u(t-n)\right]^{\mathrm{T}} \in \mathbb{R}^{n^2+n},$$
$$\hat{\vartheta}(t) := \left[\hat{\boldsymbol{b}}_1(t), \hat{\boldsymbol{b}}_2(t), \cdots, \hat{\boldsymbol{b}}_{n-2}(t), \widehat{\boldsymbol{a}+\boldsymbol{b}_{n-1}}(t), \hat{\boldsymbol{b}}_n(t), \hat{\boldsymbol{f}}^{\mathrm{T}}(t)\right]^{\mathrm{T}} \in \mathbb{R}^{n^2+n}.$$

We can obtain the following state estimation-based recursive least squares (SE-RLS) parameter identification algorithm:

$$\hat{\vartheta}(t) = \hat{\vartheta}(t-1) + \boldsymbol{P}(t)\hat{\varphi}(t)\left[y(t) - \hat{\varphi}^{\mathrm{T}}(t)\hat{\vartheta}(t-1)\right], \tag{1.27}$$

$$\boldsymbol{P}^{-1}(t) = \boldsymbol{P}^{-1}(t-1) + \hat{\varphi}(t)\hat{\varphi}^{T}(t), \qquad \boldsymbol{P}(0) = p_0\boldsymbol{I}, \tag{1.28}$$

$$\hat{\varphi}(t) = \left[\hat{\boldsymbol{x}}^{\mathrm{T}}(t-2), \hat{\boldsymbol{x}}^{\mathrm{T}}(t-3), \cdots, \hat{\boldsymbol{x}}^{\mathrm{T}}(t-n-1), u(t-1), u(t-1), u(t-2), \cdots, u(t-n)\right]^{\mathrm{T}}. \tag{1.29}$$

Using the parameter estimates and based on the canonical state-space model in (1.23) and (1.24) and the Kalman filtering theory, we can obtain the following state estimation algorithm:

$$\hat{\boldsymbol{x}}(t+1) = \hat{\boldsymbol{A}}(t)\hat{\boldsymbol{x}}(t) + \hat{\boldsymbol{B}}(t)\hat{\boldsymbol{x}}(t-1) + \hat{\boldsymbol{f}}(t)u(t) + \boldsymbol{L}_2(t)[y(t) - c\hat{\boldsymbol{x}}(t)], \tag{1.30}$$

$$\boldsymbol{L}_2(t) = \hat{\boldsymbol{A}}(t)\boldsymbol{P}_2(t)\boldsymbol{c}^{\mathrm{T}}\left[1 + \boldsymbol{c}\boldsymbol{P}_2(t)\boldsymbol{c}^{\mathrm{T}}\right]^{-1}, \tag{1.31}$$

$$\boldsymbol{P}_2(t+1) = \hat{\boldsymbol{A}}(t)\boldsymbol{P}_2(t)\hat{\boldsymbol{A}}^{\mathrm{T}}(t) - \hat{\boldsymbol{A}}(t)\boldsymbol{P}_2(t)\boldsymbol{c}^{\mathrm{T}}\left[1 + \boldsymbol{c}\boldsymbol{P}_2(t)\boldsymbol{c}^{\mathrm{T}}\right]^{-1}\boldsymbol{c}\boldsymbol{P}_2(t)\hat{\boldsymbol{A}}^{\mathrm{T}}(t), \tag{1.32}$$

$$\hat{\boldsymbol{A}}(t) = \begin{bmatrix} 0 & 1 & 0 & \cdots & 0 \\ 0 & 0 & 0 & \ddots & \vdots \\ \vdots & \vdots & \ddots & \ddots & 0 \\ 0 & 0 & \cdots & 0 & 1 \\ \hat{a}_n(t) & \hat{a}_{n-1}(t) & \hat{a}_{n-2}(t) & \cdots & \hat{a}_1(t) \end{bmatrix} \tag{1.33}$$

$$\hat{\boldsymbol{B}}(t)=\begin{bmatrix}\hat{\boldsymbol{b}}_1(t)\\ \hat{\boldsymbol{b}}_2(t)\\ \vdots\\ \hat{\boldsymbol{b}}_1(t)\end{bmatrix},\qquad \hat{f}(t)=\begin{bmatrix}\hat{f}_1(t)\\ \hat{f}_2(t)\\ \vdots\\ \hat{f}_n(t)\end{bmatrix}. \tag{1.34}$$

The parameter estimation algorithm in (1.27)−(1.29) and the state estimation algorithm in (1.30)−(1.34) perform a hierarchical computation process or interactive computation process and form a combined parameter and state estimation algorithm. The initial values are taken to be $\hat{\vartheta}(0)=1_{n^2+n/p_0}$ and $P(0)=I/p_0$, that is, $p_0=10^6$.

Regarding the convergence of the parameter estimation, we have the following theorem.

**Theorem 1.1**: For the system in (1.26) and the algorithm in (1.27) and (1.28), assume that $\{v(t)\}$ is a white noise sequence with zero mean and variance $\sigma^2$, i.e., (A1) $\mathrm{E}[v(t)]=0$ and (A2) $\mathrm{E}[v^2(t)]=\sigma^2$, and that there exist positive constants $0<p\leq q<\infty$ and integer $t_0$ such that for $t\geq t_0$, the persistent excitation condition holds: $pI\leq\frac{1}{t}\sum_{j=1}^{t}\hat{\phi}(j)\hat{\phi}^{\mathrm{T}}(j)\leq qI$. Then, the parameter estimation error $\hat{\vartheta}(t)-\vartheta$ converges to zero.

**Proof** Define the parameter estimation error vector $\tilde{\vartheta}(t):=\hat{\vartheta}(t)-\vartheta$. Using Eqs. (1.26) and (1.27) gives

$$\tilde{\vartheta}(t)=\tilde{\vartheta}(t-1)+\boldsymbol{P}(t)\hat{\varphi}(t)[-\tilde{y}(t)+\beta(t)+v(t)], \tag{1.35}$$

where

$$\begin{aligned}\tilde{y}(t)&:=\hat{\varphi}^{\mathrm{T}}(t)\tilde{\vartheta}(t-1),\\ \beta(t)&:=[\varphi(t)-\hat{\varphi}(t)]^{T}\vartheta\\ &=\boldsymbol{a}[\boldsymbol{x}(t-n)-\hat{\boldsymbol{x}}(t-n)]+\boldsymbol{b}_1[\boldsymbol{x}(t-3)-\hat{\boldsymbol{x}}(t-3)]+\boldsymbol{b}_2[\boldsymbol{x}(t-4)-\hat{\boldsymbol{x}}(t-4)]\\ &\quad+\cdots+\boldsymbol{b}_{n-1}[\boldsymbol{x}(t-n-1)-\hat{\boldsymbol{x}}(t-n-1)]+\boldsymbol{b}_n[\boldsymbol{x}(t-n-2)-\hat{\boldsymbol{x}}(t-n-2)].\end{aligned}$$

Using the persistent excitation condition and from Eq. (1.28), we have

$$\begin{aligned}pt\boldsymbol{I}&\leq\boldsymbol{P}^{-1}(t)=\sum_{j=1}^{t}\hat{\varphi}(j)\hat{\varphi}^{\mathrm{T}}(j)+\boldsymbol{P}^{-1}(0)\leq(qt+1)\boldsymbol{I},\\ (n^2+n)pt&\leq|\boldsymbol{P}^{-1}(t)|\leq(n^2+n)(qt+1).\end{aligned}$$

According to Eq. (1.28), we have

$$
\begin{aligned}
S(t) &:= \sum_{j=1}^{t} \hat{\varphi}^{\mathrm{T}}(j)\boldsymbol{P}(j)\hat{\varphi}(j) \leq \ln |\boldsymbol{P}^{-1}(t)| + \left(n^2+n\right)\ln p_0 \\
&\leq \ln\left[\left(n^2+n\right)(qt+1)\right] + \left(n^2+n\right)\ln p_0.
\end{aligned}
$$

From Eqs. (1.28) and (1.35), we have

$$
\begin{aligned}
\tilde{\vartheta}^{\mathrm{T}}(t)\boldsymbol{P}^{-1}(t)\tilde{\vartheta}(t) &= \left\{\tilde{\vartheta}(t-1)+\boldsymbol{P}(t)\hat{\varphi}(t)[-\tilde{y}(t)+\beta(t)+v(t)]\right\}^{\mathrm{T}}\boldsymbol{P}^{-1}(t) \\
&\quad \times \left\{\tilde{\vartheta}(t-1)+\boldsymbol{P}(t)\hat{\varphi}(t)[-\tilde{y}(t)+\beta(t)+v(t)]\right\} \\
&= \tilde{\vartheta}^{\mathrm{T}}(t-1)\boldsymbol{P}^{-1}(t-1)\tilde{\vartheta}(t-1) + 2\left[1-\hat{\varphi}^{\mathrm{T}}(t)\boldsymbol{P}(t)\hat{\varphi}(t)\right]\tilde{y}(t)[\beta(t)+v(t)] \\
&\quad + \hat{\varphi}^{\mathrm{T}}(t)P(t)\hat{\varphi}(t)\left[v^2(t)+\beta^2(t)+2\beta(t)v(t)\right].
\end{aligned}
\tag{1.36}
$$

Note that $1-\hat{\phi}^{\mathrm{T}}(t)P(t)\hat{\phi}(t) = [1+\hat{\phi}^{\mathrm{T}}(t)P(t-1)\hat{\phi}(t)]^{-1} \geq 0$ and $\{v(t)\}$ is a white noise with zero mean independent of the input signal $\{u(t)\}$. Assume that $\beta(t)$ is bounded with $\beta^2(t) \leq \varepsilon$. Let $V(t) := \mathrm{E}[\tilde{\vartheta}^{\mathrm{T}}(t)P^{-1}(t)\tilde{\vartheta}(t)]$. Taking the expectation of both sides of Eq. (1.36) and using (A1) and (A2) give

$$
\begin{aligned}
V(t) &\leq V(t-1) + 0 + \mathrm{E}\left\{\hat{\varphi}^{\mathrm{T}}(t)\boldsymbol{P}(t)\hat{\varphi}(t)\left[v^2(t)+\beta^2(t)\right]\right\} \\
&\leq V(t-1) + \mathrm{E}\left[\hat{\varphi}^{\mathrm{T}}(t)\boldsymbol{P}(t)\hat{\varphi}(t)\right]\left(\sigma^2+\varepsilon\right) \\
&\leq V(0) + \mathrm{E}[S(t)]\left(\sigma^2+\varepsilon\right) \\
&\leq V(0) + \left\{\ln\left[\left(n^2+n\right)(qt+1)\right] + \left(n^2+n\right)\ln p_0\right\}\left(\sigma^2+\varepsilon\right).
\end{aligned}
$$

Note that $V(0) < \infty$. Thus, we have

$$
\begin{aligned}
\lim_{t\to\infty} \mathrm{E}\left[\|\tilde{\vartheta}(t)\|^2\right] \leq \lim_{t\to\infty}\frac{1}{pt}V(t) \leq \lim_{t\to\infty}\frac{1}{pt}\Big(V(0) + \Big\{\ln\left[\left(n^2+n\right)(qt+1)\right] \\
+ \left(n^2+n\right)\ln p_0\Big\}\left(\sigma^2+\varepsilon\right)\Big) = 0.
\end{aligned}
$$

This shows that the estimation errors $\|\hat{\vartheta}(t)-\vartheta\|$ converge to zero. This proves Theorem 1.1.

### 1.2.4 Examples

**Example 1.1**: Consider the following state-space system with one-step delay:

$$
\begin{aligned}
\boldsymbol{x}(t+1) &= \begin{bmatrix} a_1 & a_2 \\ a_3 & a_4 \end{bmatrix}\boldsymbol{x}(t) + \begin{bmatrix} b_1 & b_2 \\ b_3 & b_4 \end{bmatrix}\boldsymbol{x}(t-1) + \begin{bmatrix} f_1 \\ f_2 \end{bmatrix}u(t) \\
&= \begin{bmatrix} 0 & 1 \\ -0.45 & -0.80 \end{bmatrix}\boldsymbol{x}(t) + \begin{bmatrix} 0.20 & -0.15 \\ 0.15 & -0.20 \end{bmatrix}\boldsymbol{x}(t-1) + \begin{bmatrix} 1.00 \\ -1.00 \end{bmatrix}u(t),
\end{aligned}
$$

$$
y(t) = [1, \quad 0]\boldsymbol{x}(t) + v(t).
$$

The parameter vector to be identified is

$$\begin{aligned}\vartheta &= \left[a_2+b_{11},\ a_1+b_{12},\ b_{21},\ b_{22},\ f_1,\ f_2\right]^{\mathrm{T}} \\ &= [-0.25,\ -0.95,\ 0.15,\ -0.20,\ 1.00,\ -1.00]^{\mathrm{T}}.\end{aligned}$$

In simulation, the input $\{u(t)\}$ is taken as an uncorrelated persistent excitation signal sequence with zero mean and unit variance and $\{v(t)\}$ as a white noise sequence with zero mean and variance $\sigma^2 = 0.50^2$ and $\sigma^2 = 1.00^2$, respectively. Apply the least squares parameter estimation algorithm with the known states to estimate the parameters of this example system. The parameter estimates and their estimation errors are shown in Table 1.2 with $\sigma^2 = 0.50^2$ and $\sigma^2 = 1.00^2$, and the parameter estimation errors $\delta$ versus $t$ are shown in Fig. 1.3.

From Table 1.1 and Fig. 1.3, we can draw the following conclusions: (1) the parameter estimation errors $\delta$ (in general) become smaller with the increasing of —see Fig. 1.3; (2) the parameter estimation errors under the same data lengths become smaller for low noise levels—see Table 1.1 and Fig. 1.3; (3) the parameter estimation accuracy becomes higher as the data length $t$ increases—see Table 1.1; and (4) a lower noise level leads to a faster rate of convergence of the parameter estimates to their true values—see Fig. 1.3.

**Example 1.2**: Consider the following state-space system with one-step delay:

$$\begin{aligned}\boldsymbol{x}(t+1) &= \begin{bmatrix} 0 & 1 \\ 0 & 0.35 \end{bmatrix}\boldsymbol{x}(t) + \begin{bmatrix} 0.28 & 0 \\ 0.52 & -0.20 \end{bmatrix}\boldsymbol{x}(t-1) + \begin{bmatrix} 1.00 \\ -1.00 \end{bmatrix}u(t), \\ y(t) &= \begin{bmatrix} 1, & 0 \end{bmatrix}\boldsymbol{x}(t) + v(t).\end{aligned}$$

The parameter vector to be identified is

$$\begin{aligned}\vartheta &= \left[a_2+b_{11},\ a_1+b_{12},\ b_{21},\ b_{22},\ f_1,\ f_2\right]^{\mathrm{T}} \\ &= [0.28,\ 0.35,\ 0.52,\ -0.20,\ 1.00,\ -1.00]^{\mathrm{T}}.\end{aligned}$$

The simulation conditions are as in Example 1.1, and the noise variance is $\sigma^2 = 0.50^2$; the corresponding noise-to-signal ratio is $\delta_{ns} = 33.29\%$. Apply the algorithm with the known states to estimate the parameters of this example system. The parameter estimates and their estimation errors are shown in Table 1.3.

**Table 1.2** The parameter estimates and errors with $\sigma^2 = 0.50^2$ and $\sigma^2 = 1.00^2$.

| $\sigma^2$ | $t$ | $a_2 + b_{11}$ | $a_1 + b_{12}$ | $b_{21}$ | $b_{22}$ | $f_1$ | $f_2$ | $\delta(\%)$ |
|---|---|---|---|---|---|---|---|---|
| $0.50^2$ | 100 | −0.13924 | −0.86780 | 0.25629 | −0.22918 | 1.00521 | −0.96610 | 10.33719 |
| | 200 | −0.19649 | −0.90985 | 0.20242 | −0.22499 | 1.02959 | −0.96712 | 5.69071 |
| | 500 | −0.28061 | −0.96370 | 0.13165 | −0.18307 | 1.03063 | −0.98191 | 3.15485 |
| | 1000 | −0.27817 | −0.96864 | 0.12634 | −0.20335 | 1.03317 | −0.99633 | 3.05540 |
| | 2000 | −0.24721 | −0.92783 | 0.16467 | −0.19332 | 1.02186 | −0.99018 | 2.09852 |
| | 3000 | −0.24719 | −0.93520 | 0.16110 | −0.19479 | 1.00426 | −1.00833 | 1.23899 |
| $1.00^2$ | 100 | −0.00858 | −0.76816 | 0.38167 | −0.25873 | 0.99626 | −0.93816 | 22.42908 |
| | 200 | −0.13314 | −0.86123 | 0.26428 | −0.25083 | 1.05237 | −0.93740 | 12.03377 |
| | 500 | −0.30735 | −0.97424 | 0.11688 | −0.16660 | 1.05859 | −0.96501 | 5.95758 |
| | 1000 | −0.30458 | −0.98585 | 0.10429 | −0.20693 | 1.06507 | −0.99320 | 5.94017 |
| | 2000 | −0.24350 | −0.90493 | 0.18016 | −0.18677 | 1.04305 | −0.98063 | 4.21822 |
| | 3000 | −0.24379 | −0.91993 | 0.17274 | −0.18965 | 1.00810 | −1.01683 | 2.51540 |
| True values | | −0.25000 | −0.95000 | 0.15000 | −0.20000 | 1.00000 | −1.00000 | |

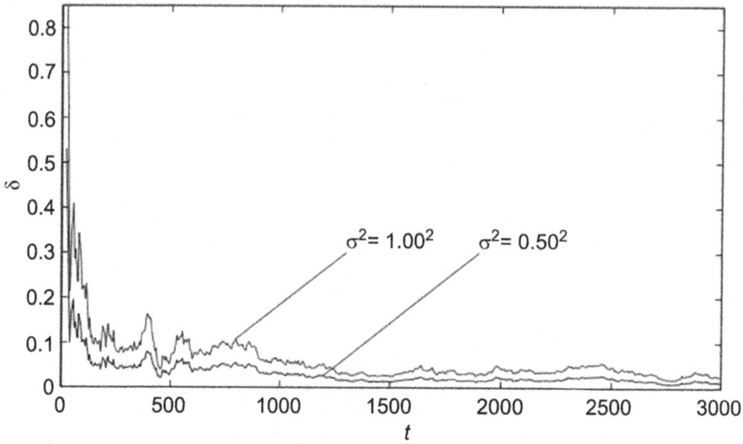


**Figure 1.3** The parameter estimation errors $\delta$ versus $t$ with $\sigma^2 = 0.50^2$ and $\sigma^2 = 1.00^2$.

For comparison, we transform the state-space system into the transfer function representation:

$$
\begin{aligned}
y(t) &= \frac{\beta(z)}{\alpha(z)} u(t) + v(t) + v(t) \\
&= \frac{\beta_1 z^{-1} + \beta_2 z^{-2} + \beta_3 z^{-3}}{1 + \alpha_1 z^{-1} + \alpha_2 z^{-2} + \alpha_3 z^{-3} + \alpha_4 z^{-4}} u(t) + v(t) \\
&= \frac{z^{-1} - 1.35 z^{-2} - 0.20 z^{-3}}{1 - 0.35 z^{-1} - 0.08 z^{-2} - 0.422 z^{-3} - 0.56 z^{-4}} u(t) + v(t),
\end{aligned}
$$

$$
\begin{aligned}
\boldsymbol{\theta} &= \left[\alpha_1,\ \alpha_2,\ \alpha_3,\ \alpha_4,\ \beta_1,\ \beta_2,\ \beta_3\right]^{\mathrm{T}} \\
&= [-0.35000,\ -0.08000,\ -0.42200,\ -0.05600,\ 1.00000,\ -1.35000,\ 0.20000]^{\mathrm{T}}.
\end{aligned}
$$

Based on the estimated parameters of the state-space model in Table 1.3, we obtain the corresponding the parameter estimates of $(\alpha_i, \beta_i)$ as shown in Table 1.4. The parameter estimates and errors using the auxiliary model-based recursive least squares algorithm are shown in Table 1.5.

From the estimation errors in Tables 1.4 and 1.5, it is clear that the proposed algorithm is superior to the auxiliary model-based recursive least squares algorithm.

## 1.3 Conclusions

This chapter discusses the identification problems for linear systems based on the canonical state-space systems with one-unit state delay and presents a recursive least squares parameter estimation algorithm using the

**Table 1.3** The parameter estimates and errors.

| $t$ | $a_2 + b_1$ | $a_1 + b_{12}$ | $b_{21}$ | $b_{22}$ | $f_1$ | $f_2$ | $\delta(\%)$ |
|---|---|---|---|---|---|---|---|
| 100 | 0.05589 | 0.08732 | 0.31546 | −0.13986 | 0.99809 | 0.95563 | 25.76053 |
| 200 | 0.18337 | 0.24815 | 0.45978 | −0.14566 | 1.02009 | 0.96256 | 10.57711 |
| 500 | 0.32564 | 0.41758 | 0.55735 | −0.20207 | 1.02685 | 0.98253 | 6.01377 |
| 1000 | 0.33400 | 0.41751 | 0.57860 | −0.22315 | 1.02959 | 0.99687 | 7.00657 |
| 2000 | 0.29168 | 0.38157 | 0.54766 | −0.20907 | 1.02044 | 0.98963 | 3.15861 |
| 3000 | 0.26023 | 0.34250 | 0.51855 | −0.18177 | 1.00386 | 1.00821 | 1.85451 |
| True values | 0.28000 | 0.35000 | 0.52000 | −0.20000 | 1.00000 | 1.00000 | |

**Table 1.4** The parameter estimates and errors for Example 1.2.

| $t$ | $\alpha_1$ | $\alpha_2$ | $\alpha_3$ | $\alpha_4$ | $\beta_1$ | $\beta_2$ | $\beta_3$ | $\delta(\%)$ |
|---|---|---|---|---|---|---|---|---|
| 100 | −0.08732 | 0.08397 | −0.31058 | −0.00782 | 0.99809 | −1.04279 | 0.13959 | 25.64515 |
| 200 | −0.24815 | −0.03771 | −0.41427 | −0.02671 | 1.02009 | −1.21570 | 0.14859 | 10.37691 |
| 500 | −0.41758 | −0.12657 | −0.42197 | −0.06580 | 1.02685 | −1.41109 | 0.20750 | 5.90705 |
| 1000 | −0.41751 | −0.11085 | −0.43915 | −0.07453 | 1.02959 | −1.42673 | 0.22975 | 6.59422 |
| 2000 | −0.38157 | −0.08260 | −0.43636 | −0.06098 | 1.02044 | −1.37900 | 0.21334 | 2.90156 |
| 3000 | −0.34250 | −0.07846 | −0.42942 | −0.04730 | 1.00386 | −1.35204 | 0.18247 | 1.27465 |
| True values | −0.35000 | −0.08000 | −0.42200 | −0.05600 | 1.00000 | −1.35000 | 0.20000 | |

**Table 1.5** The parameter estimates and errors for Example 1.2.

| $t$ | $\alpha_1$ | $\alpha_2$ | $\alpha_3$ | $\alpha_4$ | $\beta_1$ | $\beta_2$ | $\beta_3$ | $\delta(\%)$ |
|---|---|---|---|---|---|---|---|---|
| 100 | −0.14553 | −0.36393 | −0.06071 | 0.04933 | 0.99994 | −0.78765 | 0.23496 | 53.31184 |
| 200 | −0.02763 | −0.18127 | −0.20215 | 0.03021 | 1.02919 | −0.97321 | 0.17179 | 31.22246 |
| 500 | −0.39977 | −0.08799 | −0.40067 | −0.02517 | 1.03209 | −1.38508 | 0.21087 | 4.46499 |
| 1000 | −0.24218 | −0.11380 | −0.44229 | −0.09711 | 1.03399 | −1.24077 | 0.04079 | 12.96209 |
| 2000 | −0.28332 | −0.10198 | −0.44152 | −0.08536 | 1.02125 | −1.27871 | 0.08530 | 8.85242 |
| 3000 | −0.31884 | −0.07457 | −0.41318 | −0.04696 | 1.00437 | −1.33271 | 0.15820 | 3.18849 |
| True values | −0.35000 | −0.08000 | −0.42200 | −0.05600 | 1.00000 | −1.35000 | 0.20000 | |

measured states and an estimated state-based recursive least squares algorithm. Compared with the recursive least squares methods, the proposed algorithms can not only identify the system parameters, but also estimate the system states. Finally, we compare the convergence properties of the proposed algorithm and the auxiliary model-based recursive least squares algorithm.

## References


Ashiba, H.I., Awadalla, K.H., El-Halfawy, S.M., 2011. Adaptive least squares interpolation of infrared images. Circuits, Systems and Signal Processing 30, 543–551.

Devakar, M., Lyengar, T.K.V., 2011. Run up flow of an incompressible micropolar fluid between parallel plates-A state space approach. Applied Mathematical Modelling 35, 1751–1764.

Ding, F., 2013. System Identification – New Theory and Methods. Science Press, Beijing.

Ding, M.F., Liu, Y.J., Bao, B., 2012. Gradient based and least squares based iterative estimation algorithms for multi-input multi-output systems. Proceedings of the Institution of Mechanical Engineers, Part I: Journal of Systems and Control Engineering 226, 43–55.

Shi, Y., Fang, H.Z., 2010. Kalman filter-based identification for systems with randomly missing measurements in a network environment. International Journal of Control 83, 538–551.

CHAPTER 2

# D-step state-delay identification

## 2.1 State filtering and parameter estimation for d-step state delay

The state-space model can describe the law of motion with differential/difference equations which can be transformed into a set of the first-order differential/difference equations (Hmida et al., 2012). The state-space model involves the system inputs, outputs, and state variables representing the internal behavior of a system and can be used for system identification, adaptive control, and system analysis (Ding, 2014).

System identification contains the structure or order determination and parameter estimation. Parameter estimation is basic for system modeling, signal filtering, and adaptive control. In the field of system modeling and identification, the recursive least squares methods are popular and can be used to estimate the parameters of linear and nonlinear systems. Another basic estimation method for system identification is the stochastic gradient method. Others are the auxiliary model-based algorithms, the iterative identification algorithms, and the hierarchical identification algorithms.

Some work discussed the stability and stabilization of control systems with time-varying state delay or random delays. Identification of the time-delay systems which widely exist in industrial processes has drawn a great deal of attention from many researchers in system control and system analysis.

It presents new recursive least squares identification methods for joint estimating system parameters and states. The main contributions of this paper are as follows.

1. By using the property of the shift operator, this paper transforms an observability state-space model with d-step state delay into an input–output representation.
2. By replacing the state variables in the information vector with their estimates, this paper presents a parameter and state estimation algorithm for identifying the state-delay systems from given input–output data.
3. By using a numerical example, this paper demonstrates the performances of the proposed algorithm, including the estimation errors of the recursive least squares algorithm for finite measurement data.

*State Space Systems With Time-Delays Analysis, Identification, and Applications*
DOI: https://doi.org/10.1016/B978-0-323-91768-1.00004-6

### 2.1.1 The system description and identification model

Let us introduce some notation. "$A=:X$" or "$X=:A$" stands for "$A$ is defined as $X$"; the symbol $I(I_n)$ stands for an identity matrix of appropriate size ($n \times n$); the superscript T denotes the matrix transpose; the norm of a matrix $X$ is defined by $||X||^2:=\mathrm{tr}[XX^{\mathrm{T}}]$; and $\hat{\theta}(t)$ denotes the estimate of $\theta$ at time $t$.

Consider the following state-space system with d-step state delay,

$$\boldsymbol{x}(t+1)=\boldsymbol{A}\boldsymbol{x}(t)+\boldsymbol{B}\boldsymbol{x}(t-d)+\boldsymbol{f}u(t), \tag{2.1}$$

$$y(t)=\boldsymbol{c}\boldsymbol{x}(t)+v(t), \tag{2.2}$$

$$\boldsymbol{A}:=\begin{bmatrix} 0 & 1 & 0 & \cdots & 0 \\ 0 & 0 & 1 & \ddots & \vdots \\ \vdots & \vdots & & \ddots & 0 \\ 0 & 0 & \cdots & 0 & 1 \\ -a_n & -a_{n-1} & -a_{n-2} & \cdots & -a_n \end{bmatrix} \in \mathbb{R}^{n\times n},$$

$$\boldsymbol{B}:=\begin{bmatrix} \boldsymbol{b}_1 \\ \boldsymbol{b}_2 \\ \vdots \\ \boldsymbol{b}_n \end{bmatrix} \in \mathbb{R}^{n\times n}, \quad \boldsymbol{b}_i \in \mathbb{R}^{1\times n},$$

$$\boldsymbol{f}:=\begin{bmatrix} f_1 \\ f_2 \\ \vdots \\ f_n \end{bmatrix} \in \mathbb{R}^{n}, \quad \boldsymbol{c}:=[1, \ 0, \ 0, \ldots, 0] \in \mathbb{R}^{1\times n},$$

where $x(t) \in R^n$ is the state vector, $u(t) \in R$ is the system input, $y(t) \in R$ is the system output, $v(t) \in R$ is a random noise with zero mean, and $A \in R^{n\times n}$, $B \in R^{n\times n}$, $f \in R^n$, and $c \in R^{1\times n}$ are vectors. On the basis of/ vectors. Assume that $(c, A)$ is observable, and $u(t)=0, y(t)=0$, and $v(t)=0$ for $t \leq 0$.

The system in (2.1) and (2.2) is an observability canonical form, and its transformation matrix T is a nonsingular matrix, i.e.,

$$\boldsymbol{T}:=\left[\boldsymbol{c}, \ \boldsymbol{c}\boldsymbol{A}, \ \cdots, \ \boldsymbol{c}\boldsymbol{A}^{n-1}\right]^{\mathrm{T}}=\boldsymbol{I}_n \tag{2.3}$$

Since there exists the unknown state variable in (2.1) and (2.2), it is required to derive a new identification expression that only involves the available measurement input–output data $\{u(t), y(t)\}$. The following transforms the time-delay state-space model in (2.1–2.2) into an input–output representation and gives its identification model.

From (2.1) and (2.2), we have

$$
\begin{aligned}
y(t+i) = {} & \boldsymbol{c}\boldsymbol{A}^{i}\boldsymbol{x}(t) + \boldsymbol{c}\boldsymbol{A}^{i-1}\boldsymbol{B}\boldsymbol{x}(t-d) + \boldsymbol{c}\boldsymbol{A}^{i-2}\boldsymbol{B}\boldsymbol{x}(t-d+1) + \cdots + \boldsymbol{c}\boldsymbol{B}\boldsymbol{x}(t-d+i-1) \\
& + \boldsymbol{c}\boldsymbol{A}^{i-1}\boldsymbol{f}u(t) + \boldsymbol{c}\boldsymbol{A}^{i-2}\boldsymbol{f}u(t+1) + \cdots + \boldsymbol{c}\boldsymbol{f}u(t+i-1) \\
& + v(t+i), \quad i = 0, \ 1, \ldots, n-1,
\end{aligned} \tag{2.4}
$$

$$
\begin{aligned}
y(t+n) = {} & \boldsymbol{c}\boldsymbol{A}^{n}\boldsymbol{x}(t) + \boldsymbol{c}\boldsymbol{A}^{n-1}\boldsymbol{B}\boldsymbol{x}(t-d) + \boldsymbol{c}\boldsymbol{A}^{n-2}\boldsymbol{B}\boldsymbol{x}(t-d+1) + \cdots + \boldsymbol{c}\boldsymbol{B}\boldsymbol{x}(t-d+n-1) \\
& + \boldsymbol{c}\boldsymbol{A}^{n-1}\boldsymbol{f}u(t) + \boldsymbol{c}\boldsymbol{A}^{n-2}\boldsymbol{f}u(t+1) + \cdots + \boldsymbol{c}\boldsymbol{f}u(t+n-1) + v(t+n).
\end{aligned} \tag{2.5}
$$

Define some vectors/matrices,

$$
\boldsymbol{Y}(t+n) := \begin{bmatrix} y(t) \\ y(t+1) \\ \vdots \\ y(t+n-1) \end{bmatrix} \in \mathbb{R}^{n}, \quad \boldsymbol{U}(t+n) := \begin{bmatrix} u(t) \\ u(t+1) \\ \vdots \\ u(t+n-1) \end{bmatrix} \in \mathbb{R}^{n},
$$

$$
\boldsymbol{X}(t-d+n) := \begin{bmatrix} \boldsymbol{x}(t-d) \\ \boldsymbol{x}(t-d+1) \\ \vdots \\ \boldsymbol{x}(t-d+n-1) \end{bmatrix} \in \mathbb{R}^{n^2}, \quad \boldsymbol{V}(t+n) := \begin{bmatrix} v(t) \\ v(t+1) \\ \vdots \\ v(t+n-1) \end{bmatrix} \in \mathbb{R}^{n},
$$

$$
\boldsymbol{Q} := \begin{bmatrix} 0 & 0 & \cdots & 0 & 0 \\ \boldsymbol{c}\boldsymbol{f} & 0 & \cdots & 0 & 0 \\ \boldsymbol{c}\boldsymbol{A}\boldsymbol{f} & \boldsymbol{c}\boldsymbol{f} & \ddots & \vdots & \vdots \\ \vdots & \vdots & \ddots & 0 & 0 \\ \boldsymbol{c}\boldsymbol{A}^{n-2}\boldsymbol{f} & \boldsymbol{c}\boldsymbol{A}^{n-3}\boldsymbol{f} & \cdots & \boldsymbol{c}\boldsymbol{f} & 0 \end{bmatrix} \in \mathbb{R}^{n \times n},
$$

$$
\boldsymbol{M} := \begin{bmatrix} 0 & 0 & \cdots & 0 & 0 \\ \boldsymbol{c}\boldsymbol{B} & 0 & \cdots & 0 & 0 \\ \boldsymbol{c}\boldsymbol{A}\boldsymbol{B} & \boldsymbol{c}\boldsymbol{B} & \ddots & \vdots & \vdots \\ \vdots & \vdots & \ddots & 0 & 0 \\ \boldsymbol{c}\boldsymbol{A}^{n-2}\boldsymbol{B} & \boldsymbol{c}\boldsymbol{A}^{n-3}\boldsymbol{B} & \cdots & \boldsymbol{c}\boldsymbol{B} & 0 \end{bmatrix} \in \mathbb{R}^{n \times n^2}, \tag{2.6}
$$

From Eq. (2.4) and the above definitions, we have

$$\begin{aligned}\boldsymbol{Y}(t+n) &= \boldsymbol{T}\boldsymbol{x}(t) + \boldsymbol{M}\boldsymbol{X}(t-d+n) + \boldsymbol{Q}\boldsymbol{U}(t+n) + \boldsymbol{V}(t+n)\\ &= \boldsymbol{x}(t) + \boldsymbol{M}\boldsymbol{X}(t-d+n) + \boldsymbol{Q}\boldsymbol{U}(t+n) + \boldsymbol{V}(t+n),\end{aligned}$$

or

$$\boldsymbol{x}(t) = \boldsymbol{Y}(t+n) - \boldsymbol{M}\boldsymbol{X}(t-d+n) - \boldsymbol{Q}\boldsymbol{U}(t+n) - \boldsymbol{V}(t+n) \tag{2.7}$$

Define the information vector $\phi(t)$ and the parameter vector $\theta$ as

$$\varphi(t+n) := \begin{bmatrix} \boldsymbol{Y}(t+n) - \boldsymbol{V}(t+n) \\ \boldsymbol{X}(t-d+n) \\ \boldsymbol{U}(t+n) \end{bmatrix} \in \mathbb{R}^{2n+n^2}, \quad \theta := \begin{bmatrix} \theta_a \\ \theta_b \\ \theta_c \end{bmatrix} \in \mathbb{R}^{2n+n^2},$$

$$\boldsymbol{\theta}_a := [\boldsymbol{c}\boldsymbol{A}^n]^T \in \mathbb{R}^n, \quad \boldsymbol{\theta}_b := \left[-\boldsymbol{c}\boldsymbol{A}^n\boldsymbol{M} + \left[\boldsymbol{c}\boldsymbol{A}^{n-1}\boldsymbol{B},\ \boldsymbol{c}\boldsymbol{A}^{n-2}\boldsymbol{B},\ \cdots,\ \boldsymbol{c}\boldsymbol{B}\right]\right]^{\mathrm{T}} \in \mathbb{R}^{n^2},$$

$$\boldsymbol{\theta}_c := \left[-\boldsymbol{c}\boldsymbol{A}^n\boldsymbol{Q} + \left[\boldsymbol{c}\boldsymbol{A}^{n-1}\boldsymbol{f},\ \boldsymbol{c}\boldsymbol{A}^{n-2}\boldsymbol{f},\ \cdots,\ \boldsymbol{c}\boldsymbol{f}\right]\right]^{\mathrm{T}} \in \mathbb{R}^n.$$

Substituting (2.7) into (2.5) gives

$$\begin{aligned}y(t+n) &= \boldsymbol{c}\boldsymbol{A}^n[\boldsymbol{Y}(t+n) - \boldsymbol{M}\boldsymbol{X}(t-d+n) - \boldsymbol{Q}\boldsymbol{U}(t+n) - \boldsymbol{V}(t+n)]\\ &+ \boldsymbol{c}\boldsymbol{A}^{n-1}\boldsymbol{B}\boldsymbol{x}(t-d) + \boldsymbol{c}\boldsymbol{A}^{n-2}\boldsymbol{B}\boldsymbol{x}(t-d+1) + \cdots + \boldsymbol{c}\boldsymbol{B}\boldsymbol{x}(t-d+n-1)\\ &+ \boldsymbol{c}\boldsymbol{A}^{n-1}\boldsymbol{f}u(t) + \boldsymbol{c}\boldsymbol{A}^{n-2}\boldsymbol{f}u(t+1) + \cdots + \boldsymbol{c}\boldsymbol{f}u(t+n-1) + v(t+n)\\ &= \boldsymbol{c}\boldsymbol{A}^n[\boldsymbol{Y}(t+n) - \boldsymbol{M}\boldsymbol{X}(t-d+n) - \boldsymbol{Q}\boldsymbol{U}(t+n) - \boldsymbol{V}(t+n)]\\ &+ \left[\boldsymbol{c}\boldsymbol{A}^{n-1}\boldsymbol{B},\ \boldsymbol{c}\boldsymbol{A}^{n-2}\boldsymbol{B},\ \cdots,\ \boldsymbol{c}\boldsymbol{B}\right]\begin{bmatrix} x(t-d) \\ x(t-d+1) \\ \vdots \\ x(t-d+n-1) \end{bmatrix}\\ &+ \left[\boldsymbol{c}\boldsymbol{A}^{n-1}\boldsymbol{f},\ \boldsymbol{c}\boldsymbol{A}^{n-2}\boldsymbol{f},\ \cdots,\ \boldsymbol{c}\boldsymbol{f}\right]\begin{bmatrix} u(t) \\ u(t+1) \\ \vdots \\ u(t+n-1) \end{bmatrix} + v(t+n)\\ &= \boldsymbol{c}\boldsymbol{A}^n[\boldsymbol{Y}(t+n) - \boldsymbol{V}(t+n)] - \boldsymbol{c}\boldsymbol{A}^n\boldsymbol{M}\boldsymbol{X}(t-d+n) - \boldsymbol{c}\boldsymbol{A}^n\boldsymbol{Q}\boldsymbol{U}(t+n)\\ &+ \left[\boldsymbol{c}\boldsymbol{A}^{n-1}\boldsymbol{B},\ \boldsymbol{c}\boldsymbol{A}^{n-2}\boldsymbol{B},\ \cdots,\ \boldsymbol{c}\boldsymbol{B}\right]\boldsymbol{X}(t-d+n)\\ &+ \left[\boldsymbol{c}\boldsymbol{A}^{n-1}\boldsymbol{f},\ \boldsymbol{c}\boldsymbol{A}^{n-2}\boldsymbol{f},\ \cdots,\ \boldsymbol{c}\boldsymbol{f}\right]\boldsymbol{U}(t+n) + v(t+n)\\ &= \left[\boldsymbol{Y}^{\mathrm{T}}(t+n) - \boldsymbol{V}^{\mathrm{T}}(t+n),\ \boldsymbol{X}^{\mathrm{T}}(t-d+n),\ \boldsymbol{U}^{\mathrm{T}}(t+n)\right]\begin{bmatrix} \boldsymbol{\theta}_a \\ \boldsymbol{\theta}_b \\ \boldsymbol{\theta}_c \end{bmatrix} + v(t+n)\\ &= \boldsymbol{\varphi}^{\mathrm{T}}(t+n)\boldsymbol{\theta} + v(t+n)\end{aligned} \tag{2.8}$$

Replacing $t$ in (2.8) with $t-n$ yields

$$y(t)=\boldsymbol{\varphi}^{\mathrm{T}}(t)\boldsymbol{\theta}+v(t) \tag{2.9}$$

This is the identification model of the state-space system with d-step state delay.

### 2.1.2 The parameter estimation algorithm

Since the information vector $\phi(t)$ in (2.9) contains the unknown noise item $v(t-i)$ and the state vector $x(t-d-i)$, this section adopts the basic idea of replacing the unknown noise item $v(t-i)$ and the state vector $x(t-d-i)$ in $\phi(t)$ with the estimated residual $\hat{v}(t-i)$ and the estimated state vector $\hat{x}(t-d-i)$. Define

$$\hat{\boldsymbol{\varphi}}(t):=\begin{bmatrix}\boldsymbol{Y}(t)-\hat{\boldsymbol{V}}(t)\\ \hat{\boldsymbol{X}}(t-d)\\ \boldsymbol{U}(t)\end{bmatrix}\in\mathbb{R}^{2n+n^2},\quad \hat{\boldsymbol{X}}(t-d):=\begin{bmatrix}\hat{\boldsymbol{x}}(t-n-d)\\ \hat{\boldsymbol{x}}(t-n-d+1)\\ \vdots\\ \hat{\boldsymbol{x}}(t-d-1)\end{bmatrix}\in\mathbb{R}^{n^2},$$

$$\hat{\boldsymbol{V}}(t):=\begin{bmatrix}\hat{v}(t-n)\\ \hat{v}(t-n+1)\\ \vdots\\ \hat{v}(t-1)\end{bmatrix}\in\mathbb{R}^{n}.$$

Let $\hat{\theta}(t):=\begin{bmatrix}\hat{\theta}_a(t)\\ \hat{\theta}_b(t)\\ \hat{\theta}_c(t)\end{bmatrix}$ be the estimate of $\theta=\begin{bmatrix}\theta_a\\ \theta_b\\ \theta_c\end{bmatrix}$ at time $t$. According to (2.9), we have

$$\hat{v}(t)=y(t)-\hat{\boldsymbol{\varphi}}^{\mathrm{T}}(t)\hat{\boldsymbol{\theta}}(t).$$

Thus, we can obtain the following state estimation-based recursive least squares algorithm:

$$\hat{\boldsymbol{\theta}}(t)=\hat{\boldsymbol{\theta}}(t-1)+\boldsymbol{L}(t)\left[y(t)-\hat{\boldsymbol{\varphi}}^{\mathrm{T}}(t)\hat{\boldsymbol{\theta}}(t-1)\right], \tag{2.10}$$

$$\boldsymbol{L}(t)=\boldsymbol{P}(t)\hat{\boldsymbol{\varphi}}(t)=\frac{P(t-1)\hat{\boldsymbol{\varphi}}(t)}{1+\hat{\boldsymbol{\varphi}}^{\mathrm{T}}(t)\boldsymbol{P}(t-1)\hat{\boldsymbol{\varphi}}(t)}, \tag{2.11}$$

$$\boldsymbol{P}(t)=\left[\boldsymbol{I}-\boldsymbol{L}(t)\hat{\boldsymbol{\varphi}}^{\mathrm{T}}(t)\right]\boldsymbol{P}(t-1),\quad \boldsymbol{P}(0)=p_0\boldsymbol{I}, \tag{2.12}$$

$$\hat{v}(t)=y(t)-\hat{\boldsymbol{\varphi}}^{\mathrm{T}}(t)\hat{\boldsymbol{\theta}}(t), \tag{2.13}$$

$$\begin{aligned}\hat{\boldsymbol{\varphi}}(t)=[&y(t-n)-\hat{v}(t-n),\ y(t-n+1)-\hat{v}(t-n+1),\ \cdots,\ y(t-1)-\hat{v}(t-1),\\ &\hat{x}(t-n-d),\ \hat{x}(t-n-d+1),\ \cdots,\ \hat{x}(t-d-1),\ u(t-n),\\ &u(t-n+1),\ \cdots,\ u(t-1)]^{\mathrm{T}}\end{aligned} \tag{2.14}$$

### 2.1.3 The state estimation algorithm

For the case with unmeasurable states, the basic idea is to estimate the system states by using the parameter estimates and available input–output data. The details are as follows.

Post-multiplying (2.3) by $B$ gives

$$\begin{bmatrix}\boldsymbol{c}\boldsymbol{B}\\ \boldsymbol{c}\boldsymbol{A}\boldsymbol{B}\\ \vdots\\ \boldsymbol{c}\boldsymbol{A}^{n-1}\boldsymbol{B}\end{bmatrix}=\boldsymbol{B} \tag{2.15}$$

From Eq. (2.15) and the definition of $B$, we have

$$\boldsymbol{c}\boldsymbol{A}^{k-1}\boldsymbol{B}=b_k,\quad k=1,\ 2,\ \cdots,\ n.$$

Using Eqs. (2.6) and (2.15), the matrix $M$ can be expressed as

$$\boldsymbol{M}=\begin{bmatrix}0 & 0 & \cdots & 0 & 0\\ b_1 & 0 & \cdots & 0 & 0\\ b_2 & b_1 & \ddots & \vdots & \vdots\\ \vdots & \vdots & \ddots & 0 & 0\\ b_{n-1} & b_{n-2} & \cdots & b_1 & 0\end{bmatrix} \tag{2.16}$$

Post-multiplying (2.3) by $A$ gives

$$\left[\boldsymbol{c}\boldsymbol{A},\ \boldsymbol{c}\boldsymbol{A}^2,\ \cdots,\ \boldsymbol{c}\boldsymbol{A}^n\right]^{\mathrm{T}}=\boldsymbol{A}.$$

According to the definition of $A$, the vector $cA^n$ can be expressed as

$$\boldsymbol{c}\boldsymbol{A}^n = [-a_n,\ -a_{n-1},\ \cdots,\ -a_1].$$

Postmultiplying both sides of the above equation by $M$ in (2.16) gives

$$\boldsymbol{c}\boldsymbol{A}^n\boldsymbol{M} = \begin{bmatrix} -(\boldsymbol{b}_1^{\mathrm{T}}a_{n-1} + \boldsymbol{b}_2^{\mathrm{T}}a_{n-2} + \cdots + \boldsymbol{b}_{n-1}^{\mathrm{T}}a_1) \\ -(\boldsymbol{b}_1^{\mathrm{T}}a_{n-2} + \boldsymbol{b}_2^{\mathrm{T}}a_{n-3} + \cdots + \boldsymbol{b}_{n-2}^{\mathrm{T}}a_1) \\ \vdots \\ -\boldsymbol{b}_1^{\mathrm{T}}a_1 \\ \boldsymbol{0} \end{bmatrix}$$

Similarly, we can get $cA^nQ$ in the form of

$$\boldsymbol{c}\boldsymbol{A}^n\boldsymbol{Q} = \begin{bmatrix} -(f_1a_{n-1} + f_2a_{n-2} + \cdots + f_{n-1}a_1) \\ -(f_1a_{n-2} + f_2a_{n-3} + \cdots + f_{n-2}a_1) \\ \vdots \\ -f_1a_1 \\ 0 \end{bmatrix}$$

Using the above equations, the parameter vectors $\theta_a$, $\theta_b$, and $\theta_c$ can be expressed as

$$\boldsymbol{\theta}_a = [\boldsymbol{c}\boldsymbol{A}^n]^{\mathrm{T}} = \begin{bmatrix} -a_n \\ -a_{n-1} \\ \vdots \\ -a_1 \end{bmatrix}, \tag{2.17}$$

$$\boldsymbol{\theta}_b = \left[-\boldsymbol{c}\boldsymbol{A}^n\boldsymbol{M} + \left[\boldsymbol{c}\boldsymbol{A}^{n-1}\boldsymbol{B},\ \boldsymbol{c}\boldsymbol{A}^{n-2}\boldsymbol{B},\ \cdots,\ \boldsymbol{c}\boldsymbol{B}\right]\right]^{\mathrm{T}}$$

$$\begin{bmatrix} \boldsymbol{b}_1^{\mathrm{T}}a_{n-1} + \boldsymbol{b}_2^{\mathrm{T}}a_{n-2} + \cdots + \boldsymbol{b}_{n-1}^{\mathrm{T}}a_1 + \boldsymbol{b}_n^{\mathrm{T}} \\ \boldsymbol{b}_1^{\mathrm{T}}a_{n-2} + \boldsymbol{b}_2^{\mathrm{T}}a_{n-3} + \cdots + \boldsymbol{b}_{n-2}^{\mathrm{T}}a_1 + \boldsymbol{b}_{n-1}^{\mathrm{T}} \\ \vdots \\ \boldsymbol{b}_1^{\mathrm{T}}a_1 + \boldsymbol{b}_2^{\mathrm{T}} \\ \boldsymbol{b}_1^{\mathrm{T}} \end{bmatrix} = \begin{bmatrix} b_1^{\mathrm{T}} & b_2^{\mathrm{T}} & \cdots & b_{n-1}^{\mathrm{T}} & b_n^{\mathrm{T}} \\ 0 & b_1^{\mathrm{T}} & \cdots & b_{n-2}^{\mathrm{T}} & b_{n-1}^{\mathrm{T}} \\ \vdots & \vdots & & \vdots & \vdots \\ 0 & 0 & \cdots & b_1^{\mathrm{T}} & b_2^{\mathrm{T}} \\ 0 & 0 & \cdots & 0 & b_1^{\mathrm{T}} \end{bmatrix} \begin{bmatrix} a_{n-1} \\ a_{n-2} \\ \vdots \\ a_1 \\ 1 \end{bmatrix} \tag{2.18}$$

$$\boldsymbol{\theta}_c = \left[-\boldsymbol{c}\boldsymbol{A}^n\boldsymbol{Q} + \left[\boldsymbol{c}\boldsymbol{A}^{n-1}\boldsymbol{f},\ \boldsymbol{c}\boldsymbol{A}^{n-2}\boldsymbol{f},\ \cdots,\ \boldsymbol{c}\boldsymbol{f}\right]\right]^{\mathrm{T}}$$

$$\begin{bmatrix} f_1a_{n-1} + f_2a_{n-2} + \cdots + f_{n-1}a_1 + f_n \\ f_1a_{n-2} + f_2a_{n-3} + \cdots + f_{n-2}a_1 + f_{n-1} \\ \vdots \\ f_1a_1 + f_2 \\ f_1 \end{bmatrix} = \begin{bmatrix} a_{n-1} & a_{n-2} & \cdots & a_1 & 1 \\ a_{n-2} & a_{n-3} & \cdots & 1 & 0 \\ \vdots & \vdots & & \vdots & \vdots \\ a_1 & 1 & \cdots & 0 & 0 \\ 1 & 0 & \cdots & 0 & 0 \end{bmatrix} \begin{bmatrix} f_1 \\ f_2 \\ \vdots \\ f_{n-1} \\ f_n \end{bmatrix} \tag{2.19}$$

Let $\hat{\theta}_b(t)$ be the estimate of $\theta_b$ at time $t$, $\hat{\theta}_c(t)$ be the estimate of $\theta_c$ at time $t$, and the estimates of $f$ and $\theta_a$ be

$$\hat{\boldsymbol{f}}(t)=\begin{bmatrix}\hat{f}_1(t)\\ \hat{f}_2(t)\\ \vdots\\ \hat{f}_n(t)\end{bmatrix}\in\mathbb{R}^n,\quad \hat{\boldsymbol{\theta}}_a(t)=\begin{bmatrix}-\hat{a}_n(t)\\ -\hat{a}_{n-1}(t)\\ \vdots\\ -\hat{a}_1(t)\end{bmatrix}\in\mathbb{R}^n. \tag{2.20}$$

Once the estimate $\hat{\theta}(t)=\begin{bmatrix}\hat{\theta}_a(t)\\ \hat{\theta}_b(t)\\ \hat{\theta}_c(t)\end{bmatrix}$ of the parameter vector $\theta$ is obtained by using the algorithm in (2.10–2.14), we can obtain the estimate $\hat{a}_i(t)$ of $a_i$ from (2.20). According to (2.18) and (2.19), we can use the estimates $\hat{a}_i(t)$, $\hat{b}_i(t)$, and $\hat{f}_i(t)$ to establish the equations,

$$\begin{bmatrix}\hat{\boldsymbol{b}}_1^{\mathrm{T}}(t) & \hat{\boldsymbol{b}}_2^{\mathrm{T}}(t) & \cdots & \hat{\boldsymbol{b}}_{n-1}^{\mathrm{T}}(t) & \hat{\boldsymbol{b}}_n^{\mathrm{T}}(t)\\ 0 & \hat{\boldsymbol{b}}_1^{\mathrm{T}}(t) & \cdots & \hat{\boldsymbol{b}}_{n-2}^{\mathrm{T}}(t) & \hat{\boldsymbol{b}}_{n-1}^{\mathrm{T}}(t)\\ \vdots & \vdots & & \vdots & \vdots\\ 0 & 0 & \cdots & \hat{\boldsymbol{b}}_1^{\mathrm{T}}(t) & \hat{\boldsymbol{b}}_2^{\mathrm{T}}(t)\\ 0 & 0 & \cdots & 0 & \hat{\boldsymbol{b}}_1^{\mathrm{T}}(t)\end{bmatrix}\begin{bmatrix}\hat{a}_{n-1}(t)\\ \hat{a}_{n-2}(t)\\ \vdots\\ \hat{a}_1(t)\\ 1\end{bmatrix}=\hat{\boldsymbol{\theta}}_b(t),$$

and

$$\begin{bmatrix}\hat{a}_{n-1}(t) & \hat{a}_{n-2}(t) & \cdots & \hat{a}_1(t) & 1\\ \hat{a}_{n-2}(t) & \hat{a}_{n-3}(t) & \cdots & 1 & 0\\ \vdots & \vdots & & \vdots & \vdots\\ \hat{a}_1(t) & 1 & \cdots & 0 & 0\\ 1 & 0 & \cdots & 0 & 0\end{bmatrix}\begin{bmatrix}\hat{f}_1(t)\\ \hat{f}_2(t)\\ \vdots\\ \hat{f}_{n-1}(t)\\ \hat{f}_n(t)\end{bmatrix}=\hat{\boldsymbol{\theta}}_c(t).$$

Thus, we have

$$\hat{f}(t)=\begin{bmatrix}\hat{f}_1(t)\\ \hat{f}_2(t)\\ \vdots\\ \hat{f}_{n-1}(t)\\ \hat{f}_n(t)\end{bmatrix}=\begin{bmatrix}\hat{a}_{n-1}(t) & \hat{a}_{n-2}(t) & \cdots & \hat{a}_1(t) & 1\\ \hat{a}_{n-2}(t) & \hat{a}_{n-3}(t) & \cdots & 1 & 0\\ \vdots & \vdots & & \vdots & \vdots\\ \hat{a}_1(t) & 1 & \cdots & 0 & 0\\ 1 & 0 & \cdots & 0 & 0\end{bmatrix}^{-1}\hat{\boldsymbol{\theta}}_c(t). \tag{2.21}$$

Substituting $t$ in (2.7) with $t-n$ gives

$$\boldsymbol{x}(t-n)=\boldsymbol{Y}(t)-\boldsymbol{MX}(t-d)-\boldsymbol{QU}(t)-\boldsymbol{V}(t).$$

Replacing the unknown $M$, $Q$, and $V(t)$ with their estimates $\hat{M}(t)$, $\hat{Q}(t)$, and $\hat{V}(t)$ gives the state estimation vector,

$$\hat{\boldsymbol{x}}(t-n) = \boldsymbol{Y}(t) - \hat{\boldsymbol{M}}(t)\hat{\boldsymbol{X}}(t-d) - \hat{\boldsymbol{Q}}(t)\hat{\boldsymbol{U}}(t) - \hat{\boldsymbol{V}}(t). \tag{2.22}$$

From (2.16) and (2.20–2.22), we can summarize the state estimation algorithm as

$$\hat{\boldsymbol{x}}(t-n) = \boldsymbol{Y}(t) - \hat{\boldsymbol{M}}(t)\hat{\boldsymbol{X}}(t-d) - \hat{\boldsymbol{Q}}(t)\hat{\boldsymbol{U}}(t) - \hat{\boldsymbol{V}}(t),$$

$$\boldsymbol{Y}(t) = [y(t-n),\ y(t-n+t),\ \cdots,\ y(t-1)]^{\mathrm{T}},$$

$$\hat{\boldsymbol{X}}(t-d) = [\hat{\boldsymbol{x}}(t-d-n),\ \hat{\boldsymbol{x}}(t-d-n+1),\ \cdots,\ \hat{\boldsymbol{x}}(t-d-1)]^{\mathrm{T}},$$

$$\boldsymbol{U}(t) = [u(t-n),\ u(t-n+1),\ \cdots,\ u(t-1)]^{\mathrm{T}},$$

$$\hat{\boldsymbol{V}}(t) = [\hat{v}(t-n),\ \hat{v}(t-n+1),\ \cdots,\ \hat{v}(t-1)]^{\mathrm{T}},$$

$$\hat{\boldsymbol{M}}(t) = \begin{bmatrix} \mathbf{0} & \mathbf{0} & \cdots & \mathbf{0} & \mathbf{0} \\ \hat{\boldsymbol{b}}_1(t) & \mathbf{0} & \cdots & \mathbf{0} & \mathbf{0} \\ \hat{\boldsymbol{b}}_2(t) & \hat{\boldsymbol{b}}_1(t) & \ddots & \vdots & \vdots \\ \vdots & \vdots & \ddots & \mathbf{0} & \mathbf{0} \\ \hat{\boldsymbol{b}}_{n-1(t)} & \hat{\boldsymbol{b}}_{n-2}(t) & \cdots & \hat{\boldsymbol{b}}_1(t) & \mathbf{0} \end{bmatrix},$$

$$\hat{\theta}_b(t) = \begin{bmatrix} \hat{\boldsymbol{b}}_1^{\mathrm{T}}(t) & \hat{\boldsymbol{b}}_2^{\mathrm{T}}(t) & \cdots & \hat{\boldsymbol{b}}_{n-1}^{\mathrm{T}}(t) & \hat{\boldsymbol{b}}_n^{\mathrm{T}}(t) \\ \mathbf{0} & \hat{\boldsymbol{b}}_1^{\mathrm{T}}(t) & \cdots & \hat{\boldsymbol{b}}_{n-2}^{\mathrm{T}}(t) & \hat{\boldsymbol{b}}_{n-1}^{\mathrm{T}}(t) \\ \vdots & \vdots & & \vdots & \vdots \\ \mathbf{0} & \mathbf{0} & \cdots & \hat{\boldsymbol{b}}_1^{\mathrm{T}}(t) & \hat{\boldsymbol{b}}_2^{\mathrm{T}}(t) \\ \mathbf{0} & \mathbf{0} & \cdots & \mathbf{0} & \hat{\boldsymbol{b}}_1^{\mathrm{T}}(t) \end{bmatrix} \begin{bmatrix} \hat{a}_{n-1}(t) \\ \hat{a}_{n-2}(t) \\ \vdots \\ \hat{a}_1(t) \\ 1 \end{bmatrix}$$

$$\hat{\boldsymbol{Q}}(t) = \begin{bmatrix} 0 & 0 & \cdots & 0 & 0 \\ \hat{f}_1(t) & 0 & \cdots & 0 & 0 \\ \hat{f}_2(t) & \hat{f}_1(t) & \ddots & \vdots & \vdots \\ \vdots & \vdots & \ddots & 0 & 0 \\ \hat{f}_{n-1}(t) & \hat{f}_{n-2}(t) & \cdots & \hat{f}_1(t) & 0 \end{bmatrix}$$

$$\hat{\theta}_c(t) = \begin{bmatrix} \hat{a}_{n-1}(t) & \hat{a}_{n-2}(t) & \cdots & \hat{a}_1(t) & 1 \\ \hat{a}_{n-2}(t) & \hat{a}_{n-3}(t) & \cdots & 1 & 0 \\ \vdots & \vdots & & \vdots & \vdots \\ \hat{a}_1(t) & 1 & \cdots & 0 & 0 \\ 1 & 0 & \cdots & 0 & 0 \end{bmatrix} \begin{bmatrix} \hat{f}_1(t) \\ \hat{f}_2(t) \\ \vdots \\ \hat{f}_{n-1}(t) \\ \hat{f}_n(t) \end{bmatrix}$$

$$\hat{\boldsymbol{\theta}}(t) = \begin{bmatrix} \hat{\theta}_a(t) \\ \hat{\theta}_b(t) \\ \hat{\theta}_c(t) \end{bmatrix},$$

$$\hat{\boldsymbol{\theta}}_a(t) = [-\hat{a}_n(t), -\hat{a}_{n-1}(t),\ \cdots, -\hat{a}_1(t)]^{\mathrm{T}}.$$

Let $t = 1$, and set the initial values $\hat{\theta}(0) = 1_n/p_0$, $P(0) = p_0 I$, $p_0 = 10^6$, $u(i) = 0$, $y(i) = 0$, and $\hat{v}(i) = 1/p_0$ for $i \leq 0$.

### 2.1.4 Example

Consider the following state-space system with two-step state delay:

$$\boldsymbol{x}(t+1) = \begin{bmatrix} 0 & 1 \\ -0.25 & 0.35 \end{bmatrix} \boldsymbol{x}(t) + \begin{bmatrix} 0.30 & -0.20 \\ 0.54 & -0.22 \end{bmatrix} \boldsymbol{x}(t-2) + \begin{bmatrix} 0.59 \\ -1.22 \end{bmatrix} u(t),$$
$$y(t) = [1, \ 0]\boldsymbol{x}(t) + v(t).$$

The parameter vector to be identified is

$$\boldsymbol{\theta} = [\theta_1, \ \theta_2, \ \theta_3, \ \theta_4, \ \theta_5, \ \theta_6, \ \theta_7, \ \theta_8]^{\mathrm{T}}$$
$$[-0.2500, \ 0.3500, \ 0.4350, -0.1500, \ 0.3000, -0.2000, -1.4265, \ 0.5900]^{\mathrm{T}}$$

In simulation, the input $\{u(t)\}$ is taken as an uncorrelated persistent excitation signal sequence with zero mean and unit variance and $\{v(t)\}$ as a white noise sequence with zero mean and variance $\sigma^2 = 0.50^2$. Apply the least squares parameter estimation algorithm to estimate the parameter vector $\theta$ and the state estimation algorithm to estimate the state vector $x(t)$. The parameter estimates and their estimation errors are shown in Table 2.1, the parameter estimation errors $\delta$ versus $t$ are shown in Fig. 2.1, where $\delta := ||\hat{\theta}(t) - \theta||/||\theta||$, and the state estimates $\hat{x}_1(t)$ and $\hat{x}_2(t)$ versus $t$ are shown in Figs. 2.2–2.3.

From Table 2.1 and Figs. 2.1–2.3, we can draw the following conclusions: (1) the parameter estimation errors become smaller (in general) with the increasing of $t$—see Fig. 2.1; (2) the parameter estimation accuracy becomes higher as the data length $t$ increases $t$—see Table 2.1; (3) the state estimates are close to their true values with the increasing of $tt$—see Figs. 2.2–2.3.

### 2.1.5 Conclusions

This chapter discusses the identification algorithm for estimating the parameters and states for linear systems with d-step state delay based on the canonical state-space model. Compared with the recursive least squares methods, the proposed algorithms can identify the system parameters and the system states. A simulation example is given to show the effectiveness of the proposed algorithm. The convergence properties of the proposed algorithm can be analyzed by means of the martingale convergence theorem.

**Table 2.1** The parameter estimates and errors.

| $t$ | $\theta_1$ | $\theta_2$ | $\theta_3$ | $\theta_4$ | $\theta_5$ | $\theta_6$ | $\theta_7$ | $\theta_8$ | $\delta$(%) |
|---|---|---|---|---|---|---|---|---|---|
| 100 | −0.24418 | 0.33802 | 0.33341 | −0.00796 | 0.18324 | −0.25618 | −1.50557 | 0.69864 | 15.00531 |
| 200 | −0.25958 | 0.33339 | 0.36616 | −0.01103 | 0.16423 | −0.22221 | −1.48158 | 0.61122 | 12.68607 |
| 500 | −0.26533 | 0.32952 | 0.39244 | −0.05012 | 0.18138 | −0.19590 | −1.44998 | 0.59830 | 9.65947 |
| 1000 | −0.21932 | 0.35032 | 0.42307 | −0.11188 | 0.26864 | −0.22844 | −1.43309 | 0.59095 | 3.87713 |
| 2000 | −0.23665 | 0.35870 | 0.43495 | −0.11612 | 0.27028 | −0.21376 | −1.43377 | 0.58328 | 2.97348 |
| 3000 | −0.24387 | 0.35705 | 0.43563 | −0.13716 | 0.28801 | −0.20094 | −1.43226 | 0.57976 | 1.35633 |
| True values | −0.25000 | 0.35000 | 0.43500 | −0.15000 | 0.30000 | −0.20000 | −1.42650 | 0.59000 | |

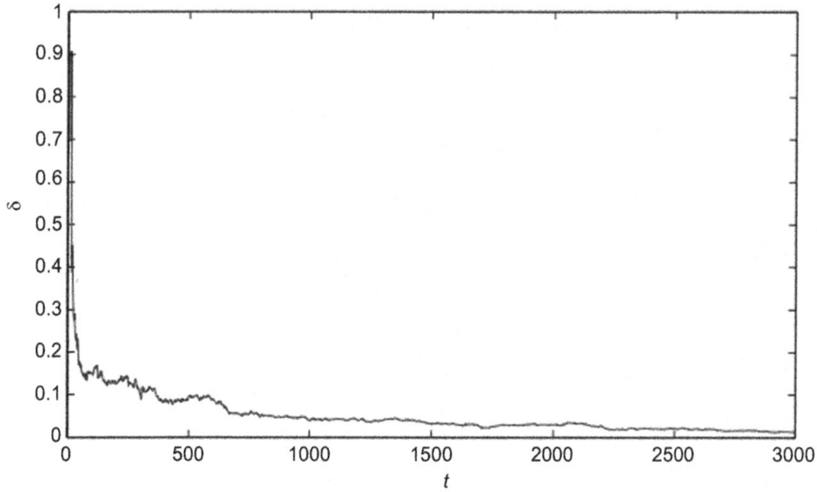


**Figure 2.1** The parameter estimation errors $\delta$ versus $t$.

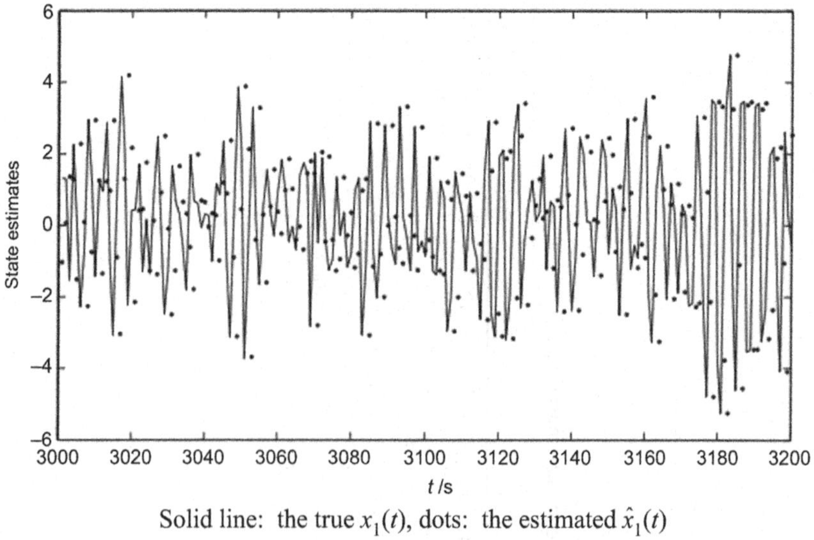


**Figure 2.2** The state and state estimate $\hat{x}_1(t)$ versus $t$.

## 2.2 Identification and U-control of dual-rate state-space models with d-step state delay

The mathematical model can represent the basic features of the system, and system identification applies the statistical method to set up the

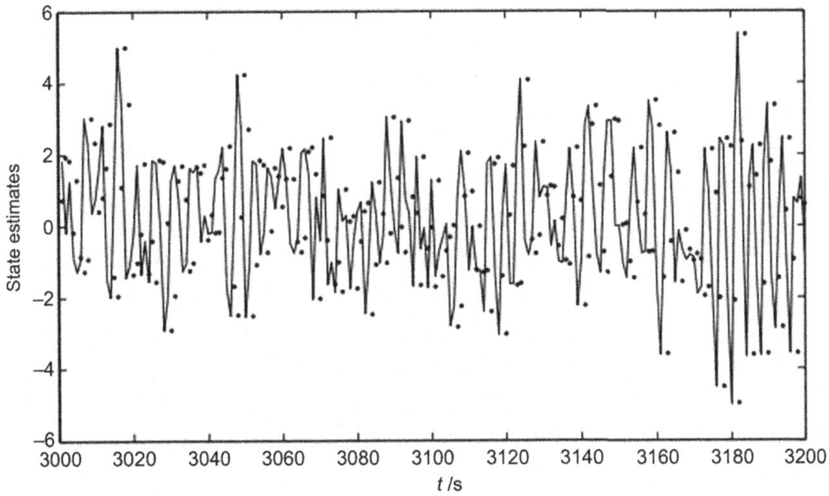


Solid line: the true $x_2(t)$, dots: the estimated $\hat{x}_2(t)$

**Figure 2.3** The state and state estimate $\hat{x}_2(t)$ versus $t$.

mathematical models of dynamic systems from available data (Ko, 2012). There exist some identification methods for a state-space model with d-step state delay, such as the recursive least squares (RLS) algorithm and the stochastic gradient (SG) algorithm. The SG algorithm is used in adaptive control because of its small computation.

In system identification, some algorithms concentrate on reducing the calculation amount and improving the recognition accuracy (Fattah et al., 2011). Since the gradient optimization method only requires calculating the first-order derivative, its computation is small, but the calculation accuracy of the gradient algorithm is low. In order to improve the calculation accuracy, many improved gradient algorithms have been proposed. Although these improved algorithms can improve the parameter estimation accuracy, the computational complexity is large. The innovation is an effective information to improve the parameter estimation accuracy. It can promote the convergence of the algorithm in the recursive process. In order to improve the estimation accuracy through using more innovation, the multi-innovation theory has been used in system recognition.

The stabilities and identification of time-delay systems have drawn a great deal of attention from many researchers in system control and system analysis. In industrial processes and control systems, time delays are difficult to avoid due to material transmission and signal interruptions.

The time delay makes it difficult for the control system to respond to the input changes in time. In addition, the time delay can cause instability and unsatisfactory performance of the controlled process. The recognition of time-delay systems has been a hot topic.

Since a foundation study, U-model based control, U-control in short, has received a certain range of attention, even not greatly. Regarding the U-control approach, it is different from those model-based design approaches in essence of separating system performance from plant models and also different from data-driven design approaches as using the plant models to determine the controller output by solving the dynamic inverse of the plant models. The merit U-control claim is that the control design on this platform is no longer classified as linear/nonlinear, polynomial/state-space model structures; second, U-control provides great simplicity/generality using linear control system design principle to nonlinear control systems, especially in specifying the system transient response and steady-state performance in a systematic formulation.

For the U-control progression, polynomial including ration (total nonlinear) model-based design has been predominantly studied with pole placement, general predictive control, and neuro control, and U-Smith predictor enhanced control with input delay. A comprehensive review of U-control has been reported. It has been noted that the critical challenge in U-control is the uncertainty in dynamic inversion. This study expands the U-control to a class of linear state-space models with state delay. Except for a recent work using U-control for dealing with nonlinear systems with input delay, U-control has not been introduced for dealing with systems with state delay.

The main contributions of the study are outlined below.

1. The input–output representation is derived from a canonical state-space model of the state-delay system for the identification through eliminating the state variables in the systems.
2. To derive a joint parameter and state estimation algorithm by means of the multi-innovation identification theory and the state observer for reducing the computational burden and improving the parameter estimation accuracy and the convergence speed.
3. U-control expansion to state-space model with state delay, which considers the problem by dynamic inversion, is different from the predictor-based and the other popular approaches.
4. Provides a simulation portfolio with model identification and U-control system design, which could be an integrated package for potential users with their ad hoc applications.

### 2.2.1 The canonical state-space model for state-delay systems

Let us introduce some symbols. The relation $A:=X$ or $X:=A$ means that $A$ is defined as $X$; $I(I_n)$ stands for an identity matrix of an appropriate size $(n \times n)$; $z$ denotes a unit forward shift operator: $zx(k)=x(k+1)$; T represents the matrix/vector transpose; $\hat{\theta}(k)$ is the estimate of $\theta$ at time $k$; $1_n$ means that an $n \times 1$ vector whose elements are all unity; E denotes the expectation operator; $adj[X]$ stands for the adjoint matrix of the square matrix $X$: $adj[X]=\det[X]X^{-1}$; and $\det[X]=|X|$ represents the determinant of the square matrix $X$.

Consider the following state-space system with d-step state delay,

$$\boldsymbol{x}(k+1)=\boldsymbol{A}\boldsymbol{x}(k)+\boldsymbol{B}\boldsymbol{x}(k-d)+\boldsymbol{f}u(k), \tag{2.23}$$

$$y(k)=\boldsymbol{c}\boldsymbol{x}(k)+w(k), \tag{2.24}$$

$$w(k)=\sum^{n_g} g_r v(k-r)+v(k), \tag{2.25}$$

$$\boldsymbol{A}:=\begin{bmatrix} 0 & 1 & 0 & \cdots & 0 \\ 0 & 0 & 1 & \ddots & \vdots \\ \vdots & \vdots & \ddots & \ddots & 0 \\ 0 & 0 & \cdots & 0 & 1 \\ a_n & a_{n-1} & a_{n-2} & \cdots & a_1 \end{bmatrix} \in \mathbb{R}^{n \times n},$$

$$\boldsymbol{B}:=\begin{bmatrix} \boldsymbol{b}_1 \\ \boldsymbol{b}_2 \\ \vdots \\ \boldsymbol{b}_n \end{bmatrix} \in \mathbb{R}^{n \times n}, \quad \boldsymbol{b}_i \in \mathbb{R}^{1 \times n},$$

$$\boldsymbol{f}:=[f_1, f_2, \quad \cdots, \quad f_n]^{\mathrm{T}} \in \mathbb{R}^n,$$

$$\boldsymbol{c}:=[1, \quad 0, \quad 0, \quad \cdots, \quad 0] \in \mathbb{R}^{1 \times n},$$

where $x(k) \in R^n$ is the system state vector, $u(k) \in R$ is the system input, $y(k) \in R$ is the system output, $v(k) \in R$ is a random noise with zero mean, and $A \in R^{n \times n}$, $B \in R^{n \times n}$, $f \in R^n$, and $c \in R^{1 \times n}$ are the system parameter matrices/vectors. Assume that $(c, A)$ is observable and $u(k)=0$ and $y(k)=0$ for $k \leq 0$. The system matrices/vector $A$, $B$, and $f$ are the unknown parameters to be estimated from the input–output data $\{u(k), y(k)\}$. If we remove $Bx(k-d)$ in Eq. (2.23), then it becomes the conventional standard state-space model.

**Remark 1**: For the system in (2.23) and (2.24), if the state vector $x(k)$ is known, the system matrix/vector $(A, b)$ is easy to identify. This chapter considers the case that the state $x(k)$ is completely unavailable. The objective is to propose new methods for jointly estimating the unknown states and parameters from the measurement data $\{u(k), y(k):k=1,2,\ldots\}$ and to study the performance of the proposed methods.

## 2.2.2 The identification model

This section derives the identification model of the canonical state-space model in (2.23–2.24). From (2.23), we have

$$x_i(k+1) = x_{i+1}(k) + \boldsymbol{b}_i\boldsymbol{x}(k-d) + f_i u(k), \tag{2.26}$$

$$i = 1,\ 2, \ldots, n-1,$$

$$x_n(k+1) = a_n x_1(k) + a_{n-1}x_2(k) + \cdots + a_1 x_n(k) + \boldsymbol{b}_n\boldsymbol{x}(k-d) + f_n u(k). \tag{2.27}$$

Let $a:=[a_n, a_{n-1}, \cdots, a_1]^{\mathrm{T}} \in R^n$, and using the properties of the shift operator $z$, multiplying Eq. (2.26) by $z^{-i}$ and (2.27) by $z^{-n}$, and adding all expressions give

$$\begin{aligned} x_1(k) &= \boldsymbol{a}\boldsymbol{x}(k-n) + \boldsymbol{b}_1\boldsymbol{x}(k-d-1) + \boldsymbol{b}_2\boldsymbol{x}(k-d-2) \\ &\quad + \cdots + \boldsymbol{b}_{n-1}\boldsymbol{x}(k-d-n+1) + \boldsymbol{b}_n\boldsymbol{x}(k-d-n) \\ &\quad + f_1u(k-1) + f_2u(k-2) + \cdots + f_nu(k-n) \\ &= \boldsymbol{a}\boldsymbol{x}(k-n) + \sum_{i=1}^{n}\boldsymbol{b}_i\boldsymbol{x}(k-d-i) + \sum_{i=1}^{n} f_iu(k-i). \end{aligned} \tag{2.28}$$

When $d \leq n-1$, define the information vector $\phi(k)$ and the parameter vector $\theta$:

$$\varphi(k) := \left[\varphi_1^{\mathrm{T}}(k),\ \ \varphi_n^{\mathrm{T}}(k)\right]^{\mathrm{T}},$$

$$\varphi_1(k) := \left[\boldsymbol{x}^{\mathrm{T}}(k-d-1),\ \ \boldsymbol{x}^{\mathrm{T}}(k-d-2),\ \ \cdots,\ \ \boldsymbol{x}^{\mathrm{T}}(k-n),\right.$$

$$\cdots,\ \ \boldsymbol{x}^{\mathrm{T}}(k-n-d),\ \ u(k-1),\ \ u(k-2),$$

$$\left.\cdots,\ \ u(k-n)\right]^{\mathrm{T}} \in \mathbb{R}^{n^2+n},$$

$$\varphi_n(k):=\left[v(k-1),\ \ v(k-2),\ldots,v(k-n_g)\right]^{\mathrm{T}}\in\mathbb{R}^{n_g},$$
$$\boldsymbol{\theta}:=\left[\boldsymbol{\theta}_1^{\mathrm{T}},\ \ \boldsymbol{\theta}_n^{\mathrm{T}}\right]^{\mathrm{T}},$$
$$\boldsymbol{\theta}_1:=\left[\boldsymbol{b}_1,\ \ \boldsymbol{b}_2,\cdots,\boldsymbol{a}+\boldsymbol{b}_{n-d},\cdots,\boldsymbol{b}_n,\ \ \boldsymbol{f}^{\mathrm{T}}\right]^{\mathrm{T}}\in\mathbb{R}^{n^2+n},$$
$$\boldsymbol{\theta}_n:=\left[g_1,\ \ g_2,\cdots,g_{n_g})\right]^{\mathrm{T}}\in\mathbb{R}^{n_g}.$$

When $d\geq n$, define the information vector $\phi(k)$ and the parameter vector $\theta$:

$$\varphi(k):\left[\varphi_1^{\mathrm{T}}(k),\ \ \varphi_n^{\mathrm{T}}(k)\right]^{\mathrm{T}},$$

$$\varphi_1(k):=\left[\boldsymbol{x}^{\mathrm{T}}(k-n),\ \ \boldsymbol{x}^{\mathrm{T}}(k-d-1),\ \ \boldsymbol{x}^{\mathrm{T}}(k-d-2),\ \ \cdots,\right.$$

$$\boldsymbol{x}^{\mathrm{T}}(k-d-n+1),\ \ \boldsymbol{x}^{\mathrm{T}}(k-d-n),$$

$$u(k-1),\ \ u(k-2),\cdots,u(k-n)]^{\mathrm{T}}\in\mathbb{R}^{n^2+2n},$$

$$\varphi_n(k):=\left[v(k-1),\ \ v(k-2),\cdots,v(k-n_g)\right]^{\mathrm{T}}\in\mathbb{R}^{n_g},$$
$$\boldsymbol{\theta}:=\left[\boldsymbol{\theta}_1^{\mathrm{T}},\ \ \boldsymbol{\theta}_n^{\mathrm{T}}\right]^{\mathrm{T}},$$
$$\boldsymbol{\theta}_1:=\left[\boldsymbol{a},\ \ \boldsymbol{b}_1,\ \ \boldsymbol{b}_2,\cdots,\boldsymbol{b}_{n-1},\ \ \boldsymbol{b}_n,\ \ \boldsymbol{f}^{\mathrm{T}}\right]^{\mathrm{T}}\in\mathbb{R}^{n^2+2n},$$
$$\boldsymbol{\theta}_n:=\left[g_1,\ \ g_2,\cdots,g_{n_g}\right]^{\mathrm{T}}\in\mathbb{R}^{n_g}.$$

From (2.24) and (2.28), we have

$$y(k)=x_1(k)+w(k)=\boldsymbol{\varphi}^{\mathrm{T}}(k)\boldsymbol{\theta}+w(k). \tag{2.29}$$

In the dual-rate scheme, the observed output is sampled by the sampler, and the sampling period is multiple of the input retention period. Assume that the sampling interval is $\tau$ ($\tau\geq 2$ is an integer), thus the measured input–output data are $\{u(k){:}k=0,1,2,\ldots\}$ at the fast rate and $\{y(k\tau){:}k=0,1,2,\ldots\}$ at the slow rate. Replacing $k$ in (2.7) with $k\tau$ gives

$$y(k\tau)=x_1(k\tau)+w(k\tau)=\boldsymbol{\varphi}^{\mathrm{T}}(k\tau)\boldsymbol{\theta}+w(k\tau). \tag{2.30}$$

**Remark 2**: This is the identification model of the dual-rate state-space system with d-step state delay. The information vector $\phi(k\tau)$ consists of the state vector $x(k\tau-i)$, the input $u(k\tau-i)$, and the correlated noise $\omega(k\tau-i)$, and the parameter vector $\theta$ consists of the parameters $a_i$, $b_i$, and $f_i$ of the state-space model in (2.23–2.24).

**Remark 3**: In what follows, a SG algorithm is derived for the state-space system with colored noise. Furthermore, a MISG algorithm is presented to reduce the computational burden and enhance the parameter estimation accuracy. A simulation example is provided to evaluate the estimation accuracy and the computational efficiency of the proposed algorithms.

### 2.2.3 The parameter and state estimation algorithm

This section derives a multi-innovation stochastic gradient algorithm to estimate the parameter vector $\theta$ in (2.30) and uses the observer to estimate the state vector $x(k\tau)$ of the system.

*The SG algorithm*

Defining and minimizing the cost function,

$$J(\boldsymbol{\theta}):=\mathrm{E}\left[||y(k\tau)-\varphi^{\mathrm{T}}(k\tau)\boldsymbol{\theta}||^2\right],$$

and using the gradient search principle, we may obtain a stochastic gradient algorithm:

$$\hat{\boldsymbol{\theta}}(k\tau)=\hat{\boldsymbol{\theta}}(k\tau-\tau)+\frac{\varphi(k\tau)}{r(k\tau)}\left[y(k\tau)-\varphi^{\mathrm{T}}(k\tau)\hat{\boldsymbol{\theta}}(k\tau-\tau)\right], \tag{2.31}$$

$$r(k\tau)=r(k\tau-\tau)+||\varphi^{\mathrm{T}}(k\tau)||^2,\quad r(0)=1 \tag{2.32}$$

Here, $1/r(k\tau)$ is the step size or convergence factor. The choice of $r(k\tau)$ guarantees that the parameter estimation error converges to zero. However, difficulties arise in that the information vector $\phi(k\tau)$ contains the unknown state vector $x(k\tau-i)(i=1+d,2+d,\ldots,n+d)$, the SG algorithm in (2.31) and (2.32) cannot compute the estimate of $\theta$ in (2.30). The approach here is to replace the unknown $x(k\tau-i)$ in $\phi(k\tau)$ with its $\hat{x}(k\tau-i)$. Based on the identification model in (2.30), we can obtain the following stochastic gradient parameter estimation algorithm for estimating $\theta$:

$$\hat{\boldsymbol{\theta}}(k\tau)=\hat{\boldsymbol{\theta}}(k\tau-\tau)+\frac{\hat{\varphi}(k\tau)}{r(k\tau)}\left[y(k\tau)-\hat{\varphi}^{\mathrm{T}}(k\tau)\hat{\boldsymbol{\theta}}(k\tau-\tau)\right], \tag{2.33}$$

$$r(k\tau)=r(k\tau-\tau)+||\hat{\varphi}(k\tau)||^2,\quad r(0)=1, \tag{2.34}$$

when $d\leq n-1$

$$\hat{\varphi}(k\tau)=\left[\hat{\varphi}_1^{\mathrm{T}}(k\tau),\ \hat{\varphi}_n^{\mathrm{T}}(k\tau)\right]^{\mathrm{T}}$$

$$\hat{\varphi}_1(k\tau) = \left[\hat{\boldsymbol{x}}^{\mathrm{T}}(k\tau - d - 1),\ \ \hat{\boldsymbol{x}}^{\mathrm{T}}(k - d - 2),\ \ \cdots,\right.$$
$$\hat{\boldsymbol{x}}^{\mathrm{T}}(k\tau - n), \cdots, \hat{\boldsymbol{x}}^{\mathrm{T}}(k\tau - n - d),$$
$$u(k\tau - 1),\ \ u(k\tau - 2), \cdots, u(k\tau - n)]^{\mathrm{T}},$$

$$\hat{\boldsymbol{\varphi}}_n(k\tau) = \left[\hat{v}(k\tau - 1),\ \ \hat{v}(k\tau - 2), \cdots, \hat{v}(k\tau - n_g)\right]^{\mathrm{T}} \in \mathbb{R}^{n_g},$$

when $d \geq n$

$$\hat{\boldsymbol{\varphi}}(k\tau) = \left[\hat{\varphi}_1^{\mathrm{T}}(k\tau),\ \ \hat{\boldsymbol{\varphi}}_n^{\mathrm{T}}(k\tau)\right]^{\mathrm{T}},$$

$$\hat{\boldsymbol{\varphi}}_1(k\tau) = \left[\hat{\boldsymbol{x}}^{\mathrm{T}}(k\tau - n),\ \ \hat{\boldsymbol{x}}^{\mathrm{T}}(k\tau - d - 1),\ \ \hat{\boldsymbol{x}}^{\mathrm{T}}(k\tau - d - 2),\right.$$

$$\cdots,\ \ \hat{\boldsymbol{x}}^{\mathrm{T}}(k\tau - d - n + 1),\ \ \hat{\boldsymbol{x}}^{\mathrm{T}}(k\tau - d - n),$$

$$u(k\tau - 1),\ \ u(k\tau - 2), \cdots, u(k\tau - n)]^{\mathrm{T}},$$

$$\hat{\boldsymbol{\varphi}}_n(k\tau) = \left[\hat{v}(k\tau - 1),\ \ \hat{v}(k\tau - 2), \cdots, \hat{v}(k\tau - n_g)\right]^{\mathrm{T}} \in \mathbb{R}^{n_g}.$$

*The MISG algorithm*

In order to improve the accuracy of the SG algorithm, we extend the SG algorithm and derive a multi-innovation stochastic gradient algorithm by expanding the innovation length.

Define an innovation vector:

$$\boldsymbol{E}(p,\ \ k\tau) = \begin{bmatrix} e(k\tau) \\ e(k\tau - 1) \\ \vdots \\ e(k\tau - p + 1) \end{bmatrix} \in \mathbb{R}^p,$$

where the positive integer $p$ represents the innovation length, and $e(k\tau - i) = y(k\tau - i) - \hat{\boldsymbol{\varphi}}^{\mathrm{T}}(k\tau - i)\hat{\boldsymbol{\theta}}(k\tau - i - \tau)$ In general, one may think that the estimate $\hat{\theta}(k\tau - \tau)$ is closer to $\theta$ than $\hat{\theta}(k\tau - i)$ at time $k\tau - i(i = 2, 3, 4, \ldots, p - 1)$. Thus, the innovation vector is taken more reasonably to be

$$\boldsymbol{E}(p,\ \ k\tau) = \begin{bmatrix} y(k\tau) - \hat{\boldsymbol{\varphi}}^{\mathrm{T}}(k\tau)\hat{\boldsymbol{\theta}}(k\tau - \tau) \\ y(k\tau - 1) - \hat{\boldsymbol{\varphi}}^{\mathrm{T}}(k\tau - 1)\hat{\boldsymbol{\theta}}(k\tau - \tau) \\ \vdots \\ y(k\tau - p + 1) - \hat{\boldsymbol{\varphi}}^{\mathrm{T}}(k\tau - p + 1)\hat{\boldsymbol{\theta}}(k\tau - \tau) \end{bmatrix} \in \mathbb{R}^p,$$

Defining the information matrix $\hat{\Phi}(p, k\tau)$ and stacking output vector $Y(p, k\tau)$ as

$$\hat{\boldsymbol{\Phi}}(p,\ k\tau) := \hat{\varphi}(k\tau),\ \hat{\varphi}(k\tau-1),\ \cdots,$$
$$\hat{\varphi}(k\tau-p+1) \in \mathbb{R}^{(n^2+n+n_g)\times p},$$

$$\boldsymbol{Y}(p,\ k\tau) := \left[y(k\tau),\ y(k\tau-1), \cdots, y(k\tau-p+1)\right]^{\mathrm{T}} \in \mathbb{R}^{p},$$

the innovation vector $E(p, k\tau)$ can be equivalently expressed as

$$\boldsymbol{E}(p, k\tau) = \boldsymbol{Y}(p, k\tau) - \hat{\boldsymbol{\Phi}}^{\mathrm{T}}(p, k\tau)\hat{\boldsymbol{\theta}}(k\tau-\tau).$$

Furthermore, we can obtain the following multi-innovation stochastic gradient algorithm with the innovation length $p$:

$$\hat{\boldsymbol{\theta}}(k\tau) = \hat{\boldsymbol{\theta}}(k\tau-\tau) + \frac{\hat{\boldsymbol{\Phi}}(p,\ k\tau)}{r(k\tau)}\boldsymbol{E}(p,\ k\tau) \tag{2.35}$$

$$\mathrm{E}(p, k\tau) = \boldsymbol{Y}(p,\ k\tau) - \hat{\boldsymbol{\Phi}}^{\mathrm{T}}(p,\ k\tau)\hat{\boldsymbol{\theta}}(k\tau-\tau), \tag{2.36}$$

$$r(k\tau) = r(k\tau-\tau) + ||\hat{\varphi}^{\mathrm{T}}(p,\ k\tau)||^2,\ r(0) = 1, \tag{2.37}$$

$$\boldsymbol{Y}(p,\ k\tau) = \left[y(k\tau),\ y(k\tau-1),\ \ldots,\ y(k\tau-p+1)\right]^{\mathrm{T}}, \tag{2.38}$$

$$\hat{\boldsymbol{\Phi}}(p,\ k\tau) = \left[\hat{\varphi}(k\tau),\ \hat{\varphi}(k\tau-1),\ \cdots,\ \hat{\varphi}(k\tau-p+1)\right], \tag{2.39}$$

when $d \leq n-1$

$$\hat{\varphi}(k\tau) = \left[\hat{\varphi}_1^{\mathrm{T}}(k\tau),\ \hat{\varphi}_n^{\mathrm{T}}(k\tau)\right]^{\mathrm{T}}, \tag{2.40}$$

$$\hat{\varphi}_1(k\tau) = \left[\hat{\boldsymbol{x}}^{\mathrm{T}}(k\tau-d-1),\ \hat{\boldsymbol{x}}^{\mathrm{T}}(k\tau-d-2),\ \cdots,\right.$$

$$\hat{\boldsymbol{x}}^{\mathrm{T}}(k\tau-n),\ \cdots,\ \hat{\boldsymbol{x}}^{\mathrm{T}}(k\tau-n-d),$$

$$u(k\tau-1),\ u(k\tau-2), \cdots, u(k\tau-n)]^{\mathrm{T}}, \tag{2.41}$$

$$\hat{\varphi}_n(k\tau) = \left[\hat{v}(k\tau-1),\ \hat{v}(k\tau-2), \cdots, \hat{v}(k\tau-n_g)\right]^{\mathrm{T}} \in \mathbb{R}^{n_g}, \tag{2.42}$$

when $d \geq n$

$$\hat{\varphi}(k\tau) = \left[\hat{\varphi}_1^{\mathrm{T}}(k\tau),\ \hat{\varphi}_n^{\mathrm{T}}(k\tau)\right]^{\mathrm{T}}, \tag{2.43}$$

$$\begin{aligned}\hat{\varphi}_1(k\tau) = \big[&\hat{\boldsymbol{x}}^{\mathrm{T}}(k\tau - n),\ \hat{\boldsymbol{x}}^{\mathrm{T}}(k\tau - d - 1),\ \hat{\boldsymbol{x}}^{\mathrm{T}}(k\tau - d - 2)\cdots,\\ &\hat{\boldsymbol{x}}^{\mathrm{T}}(k\tau - d - n + 1),\ \hat{\boldsymbol{x}}^{\mathrm{T}}(k\tau - n - d),\\ &u(k\tau - 1),\ u(k\tau - 2),\cdots,u(k\tau - n)\big]^{\mathrm{T}},\end{aligned} \tag{2.44}$$

$$\hat{\varphi}_n(k\tau) = \left[\hat{v}(k\tau - 1),\ \hat{v}(k\tau - 2),\cdots,\hat{v}(k\tau - n_g)\right]^{\mathrm{T}} \in \mathbb{R}^{n_g}, \tag{2.45}$$

When the innovation length $p = 1$, the MISG algorithm degrades to the SG algorithm.

**Theorem 1**: For the system (2.30) and the MISG algorithm (2.35)–(2.45), assume that the system noise $v(k\tau)$ is random noise with zero mean and variance $\sigma^2$ that is uncorrelated with the input $u(k\tau)$, and the mean square is bounded, that is, (A1) $E[v(k\tau)] = 0$, $E[||v(k\tau)||^2] = \sigma^2(k\tau) \leq \sigma^2$; the existence of constant $0 < \alpha \leq \beta < \infty$ and integer $p \geq n$ makes the following condition: (A2) $\alpha I \leq \frac{1}{p}\sum_{i=1}^{p} \phi(k\tau - i + 1)\phi^{\mathrm{T}}(k\tau - i + 1) \leq \beta I$, *a.s.*, $k\tau \geq p$. Then, the mean square error of the parameter estimation error given by MISG algorithm is bounded, that is, $\limsup_{k\tau \to \infty} E\left[||\hat{\theta}(k\tau) - \theta||^2\right] \leq \frac{4\beta^3\sigma^2}{\alpha^4}$.

## 2.2.4 The state estimation algorithm

Using the parameter estimation vector $\hat{\theta}(k\tau)$ to form the system matrices/vector $\hat{A}(k\tau)$, $\hat{B}(k\tau)$, and $\hat{f}(k\tau)$ and based on the canonical state-space model in (2.23)–(2.24), we can use the following observer to estimate the state vector $x(k\tau)$:

$$\hat{\boldsymbol{x}}(k\tau + \tau) = \hat{\boldsymbol{A}}(k\tau)\hat{\boldsymbol{x}}(k\tau) + \hat{\boldsymbol{B}}(k\tau)\hat{\boldsymbol{x}}(k\tau - d\tau) + \hat{\boldsymbol{f}}(k\tau)u(k\tau), \tag{2.46}$$

$$\hat{A}(k\tau) = \begin{bmatrix} 0 & 1 & \cdots & 0 \\ 0 & 0 & \ddots & \vdots \\ \vdots & \vdots & \ddots & 0 \\ 0 & 0 & 0 & 1 \\ \hat{a}_n(k\tau) & \hat{a}_{n-1}(k\tau) & \cdots & \hat{a}_n(k\tau) \end{bmatrix} \tag{2.47}$$

$$\hat{\boldsymbol{B}}(k\tau) = \begin{bmatrix} \hat{\boldsymbol{b}}_1(k\tau) \\ \hat{\boldsymbol{b}}_2(k\tau) \\ \vdots \\ \hat{\boldsymbol{b}}_n(k\tau) \end{bmatrix},\ \hat{\boldsymbol{f}}(k\tau) = \begin{bmatrix} \hat{f}_1(k\tau) \\ \hat{f}_2(k\tau) \\ \vdots \\ \hat{f}_n(k\tau) \end{bmatrix}, \tag{2.48}$$

$$\hat{\boldsymbol{\theta}}(k\tau) = \left[\hat{\boldsymbol{b}}_1(k\tau),\ \hat{\boldsymbol{b}}_2(k\tau), \cdots, \hat{\boldsymbol{a}}(k\tau) + \hat{\boldsymbol{b}}_{n-d}(k\tau), \cdots, \hat{\boldsymbol{b}}_n(k\tau),\ \hat{\boldsymbol{f}}^{\mathrm{T}}(k\tau)\right]^{\mathrm{T}},\ d \le n-1, \tag{2.49}$$

$$\hat{\boldsymbol{\theta}}(k\tau) = \left[\hat{\boldsymbol{a}}(k\tau),\ \hat{\boldsymbol{b}}_1(k\tau),\ \hat{\boldsymbol{b}}_2(k\tau), \cdots, \hat{\boldsymbol{b}}_{n-1}(k\tau),\ \hat{\boldsymbol{b}}_n(k\tau),\ \hat{\boldsymbol{f}}^{\mathrm{T}}(k\tau)\right]^{\mathrm{T}},\ d \ge n. \tag{2.50}$$

The proposed algorithms in this chapter can combine some mathematical tools and other identification methods to investigate the parameter identification methods of other linear and nonlinear systems and can be applied to other literatures such as engineering application systems.

The steps of computing the parameter estimate $\hat{\theta}(k\tau)$ in (2.35)–(2.39) and the state estimate $\hat{x}(k\tau+\tau)$ in (2.46)–(2.50) are listed in the following.

1. Let $k=1$, and set the initial values $\hat{\theta}(0) = 1_{n^2+2n}/p_0$, $r(0)=1$, $p_0 = 10^6$.
2. Collect the input–output data $u(k\tau)$ and $y(k\tau)$, and form $\hat{\phi}(k\tau)$ by (2.40) or (2.43), $\hat{\Phi}(p, k\tau)$ by (2.39), and $Y(p, k\tau)$ by (2.38), respectively.
3. Compute $E(p, k\tau)$ by (2.36) and $r(k\tau)$ by (2.37).
4. Update the parameter estimation vector $\hat{\theta}(k\tau)$ by (2.35).
5. Read $\hat{a}_i(k\tau)$, $\hat{b}_i(k\tau)$, and $\hat{f}(k\tau)$ from $\hat{\theta}(k\tau)$ according to the definition of $\hat{\theta}(k\tau)$.
6. Form $\hat{A}(k\tau)$ and $\hat{B}(k\tau)$ by (2.47) and (2.48).
7. Compute the state estimation vector $\hat{x}(k\tau+\tau)$ by (2.46).
8. Increase $k$ by 1 and go to step 2, and continue the recursive calculation.

The flowchart of computing the parameter estimation vector $\hat{\theta}(k\tau)$ and the state estimate $\hat{x}(k\tau+\tau)$ is shown in Fig. 2.4.

**Remark 4**: The flop number is used to measure the calculation efficiency (calculation amount) of a complex algorithm. The total number of four floating-point operations required by an algorithm is defined as its calculation amount. Based on this, the calculation efficiency of the algorithm is evaluated as a benchmark, and an efficient and economical algorithm is sought. The calculation method is necessary to analyze the performance of the proposed algorithm.

The computational efficiency of the MISG and the SG algorithms is shown in Tables 2.2–2.4. Total float point operation (flop) numbers of the MISG and the SG algorithms are $N_1 = (n^2+n+n_g)(4p+3)$ and

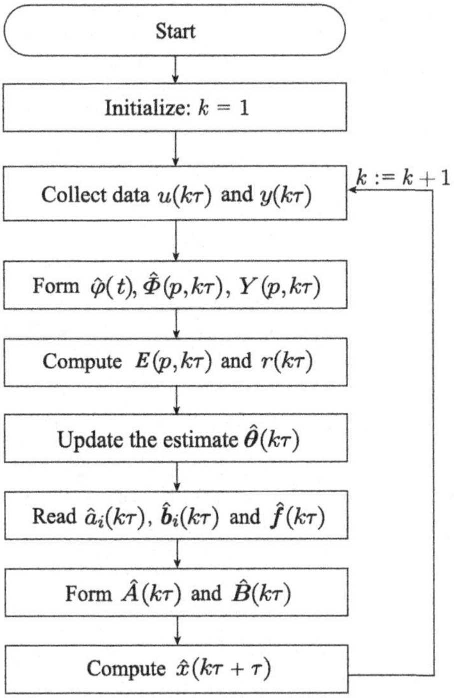


**Figure 2.4** The flowchart of computing the parameter estimate $\hat{\theta}(k\tau)$ and the state estimate $\hat{x}(k\tau + \tau)$.

**Table 2.2** The computational burden of the MISG algorithm.

| **Computational sequences** | **Number of multiplications** | **Number of additions** |
|---|---|---|
| $\hat{\theta}(k\tau) = \hat{\theta}(k\tau - \tau) + \frac{\hat{\Phi}(p,\ k\tau)}{r(k\tau)} E(p,\ k\tau)$ | $(n^2 + n + n_g)(p + 1)$ | $(n^2 + n + n_g)p$ |
| $E(p,\ k\tau) = Y(p,\ k\tau) - \hat{\Phi}^{\mathrm{T}}(p,\ k\tau)\hat{\theta}(k\tau - \tau)$ | $(n^2 + n + n_g)p$ | $(n^2 + n + n_g)p$ |
| $r(k\tau) = r(k\tau - \tau) + \|\hat{\varphi}(p,\ k\tau)\|^2$ | $n^2 + n + n_g$ | $n^2 + n + n_g$ |
| Sum | $2(n^2 + n + n_g)(p + 1)$ | $(n^2 + n + n_g)(2p + 1)$ |
| Total flops | $(n^2 + n + n_g)(4p + 3)$ | |

$N_2 = 7(n^2 + n + n_g)$, respectively. The difference between the MISG algorithm and the SG algorithm is

$$N_1 - N_2 = \left(n^2 + n + n_g\right)(4p + 3) - 7\left(n^2 + n + n_g\right)$$
$$(4p - 4)\left(n^2 + n + n_g\right) > 0.$$

**Table 2.3** The computational efficiency of the SG algorithm.

| Computational sequences | Number of multiplications | Number of additions |
|---|---|---|
| $\hat{\theta}(k\tau) = \hat{\theta}(k\tau - \tau) + \frac{\hat{\Phi}(k\tau)}{\tau(k\tau)}\left[y(k\tau) - \hat{\varphi}^{\mathrm{T}}(k\tau)\hat{\theta}(k\tau - \tau)\right]$ | $3(n^2 + n + n_g)$ | $(n^2 + n + n_g)p$ |
| $\tau(k\tau) = \tau(k\tau - \tau) + \|\|\hat{\varphi}(k\tau)\|\|^2$ | $n^2 + n + n_g$ | $n^2 + n + n_g$ |
| Sum | $4(n^2 + n + n_g)$ | $3(n^2 + n + n_g)$ |
| Total flops | $7(n^2 + n + n_g)$ | |

**Table 2.4** Comparison of the computational efficiency of the MISG and the SG algorithms.

| Algorithms | Number of multiplications | Number of additions | Total flops |
|---|---|---|---|
| MISG | $2(n^2 + n + n_g)(p + 1)$ | $(n^2 + n + n_g)(2p + 1)$ | $(n^2 + n + n_g)(4p + 3)$ |
| SG | $4(n^2 + n + n_g)$ | $3(n^2 + n + n_g)$ | $7(n^2 + n + n_g)$ |

For example, when $r = 2$, $n = 2$, and $L = 20$, $N_1 = 2023$ flops, $N_2 = 3903$ flops, $N_1 - N_2 = -1880$ flops. Thus, the SG algorithm has smaller computational efforts than the MISG algorithm.

### 2.2.5 U-model control

Consider a class of general state-space (including state-delayed) representation for single-input single-output (SISO) nonlinear discrete-time systems

$$\boldsymbol{x}(k+1) = \boldsymbol{F}(\boldsymbol{x}(k),\ \boldsymbol{x}(k-d),\ u(k)), \tag{2.51}$$

$$y(k) = h(\boldsymbol{x}(k)), \tag{2.52}$$

where $x(k) \in R^n$ is the state vector, $u(k) \in R$ is the control input, $d > 0$ is an integer denoting state delay, and $y(k) \in R$ is the system output, respectively. $F \in R^n$ is a smooth vector function to describe the model dynamics, and $h \in R$ is a smooth function to map the state and input to the output. Throughout the study, assume the system relative degree $r$ equals to the system order $n$ and has stable zero dynamics (i.e., the system model has stable inverse), the full state vector $x$ is available for measurement.

The state-space model of (2.51)−(2.52) can be converted into a multi-layer U-model as

$$x_1(k+1)=\sum_{j=0}^{M_1}\lambda_{1j}(k)f_{1j}(x_2(k)),$$
$$\vdots$$
$$x_{n-1}(k+1)=\sum_{j=0}^{M_{n-1}}\lambda_{n-1j}(k)f_{n-1j}(x_n(k)),$$
$$x_n(k+1)=\sum_{j=0}^{M_n}\lambda_{nj}(k)f_{nj}(u(k)),$$
$$y(k)=h(\mathbf{x}(k)),$$

where $f_{ij}(x_i(k))$ is a smooth function of the $i$th state.

Backstepping root-resolving routing to determine control input $u(k)$ from the U-state-space model.

The algorithm is listed below.

1. For a given desired trajectory $y_m(k)$, assign $y(k+1)=y_m(k)$.
2. Solve $x_1(k+1)=h^{-1}(y_m(k))$.
3. In backstep order, solve

$$x_2(k+1)\in x_1(k+1)-\sum_{j=0}^{M_1}\lambda_{1j}(k)f_{1j}(x_2(k))=0,$$
$$\vdots$$
$$x_n(k+1)\in x_{n-1}(k+1)$$
$$-\sum_{j=0}^{M_1}\lambda_{n-1j}(k)f_{n-1j}(x_{n-1}(k))=0.$$

4. To determine the control input $u(k)$, solve the last line of the U-state-space model by

$$u(k)\in x_n(k+1)-\sum_{j=0}^{M_n}\lambda_{nj}(k)f_{nj}(u(k))=0.$$

With reference to Fig. 2.5, the design procedure is outlined below:

1. Establish a stable linear feedback control system structured in Fig. 2.5, and assign G for the closed-loop system transfer function.
2. Specify G as a linear system with damping ratio, undamped natural frequency, and steady-state error and/or other performance indices (such as poles and zeros, and frequency response).

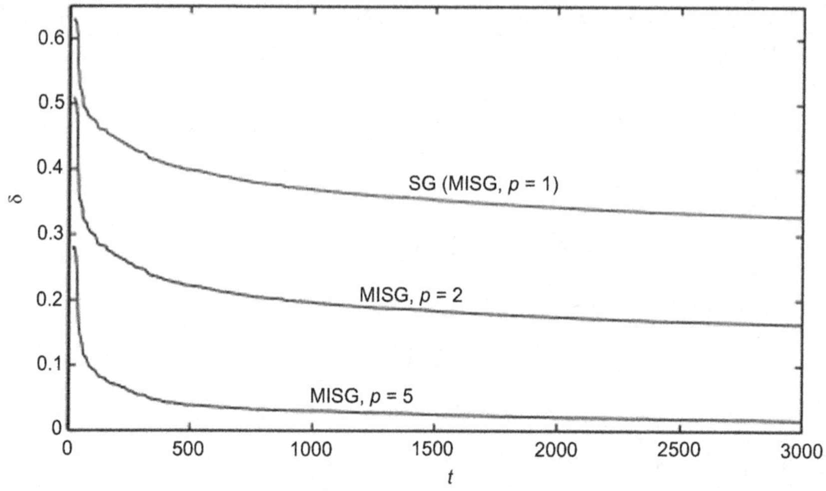


**Figure 2.5** U-model control system design framework.

3. Let the plant model be a constant unit or the virtual plant $G_{ip} = G_p^{-1} G_p = 1{:}u \rightarrow y$ have been achieved. To determine a linear invariant controller $G_{c1}$ by taking inverse of the closed-loop transfer function $G$ using $G_{c1} = \frac{G}{1-G} = (1-G)^{-1}G$. Accordingly, the desired system output is equivalently determined by the output $y_m$ of the controller $G_{c1}$.
4. Convert plant model $G_p$ into $G_p$ (U-model).
5. To achieve $G_{ip} = G_p^{-1} G_p = 1{:}u \rightarrow y$ to guarantee the desired output $y_m(t)$, determine the controller output $u(t)$ by solving an equation $y_m(t) - G_p$ (U-model) $= 0$, that is, $u(t) \in y_m(t) - G_p$ (U-model) $= 0$.
6. Locate/connect the blocks with reference to Fig. 2.5.

## 2.2.6 Examples

*Case study 1.* Model identification

Consider the following dual-rate time-delay system with $\tau = 2$:

$$\begin{aligned} \boldsymbol{x}(t+1) &= \begin{bmatrix} 0 & 1 \\ -0.01 & -0.22 \end{bmatrix} \boldsymbol{x}(t) \\ &+ \begin{bmatrix} 0.19 & -0.08 \\ 0.16 & -0.12 \end{bmatrix} \boldsymbol{x}(t-2) \\ &+ \begin{bmatrix} 0.8 \\ -0.8 \end{bmatrix} u(t), \\ y(t) &= [1, \ \ 0]\boldsymbol{x}(t) + v(t). \end{aligned}$$

The parameter vector to be identified is

$$\boldsymbol{\theta} = \left[a_2,\ a_1,\ b_{11},\ b_{12},\ b_{21},\ b_{22},\ f_1,\ f_2\right]^{\mathrm{T}}$$
$$[-0.01, -0.22,\ 0.19, -0.08,\ 0.16, -0.12,\ 0.8, -0.8]^{\mathrm{T}}.$$

In simulation, the input $\{u(t)\}$ is taken as an uncorrelated persistent excitation signal sequence with zero mean and unit variance and $\{v(t)\}$ as a white noise sequence with zero mean and variances $\sigma^2 = 0.30^2$ and $\sigma^2 = 1.00^2$. Apply the parameter estimation-based MISG algorithm to estimate the parameter vector $\theta$ and the state estimation algorithm to estimate the state vector $x(t)$ of this example system. The parameter estimates and their estimation errors are shown in Tables 2.5–2.6, the parameter estimation errors $\delta$ versus $t$ are shown in Figs. 2.6–2.7 with $p = 1, 2, 5$, respectively, and the state estimates $\hat{x}_1(t)$ and $\hat{x}_2(t)$ versus $t$ are shown in Figs. 2.8–2.9.

From Tables 2.5–2.6 and Figs. 2.6–2.9, we can draw the following conclusions:

1. The parameter estimates converge fast to their true values for large $p$—see Tables 2.5–2.6.
2. The MISG algorithm with $p \geq 2$ has higher accuracy than the SG algorithm—see Figs. 2.6, 2.7.
3. The parameter estimation errors given by the MISG algorithm become smaller with the data length $t$ and the innovation length $p$ increasing—see Tables 2.5–2.6 and Figs. 2.6, 2.7.
4. The state estimates are close to their true values with $t$ increasing—see Figs. 2.8–2.9.

*Case study 2.* U-control system design

For showing off the efficiency of the U-control system design directly, take up the plant model

$$\begin{bmatrix} x_1(k+1) \\ x_2(k+1) \end{bmatrix} = \begin{bmatrix} 0 & 1 \\ -0.01 & -0.22 \end{bmatrix} \begin{bmatrix} x_1(k) \\ x_2(k) \end{bmatrix} + \begin{bmatrix} 0.19 & -0.08 \\ 0.16 & -0.12 \end{bmatrix} \begin{bmatrix} x_1(k-2) \\ x_2(k-2) \end{bmatrix} + \begin{bmatrix} 0 \\ 1 \end{bmatrix} u(t),$$

$$y(t) = [1,\ 0] \begin{bmatrix} x_1(k) \\ x_2(k) \end{bmatrix}$$

**Table 2.5** The parameter estimates and errors with $\sigma^2 = 0.30^2$.

| Algorithms | $t$ | $a_2$ | $a_1$ | $b_{11}$ | $b_{12}$ | $b_{21}$ | $b_{22}$ | $f_1$ | $f_2$ | $\delta$(%) |
|---|---|---|---|---|---|---|---|---|---|---|
| SG (MISG, $p=1$) | 100 | −0.00912 | −0.15036 | 0.04307 | −0.06630 | 0.02505 | −0.01004 | 0.49833 | −0.38674 | 47.52683 |
| | 200 | −0.00071 | −0.15897 | 0.03430 | −0.06226 | 0.03583 | −0.01874 | 0.51905 | −0.41671 | 44.53744 |
| | 500 | −0.01158 | −0.16218 | 0.04174 | −0.07458 | 0.04926 | −0.02578 | 0.55028 | −0.45958 | 39.90966 |
| | 1000 | −0.01300 | −0.16742 | 0.04407 | −0.07779 | 0.05693 | −0.02935 | 0.57120 | −0.48809 | 36.94100 |
| | 2000 | −0.01608 | −0.17243 | 0.04797 | −0.08160 | 0.06551 | −0.03294 | 0.58943 | −0.51204 | 34.32511 |
| | 3000 | −0.01644 | −0.17616 | 0.04817 | −0.08312 | 0.07199 | −0.03621 | 0.59949 | −0.52557 | 32.86166 |
| MISG, $p=2$ | 100 | 0.00359 | −0.10678 | 0.07097 | −0.11681 | 0.07671 | −0.04656 | 0.66219 | −0.54089 | 29.99962 |
| | 200 | 0.01599 | −0.12109 | 0.06185 | −0.11177 | 0.09003 | −0.05590 | 0.68106 | −0.57759 | 26.68912 |
| | 500 | 0.00850 | −0.13309 | 0.07485 | −0.12515 | 0.10632 | −0.06220 | 0.70645 | −0.62350 | 22.16532 |
| | 1000 | 0.01018 | −0.14260 | 0.07880 | −0.12629 | 0.11409 | −0.06384 | 0.72115 | −0.65182 | 19.62474 |
| | 2000 | 0.00966 | −0.15129 | 0.08537 | −0.12916 | 0.12141 | −0.06584 | 0.73406 | −0.67410 | 17.47593 |
| | 3000 | 0.01029 | −0.15688 | 0.08764 | −0.13042 | 0.12688 | −0.06787 | 0.74038 | −0.68567 | 16.37969 |
| MISG, $p=5$ | 100 | −0.01604 | −0.16063 | 0.16648 | −0.04942 | 0.12274 | −0.11018 | 0.77804 | −0.72833 | 9.28037 |
| | 200 | −0.00941 | −0.17261 | 0.16361 | −0.05183 | 0.13136 | −0.11273 | 0.78161 | −0.75681 | 6.94835 |
| | 500 | −0.01370 | −0.18867 | 0.17725 | −0.06884 | 0.14888 | −0.11953 | 0.79311 | −0.77446 | 3.86359 |
| | 1000 | −0.00956 | −0.19552 | 0.17681 | −0.06720 | 0.15297 | −0.11685 | 0.79300 | −0.78499 | 3.00103 |
| | 2000 | −0.00865 | −0.20173 | 0.18160 | −0.07054 | 0.15557 | −0.11647 | 0.79656 | −0.79077 | 2.10426 |
| | 3000 | −0.00843 | −0.20539 | 0.18430 | −0.07311 | 0.15762 | −0.11675 | 0.79793 | −0.79282 | 1.61476 |
| True values | | −0.01000 | −0.22000 | 0.19000 | −0.08000 | 0.16000 | −0.12000 | 0.80000 | −0.80000 | |

**Table 2.6** The parameter estimates and errors with $\sigma^2 = 1.00^2$.

| Algorithms | $t$ | $a_2$ | $a_1$ | $b_{11}$ | $b_{12}$ | $b_{21}$ | $b_{22}$ | $f_1$ | $f_2$ | $\delta(\%)$ |
|---|---|---|---|---|---|---|---|---|---|---|
| SG (MISG, $p = 1$) | 100 | −0.01550 | −0.15470 | 0.04028 | −0.06265 | 0.02094 | −0.01630 | 0.52050 | −0.34565 | 49.20737 |
| | 200 | −0.00846 | −0.16114 | 0.03142 | −0.05900 | 0.02910 | −0.02302 | 0.53817 | −0.38165 | 46.04580 |
| | 500 | −0.02283 | −0.16425 | 0.04190 | −0.07513 | 0.04531 | −0.03215 | 0.57022 | −0.42744 | 41.04400 |
| | 1000 | −0.02246 | −0.16947 | 0.04133 | −0.07521 | 0.05323 | −0.03429 | 0.58828 | −0.45950 | 38.05525 |
| | 2000 | −0.02481 | −0.17522 | 0.04532 | −0.07876 | 0.06177 | −0.03769 | 0.60570 | −0.48594 | 35.31488 |
| | 3000 | −0.02492 | −0.17967 | 0.04580 | −0.08054 | 0.06861 | −0.04099 | 0.61533 | −0.50079 | 33.76525 |
| MISG, $p = 2$ | 100 | −0.00070 | −0.09901 | 0.08115 | −0.11549 | 0.07387 | −0.05378 | 0.68241 | −0.51521 | 30.88244 |
| | 200 | 0.01133 | −0.11098 | 0.06966 | −0.10975 | 0.08403 | −0.06043 | 0.69602 | −0.56265 | 27.17174 |
| | 500 | 0.00040 | −0.12474 | 0.08572 | −0.12822 | 0.10703 | −0.07177 | 0.72361 | −0.60977 | 22.19946 |
| | 1000 | 0.00665 | −0.13482 | 0.08321 | −0.12254 | 0.11560 | −0.07075 | 0.73290 | −0.64205 | 19.67043 |
| | 2000 | 0.00802 | −0.14527 | 0.08971 | −0.12475 | 0.12278 | −0.07232 | 0.74507 | −0.66653 | 17.34685 |
| | 3000 | 0.00917 | −0.15233 | 0.09273 | −0.12666 | 0.12873 | −0.07427 | 0.75100 | −0.67898 | 16.12797 |
| MISG, $p = 5$ | 100 | −0.02959 | −0.13914 | 0.19948 | −0.01531 | 0.08421 | −0.15341 | 0.76481 | −0.73279 | 12.99066 |
| | 200 | −0.01861 | −0.15341 | 0.18733 | −0.01475 | 0.09415 | −0.14951 | 0.76310 | −0.78040 | 10.55685 |
| | 500 | −0.02916 | −0.17950 | 0.20785 | −0.04599 | 0.13403 | −0.16618 | 0.79390 | −0.78371 | 6.83573 |
| | 1000 | −0.01718 | −0.18774 | 0.19516 | −0.03222 | 0.14082 | −0.15378 | 0.78631 | −0.79542 | 6.02358 |
| | 2000 | −0.01503 | −0.19804 | 0.20265 | −0.03826 | 0.14446 | −0.14983 | 0.79412 | −0.79944 | 5.03453 |
| | 3000 | −0.01509 | −0.20501 | 0.20834 | −0.04462 | 0.14803 | −0.14864 | 0.79741 | −0.80005 | 4.46176 |
| True values | | −0.01000 | −0.22000 | 0.19000 | −0.08000 | 0.16000 | −0.12000 | 0.80000 | −0.80000 | |

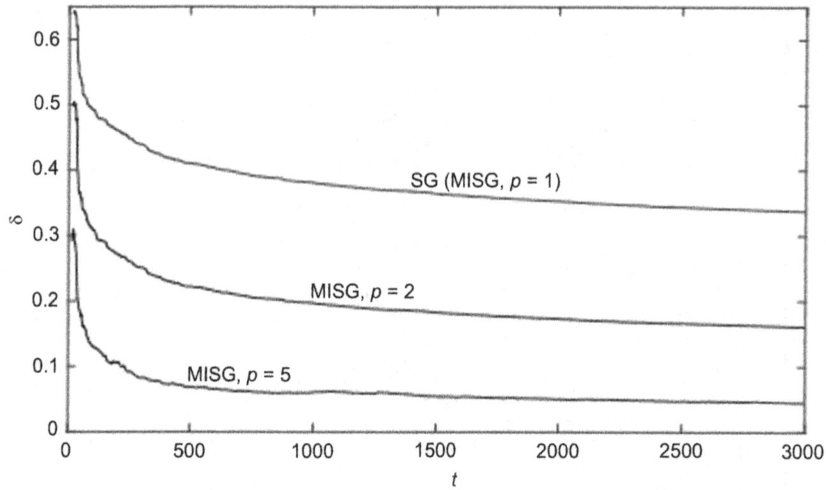


**Figure 2.6** The parameter estimation errors $\delta$ versus $t$ with $\sigma^2 = 0.30^2$.

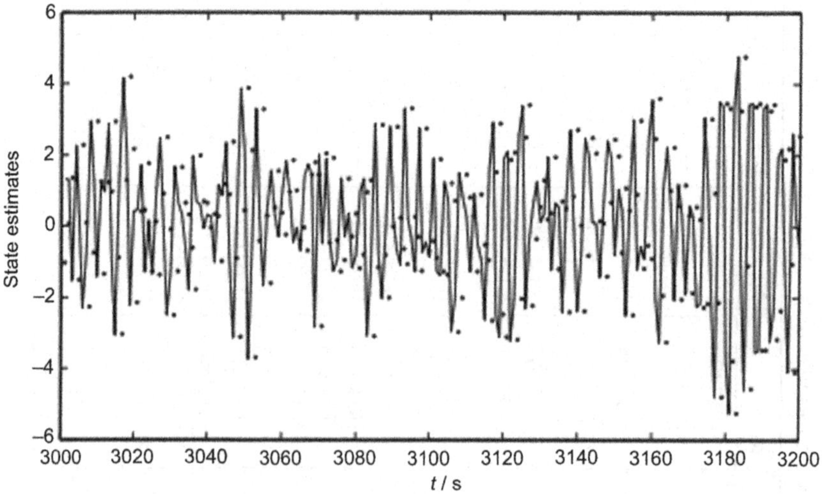


**Figure 2.7** The parameter estimation errors $\delta$ versus $t$ with $\sigma^2 = 1.00^2$.

and convert it into U-state-space model below

$$x_1(k+1) = \lambda_{10}(k) + \lambda_{11}(k)x_2(k),$$

$$x_2(k+1) = \lambda_{20}(k) + \lambda_{21}(k)u(k),$$

$$y(k) = x_1(k),$$

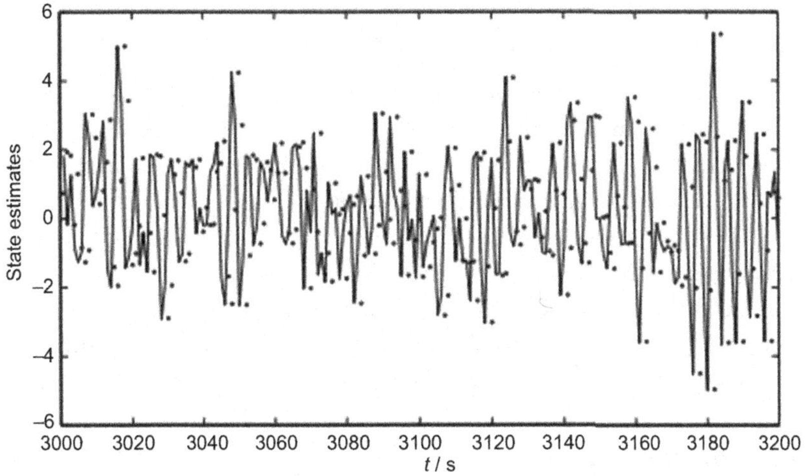


**Figure 2.8** The state and state estimate $\hat{x}_1(t)$ versus $t$, solid line: the true $x_1(t)$, dots: the estimated $\hat{x}_1(t)$.

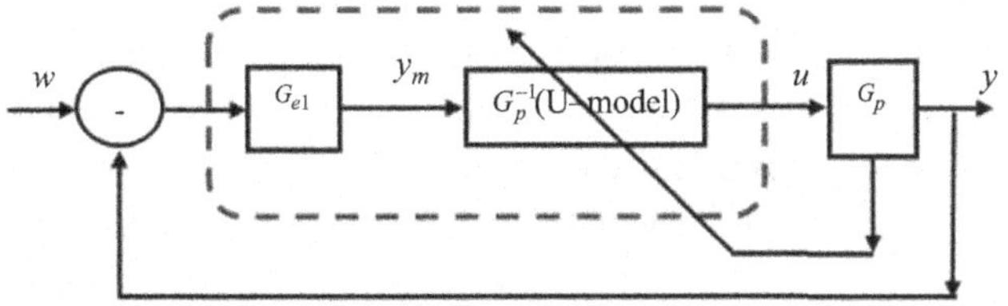


**Figure 2.9** The state and state estimate $\hat{x}_2(t)$ versus $t$, solid line: the true $x_2(t)$, dots: the estimated $\hat{x}_2(t)$.

where

$$
\begin{aligned}
\lambda_{10}(k) &= 0.21x_1(k-2) - 0.01x_2(k-2), \\
\lambda_{11}(k) &= 1, \\
\lambda_{20}(k) &= -0.01x_1(k) - 0.22x_2(k) + 0.16x_1(k-2) \\
&\quad + 0.12x_2(k-2), \\
\lambda_{21}(k) &= 1.
\end{aligned}
$$

With reference to the general U-control system design routine, make the following step-by-step design:

1. Establish a stable linear feedback control system structured in Fig. 2.5, and assign the closed-loop system transfer function with $\frac{Y(z)}{W(z)} = G(z) = \frac{0.45z^{-1}}{1-0.6z^{-1}+0.05z^{-2}}$, where $z$ is the $Z$ transform operator, the two poles are $p_1 = 0.1$, $p_2 = 0.5$, and no steady-state error to a step input.

2. To determine a linear invariant controller $G_{c1}$ by taking inverse of the closed-loop transfer function $G$ gives $G_{c1} = \frac{G}{1-G} = (1-G)^{-1}G = \frac{0.45z^{-1}}{1-1.05z^{-1}+0.05z^{-2}}$. Accordingly, the desired system output is equivalently determined by the output $y_m$ of the invariant controller $G_{c1}$.
3. To achieve $G_{ip} = G_p^{-1}G_p = 1{:}u \rightarrow y$ to guarantee the desired output $y_m(t)$, determine the controller output $u(t)$ by solving an equation $y_m(t) - G_p$ (U-model) $= 0$, that is, $u(t) \in y_m(t) - G_p$ (U-model) $= 0$. In this case, the backstepping routine is used in the root-solving.
4. The established control system is consistent with structure in Fig. 2.5.

Figs. 2.10–2.11 show the simulated responses, which confirm the specified performance and design efficiency.

### 2.2.7 Conclusions

This study has taken up a category of state-space models with state time delay as the research background and accordingly developed twofolds of solutions for the model identification (parameter and state estimation in specific) and control system design. The theoretical analysis has proved that the estimates converge to the real value under the condition of continuous excitation in modeling. The algorithms used in this paper can be applied to hybrid switched impulsive power networks and uncertain chaotic nonlinear systems with time-delay or other fields. U-control has been adopted from the authors' recent research, but

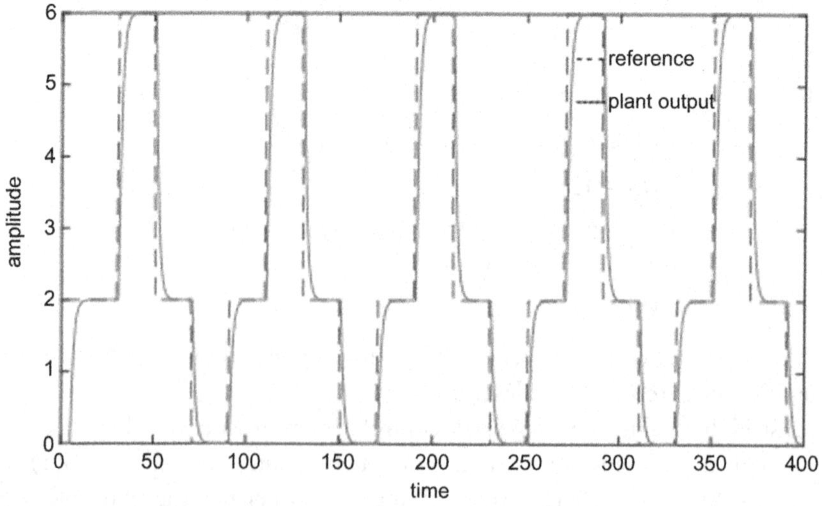


Figure 2.10 Plant output.

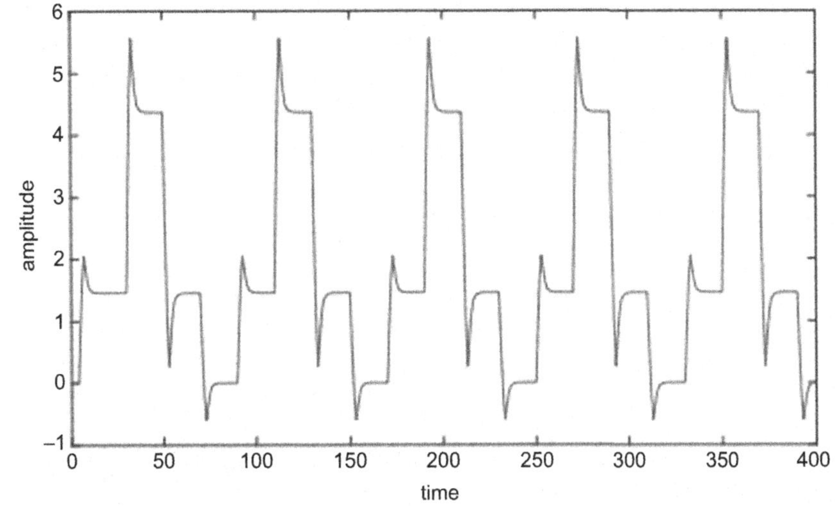


**Figure 2.11** Control input.

this study has presented a comprehensive backstepping routine for dynamic inversion for U-state-space models in such delayed control system design, which is different from those predominant approaches in this field. Hopefully, this can stimulate a new research/application direction in the future. Both side simulation case studies have demonstrated that the proposed algorithms/procedures are effective and efficient in design and implementation.

## 2.3 Parameter estimation algorithm for d-step time-delay systems

Modeling and estimation play an important role in solving many industrial problems. The state-space model derived from the description of state variables of differential equations has developed into an important tool for the identification and analysis of dynamic systems. A growing number of researchers are trying to show the advantage and universality of their applications in parameter identification, adaptive control, and system analysis. Essentially, states can define physical characteristics and represent descriptions of the internal behavior of a system. In the literature, many state-space models related papers were reported.

In comparison to the least squares method, the proposed stochastic gradient method receives more and more attention in the adaptive control.

On the contrary, the iterative recognition algorithm can lead higher faster convergence rate and precision of parameter estimation. Over the decades, time-delay analysis in the field of control has developed a great deal of literature. The time-delay influence on the performance of the control system takes great notice by many industrial process researchers.

### 2.3.1 Problem description

Consider the following mathematical model of state-space system.

$$\bar{\boldsymbol{x}}(t+1) = \overline{\boldsymbol{M}}\bar{\boldsymbol{x}}(t) + \overline{\boldsymbol{K}}\bar{\boldsymbol{x}}(t-d) + \bar{\boldsymbol{h}}u(t), \tag{2.53}$$

$$y(t) = \bar{\boldsymbol{l}}\bar{\boldsymbol{x}}(t) + v(t), \tag{2.54}$$

where $u(t) \in R$ represents the system input, $y(t) \in R$ stands for the system output, $\bar{x}(t) \in R^n$ represents the state vector, $v(t) \in R$ is a random noise with zero mean, and $\overline{M} \in R^{n \times n}$, $\overline{K} \in R^{n \times n}$, $\bar{h} \in R^n$, and $\bar{l} \in R^{1 \times n}$ are the system parameter matrices/vectors. On the basis of (2.53)–(2.54), we have

$$\begin{aligned} \boldsymbol{x}(t+1) &= \boldsymbol{T}^{-1}\bar{\boldsymbol{x}}(t+1) \\ &=: \boldsymbol{M}\boldsymbol{x}(t) + \boldsymbol{K}\boldsymbol{x}(t-d) + \boldsymbol{h}u(t), \end{aligned}$$

$$\begin{aligned} y(t) &= \bar{\boldsymbol{l}}\bar{\boldsymbol{x}}(t) + v(t) \\ &=: \boldsymbol{l}\boldsymbol{x}(t) + v(t), \end{aligned}$$

The following formula is the transformation matrix:

$$\boldsymbol{T} := \begin{bmatrix} \bar{\boldsymbol{l}} \\ \bar{\boldsymbol{l}}\overline{\boldsymbol{M}} \\ \vdots \\ \bar{\boldsymbol{l}}\overline{\boldsymbol{M}}^{n-1} \end{bmatrix}^{-1} \in \mathbb{R}^{n \times n}.$$

The following derives the expression of $M$.

For $(\bar{l}, \overline{M})$ is observable, the observable matrix $\mathrm{T}^{-1}$ is a complete rank matrix and we have

$$\boldsymbol{T}^{-1}\boldsymbol{T} = \begin{bmatrix} \bar{\boldsymbol{l}}T \\ \bar{\boldsymbol{l}}\overline{\boldsymbol{M}}T \\ \vdots \\ \bar{\boldsymbol{l}}\overline{\boldsymbol{M}}^{n-1}\boldsymbol{T} \end{bmatrix} = \boldsymbol{I}_n$$

Expanding the above matrix equation gives

$$\begin{cases} \bar{l}T = [1,\ 0,\ 0, \cdots, 0] =: e_1^{\mathrm{T}}, \\ \bar{l}\overline{M}T = [0,\ 1,\ 0, \cdots, 0] =: e_2^{\mathrm{T}}, \\ \bar{l}\overline{M}^2 T = [0,\ 0,\ 1, \cdots, 0] =: e_3^{\mathrm{T}}, \\ \vdots \\ \bar{l}\overline{M}^{n-1} T = [0,\ 0,\ 0, \cdots, 1] =: e_n^{\mathrm{T}}. \end{cases} \tag{2.55}$$

Thus, we have

$$l = \bar{l}T = [1,\ 0,\ 0,\ \cdots,\ 0] \in \mathbb{R}^{1\times n},$$

$$M = T^{-1}\overline{M}T = \begin{bmatrix} e_2^{\mathrm{T}} \\ e_3^{\mathrm{T}} \\ \\ e_n^{\mathrm{T}} \\ \bar{l}\overline{M}^n T \end{bmatrix} \in \mathbb{R}^{n\times n}. \tag{2.56}$$

The characteristic equations of the system in (2.53)–(2.54) are given by

$$\begin{aligned} p(s) &:= |sI_n - \overline{M}| \\ &= s^n - m_1 s^{n-1} - \cdots - m_{n-1}s - m_n = 0. \end{aligned} \tag{2.57}$$

The Cayley–Hamilton lemma shows that $\overline{M}$ satisfies its characteristic equation, that is,

$$\overline{M}^n - m_1\overline{M}^{n-1} - m_2\overline{M}^{n-2} - \cdots - m_n I = 0,$$

or

$$\overline{M}^n - m_1\overline{M}^{n-1} - m_2\overline{M}^{n-2} + \cdots + m_n I.$$

Pre-multiplying by $\bar{l}$, post-multiplying by T, and using (2.55) give

$$\begin{aligned} \bar{l}\overline{M}^n T &= m_1\bar{l}\overline{M}^{n-1}T + m_2\bar{l}\overline{M}^{n-2}T + \cdots + m_n\bar{l}T \\ &= [m_n,\ m_{n-1}, \cdots, m_1] =: m. \end{aligned} \tag{2.58}$$

Replacing (2.58) with (2.56) gives the expression of $M$. Therefore, we have the following expressions:

$$x(t+1) = \begin{bmatrix} 0 & 1 & \cdots & 0 \\ 0 & 0 & \ddots & \vdots \\ \vdots & \vdots & \ddots & 0 \\ 0 & 0 & 0 & 1 \\ m_n & m_{n-1} & \cdots & m_1 \end{bmatrix} x(t) + Kx(t-d) + hu(t), \tag{2.59}$$

$$y(t) = [1,\ 0,\ 0, \cdots, 0]x(t) + v(t) \tag{2.60}$$

### 2.3.2 The identification model

In the following part, the recognition model of standard state-space model in expressions (2.59)–(2.60) is derived. The detailed steps are shown below.

From (2.59), we have

$$\begin{cases} x_i(t+1) = x_{i+1}(t) + \boldsymbol{m}_i\boldsymbol{x}(t-d) + h_iu(t), \\ \qquad i = 1,\ 2, \ldots, n-1, \\ x_n(t+1) = m_nx_1(t) + m_{n-1}x_2(t) + \cdots + m_1x_n(t) \\ \qquad + \boldsymbol{k}_n\boldsymbol{x}(t-d) + h_nu(t). \end{cases} \tag{2.61}$$

When $d \leq n-1$, define $\boldsymbol{\psi}(t)$ and $\boldsymbol{\vartheta}$ as

$$\boldsymbol{\psi}(t) := \left[\boldsymbol{x}^{\mathrm{T}}(t-d-1), \cdots, \boldsymbol{x}^{\mathrm{T}}(t-n),\ \cdots,\right.$$

$$\boldsymbol{x}^{\mathrm{T}}(t-n-d),\ u(t-1),\ \cdots,$$

$$\left.u(t-n)\right]^{\mathrm{T}} \in \mathbb{R}^{n^2+n},$$

$$\boldsymbol{\vartheta}: \left[\boldsymbol{k}_1,\ \boldsymbol{k}_2, \cdots, \boldsymbol{a}+\boldsymbol{k}_{n-d}, \cdots, \boldsymbol{k}_n,\ \boldsymbol{h}^{\mathrm{T}}\right]^{\mathrm{T}} \in \mathbb{R}^{n^2+n}.$$

When $d \geq n$, define $\boldsymbol{\psi}(t)$ and $\boldsymbol{\vartheta}$ as

$$\boldsymbol{\psi}(t) := \left[\boldsymbol{x}^{\mathrm{T}}(t-n),\ \boldsymbol{x}^{\mathrm{T}}(t-d-1), \cdots, \boldsymbol{x}^{\mathrm{T}}(t-d-n),\right.$$

$$\left.u(t-1), \cdots, u(t-1)\right]^{\mathrm{T}} \in \mathbb{R}^{n^2+2n},$$

$$\boldsymbol{\vartheta}: \left[\boldsymbol{m}, \boldsymbol{k}_1, \boldsymbol{k}_2 \cdots, \boldsymbol{k}_{n-1}, \boldsymbol{k}_n, \boldsymbol{h}^{\mathrm{T}}\right]^{\mathrm{T}} \in \mathbb{R}^{n^2+2n}.$$

All expressions of (2.59) added are given

$$\begin{aligned} x_1(t) &= \boldsymbol{m}\boldsymbol{x}(t-n) + \boldsymbol{k}_1\boldsymbol{x}(t-d-1) + \cdots \\ &\quad + \boldsymbol{k}_n\boldsymbol{x}(t-d-n) + h_1u(t-1) + \cdots \\ &\quad + h_nu(t-n) \\ &= \boldsymbol{\psi}^{\mathrm{T}}(t)\boldsymbol{\vartheta}. \end{aligned}$$

From (2.60), we have

$$y(t) = x_1(t) + v(t) = \boldsymbol{\psi}^{\mathrm{T}}(t)\boldsymbol{\vartheta} + v(t) \tag{2.62}$$

The above formula is the recognition model as shown in Fig. 2.12.

**Example 1**: The following gives a second-order delay system

$$\overline{\boldsymbol{x}}(t+1) = \overline{\boldsymbol{M}}\overline{\boldsymbol{x}}(t) + \overline{\boldsymbol{K}}\overline{\boldsymbol{x}}(t-d) + \overline{\boldsymbol{h}}u(t), \tag{2.63}$$

$$y(t) = \overline{\boldsymbol{l}}\overline{\boldsymbol{x}}(t) + v(t), \tag{2.64}$$

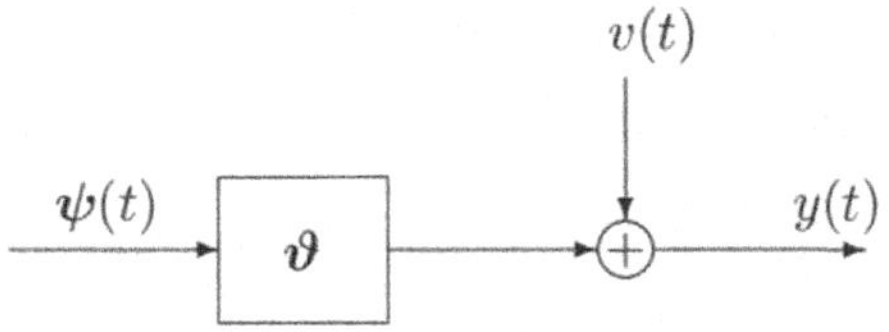


**Figure 2.12** The identification model.

where

$$\overline{\boldsymbol{M}} = \begin{bmatrix} \overline{m}_1 & \overline{m}_2 \\ \overline{m}_3 & \overline{m}_4 \end{bmatrix} \in \mathbb{R}^{2\times 2}, \quad \overline{\boldsymbol{h}} = \begin{bmatrix} \overline{h}_1 \\ \overline{h}_2 \end{bmatrix} \in \mathbb{R}^{2\times 1},$$

$$\overline{\boldsymbol{K}} = \begin{bmatrix} \overline{k}_1 & \overline{k}_2 \\ \overline{k}_3 & \overline{k}_4 \end{bmatrix} \in \mathbb{R}^{2\times 2}, \quad \overline{\boldsymbol{l}} = \begin{bmatrix} \overline{l}_1, & \overline{l}_2 \end{bmatrix} \in \mathbb{R}^{1\times 2}.$$

From Eqs. (2.59) and (2.60), we have

$$\begin{aligned} \begin{bmatrix} x_1(t+1) \\ x_2(t+1) \end{bmatrix} &= \begin{bmatrix} 0 & 1 \\ m_2 & m_1 \end{bmatrix} \begin{bmatrix} x_1(t) \\ x_2(t) \end{bmatrix} \\ &\quad + \begin{bmatrix} \boldsymbol{k}_1 \\ \boldsymbol{k}_2 \end{bmatrix} \boldsymbol{x}(t-d) + \begin{bmatrix} h_1 \\ h_2 \end{bmatrix} u(t), \\ y(t) &= [1, \quad 0] \begin{bmatrix} x_1(t) \\ x_2(t) \end{bmatrix} + v(t). \end{aligned}$$

By extending this matrix equation, we can get

$$\begin{aligned} x_1(t+1) &= x_2(t) + \boldsymbol{k}_1 \boldsymbol{x}(t-d) + h_1 u(t), \\ x_2(t+1) &= m_2 x_1(t) + m_1 x_2(t) + \boldsymbol{k}_2 \boldsymbol{x}(t-d) + h_2 u(t). \end{aligned}$$

The $i$th of the above equations multiplied by $z^{-i}$ gives

$$\begin{aligned} z^{-1} x_1(t+1) &= z^{-1} x_2(t) + z^{-1} \boldsymbol{k}_1 \boldsymbol{x}(t-d) + z^{-1} h_1 u(t), \\ z^{-2} x_2(t+1) &= z^{-2} m_2 x_1(t) + z^{-2} m_1 x_2(t) \\ &\quad + z^{-2} \boldsymbol{k}_2 \boldsymbol{x}(t-d) + z^{-2} h_2 u(t). \end{aligned}$$

Adding the above two equations gives

$$\begin{aligned} x_1(t) &= \boldsymbol{m}\boldsymbol{x}(t-2) + \boldsymbol{k}_1 \boldsymbol{x}(t-d-1) + \boldsymbol{k}_2 \boldsymbol{x}(t-d-2) \\ &\quad + h_1 u(t-1) + h_2 u(t-2). \end{aligned}$$

Its relevant input–output expression is provided by the following formula

$$\begin{aligned} y(t) &= \boldsymbol{m}\boldsymbol{x}(t-2) + \boldsymbol{k}_1 \boldsymbol{x}(t-d-1) + \boldsymbol{k}_2 \boldsymbol{x}(t-d-2) \\ &\quad + h_1 u(t-1) + h_2 u(t-2) + v(t) \end{aligned} \tag{2.65}$$

Therefore, we have the following representation:

$$y(t) = (\boldsymbol{m} + \boldsymbol{k}_1)\boldsymbol{x}(t-2) + \boldsymbol{k}_2\boldsymbol{x}(t-3) + h_1 u(t-1) + h_2 u(t-2).$$

## 2.3.3 The parameter and state estimation algorithm

In the following part, a stochastic gradient method is derived to calculate the coefficients of a state-space model with time delay, and the Kalman filter theory is applied to calculate the state of the system. Details are shown below.

Minimizing the standard features,

$$M(\vartheta) := \sum_{j=1}^{t} \left[y(j) - \boldsymbol{\psi}^{\mathrm{T}}(j)\vartheta\right]^2.$$

We can get the below recursive least square (RLS) method to calculate $\vartheta$:

$$\hat{\vartheta}(t) = \hat{\vartheta}(t-1) + \boldsymbol{E}(t)\left[y(t) - \boldsymbol{\psi}^{\mathrm{T}}(t)\hat{\vartheta}(t-1)\right], \tag{2.66}$$

$$\begin{aligned} \boldsymbol{E}(t) &= \mathbf{Q}(t)\boldsymbol{\psi}(t) \\ &= \mathbf{Q}(t-1)\boldsymbol{\psi}(t)\left[1 + \boldsymbol{\psi}^{\mathrm{T}}(t)\mathbf{Q}(t-1)\boldsymbol{\psi}(t)\right]^{-1}, \end{aligned} \tag{2.67}$$

$$\begin{aligned} \mathbf{Q}(t) &= \mathbf{Q}(t-1) \\ &- \mathbf{Q}(t-1)\boldsymbol{\psi}(t)\boldsymbol{\psi}^{\mathrm{T}}(t)\mathbf{Q}(t-1)\left[1 + \boldsymbol{\psi}^{\mathrm{T}}(t)\mathbf{Q}(t-1)\boldsymbol{\psi}(t)\right]^{-1}. \end{aligned} \tag{2.68}$$

The method here is to replace the unavailable $x(t-i)$ in $\psi(t)$ with $\hat{\psi}(t)$. The estimate of $\vartheta$ can be obtained by the following parameter estimation method:

$$\hat{\vartheta}(t) = \hat{\vartheta}(t-1) + \frac{\hat{\boldsymbol{\psi}}(t)}{\gamma(t)}\left[y(t) - \hat{\psi}^{\mathrm{T}}(t)\hat{\vartheta}(t-1)\right], \tag{2.69}$$

$$\gamma(t) = \lambda\gamma(t-1) + \|\hat{\psi}(t)\|^2, \quad 0 < \lambda < 1, \quad \gamma(0) = 1 \tag{2.70}$$

$$\begin{aligned} \hat{\boldsymbol{\psi}}(t) := \Big[&\hat{\boldsymbol{x}}^{\mathrm{T}}(t-d-1), \cdots, \hat{\boldsymbol{x}}^{\mathrm{T}}(t-n), \cdots, \\ &\hat{\boldsymbol{x}}^{\mathrm{T}}(t-n-d), \ u(t-1), \ \cdots, \\ &u(t-n)\Big]^{\mathrm{T}}, \ \text{for } d \leq n-1, \end{aligned} \tag{2.71}$$

$$\hat{\psi}(t) := \left[\hat{\boldsymbol{x}}^{\mathrm{T}}(t-n),\ \hat{\boldsymbol{x}}^{\mathrm{T}}(t-d-1), \cdots, \hat{\boldsymbol{x}}^{\mathrm{T}}(t-d-n),\right.$$
$$\left. u(t-1), \cdots, u(t-n)\right]^{\mathrm{T}}, \text{ for } d \geq n. \tag{2.72}$$

The following gives the state identification method:

$$\begin{aligned}\hat{\boldsymbol{x}}(t+1) = \hat{\boldsymbol{M}}(t)\hat{\boldsymbol{x}}(t) + \hat{\boldsymbol{M}}(t)\hat{\boldsymbol{x}}(t-1) + \hat{\boldsymbol{h}}(t)u(t) \\ + \boldsymbol{E}(t)[y(t) - \boldsymbol{l}\hat{\boldsymbol{x}}(t)],\end{aligned} \tag{2.73}$$

$$\boldsymbol{E}(t) = \hat{\boldsymbol{M}}(t)\mathbf{Q}(t)\boldsymbol{l}^{\mathrm{T}}\left[1 + \boldsymbol{l}\mathbf{Q}(t)\boldsymbol{l}^{\mathrm{T}}\right]^{-1}, \tag{2.74}$$

$$\begin{aligned}\mathbf{Q}(t+1) = \hat{\boldsymbol{M}}(t)\mathbf{Q}(t)\hat{\boldsymbol{M}}^{\mathrm{T}}(t) \\ - \hat{\boldsymbol{M}}(t)\mathbf{Q}(t)\boldsymbol{l}^{\mathrm{T}}\left[1 + \boldsymbol{l}\mathbf{Q}(t)\boldsymbol{c}^{\mathrm{T}}\right]^{-1}\boldsymbol{l}\mathbf{Q}(t)\hat{\boldsymbol{M}}^{\mathrm{T}}(t),\end{aligned} \tag{2.75}$$

$$\hat{\boldsymbol{M}}(t) = \begin{bmatrix} 0 & 1 & \cdots & 0 \\ 0 & 0 & \ddots & \vdots \\ \vdots & \vdots & \ddots & 0 \\ 0 & 0 & 0 & 1 \\ \hat{m}_n(t) & \hat{m}_{n-1}(t) & \cdots & \hat{m}_1(t) \end{bmatrix}, \tag{2.76}$$

$$\hat{\boldsymbol{M}}(t) = \begin{bmatrix} \hat{\boldsymbol{m}}_1(t) \\ \hat{\boldsymbol{m}}_2(t) \\ \vdots \\ \hat{\boldsymbol{m}}_n(t) \end{bmatrix}, \quad \hat{\boldsymbol{h}}(t) = \begin{bmatrix} \hat{h}_1(t) \\ \hat{h}_2(t) \\ \vdots \\ \hat{h}_n(t) \end{bmatrix} \tag{2.77}$$

## 2.3.4 Example

The following gives one example of state-space system with delay:

$$\begin{aligned}\boldsymbol{x}(t+1) = \begin{bmatrix} 0 & 1 \\ -0.45 & -0.80 \end{bmatrix}\boldsymbol{x}(t) \\ + \begin{bmatrix} 0.20 & -0.15 \\ 0.15 & -0.20 \end{bmatrix}\boldsymbol{x}(t-d) \\ + \begin{bmatrix} 1.00 \\ -1.00 \end{bmatrix}u(t),\end{aligned}$$

$$y(t) = [1,\ \ 0]\boldsymbol{x}(t) + v(t).$$

Here, the input $\{u(t)\}$ is seen as not related with zero mean and unit variance continuous excitation signal sequence and $\{v(t)\}$ as a white noise sequence with zero mean and variance $\sigma^2 = 0.10^2$. The parameter estimation and their estimation errors are displayed in Table 2.7 with

**Table 2.7** The parameter estimates and errors with $\sigma^2 = 0.10^2$.

| $\lambda$ | $T$ | $a_2 + b_{11}$ | $a_1 + b_{12}$ | $b_{21}$ | $b_{22}$ | $f_1$ | $f_2$ | $\delta(\%)$ |
|---|---|---|---|---|---|---|---|---|
| 0.98 | 500 | 0.03172 | −0.58665 | 0.51228 | −0.10483 | 0.87787 | −0.78205 | 36.98424 |
| | 1000 | 0.04170 | −0.60038 | 0.50042 | −0.11923 | 0.91282 | −0.83978 | 34.95480 |
| | 2000 | 0.04029 | −0.62343 | 0.48131 | −0.14711 | 0.96230 | −0.92817 | 32.00225 |
| | 5000 | 0.02134 | −0.66629 | 0.44302 | −0.17637 | 0.99389 | −0.99578 | 28.19016 |
| | 10000 | −0.02220 | −0.71607 | 0.39482 | −0.18608 | 1.00177 | −0.99537 | 23.47033 |
| 0.94 | 500 | 0.04101 | −0.60965 | 0.49387 | −0.13463 | 0.94760 | −0.89317 | 33.37353 |
| | 1000 | 0.03827 | −0.63749 | 0.47012 | −0.16547 | 0.98483 | −0.96287 | 30.73752 |
| | 2000 | 0.00421 | −0.68839 | 0.42028 | −0.18333 | 1.00297 | −0.99537 | 26.11183 |
| | 5000 | −0.07760 | −0.77418 | 0.33315 | −0.18991 | 0.99981 | −1.00486 | 17.64913 |
| | 10000 | −0.15947 | −0.85882 | 0.24658 | −0.19711 | 1.00369 | −0.99355 | 9.24977 |
| 0.88 | 500 | 0.03961 | −0.63912 | 0.46584 | −0.16647 | 0.99258 | −0.97031 | 30.53790 |
| | 1000 | 0.00974 | −0.68333 | 0.42800 | −0.18784 | 0.99880 | −0.99111 | 26.71639 |
| | 2000 | −0.06569 | −0.76067 | 0.34349 | −0.19335 | 1.00515 | −0.99885 | 18.82838 |
| | 5000 | −0.17291 | −0.87363 | 0.23004 | −0.19806 | 1.00283 | −1.00266 | 7.75336 |
| | 10000 | −0.22974 | −0.93188 | 0.17036 | −0.20224 | 1.00172 | −0.99381 | 1.99090 |
| True values | | −0.25000 | −0.95000 | 0.15000 | −0.20000 | 1.00000 | −1.00000 | |

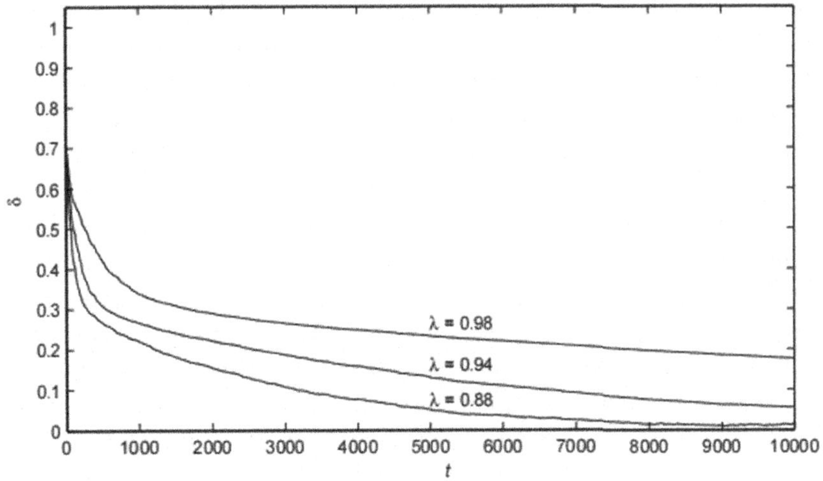


**Figure 2.13** The parameter estimation errors $\delta$ versus $t$ with $\sigma^2 = 0.10^2$.

$\lambda = 0.98, 0.94$ and $0.88$, respectively, and the errors $\delta$ of parameter estimation versus $t$ are displayed in Fig. 2.13.

We can draw some conclusions from Table 2.7 and Fig. 2.13: (1) with the increase of $t$, the parameter identification errors $\delta$ get smaller (in general)—see Fig. 2.13; (2) the accuracy of parameter estimation with the same length of data becomes higher with the increase of forgetting factor—see Table 2.7 and Fig. 2.13.

### 2.3.5 Conclusions

This chapter discusses the problem of identifying linear systems based on a standard state-space system with state delay. A SG estimation method (assuming the state is available) and a recursive least squares algorithm based on state estimation are proposed. The assumed state is available, and the estimated state is used for parameter estimation based on the RLS method of state estimation.

## 2.4 Communicative state and parameters estimation for dual-rate state-space systems with time delays

Many conventional algorithms are used to estimate system parameters, but most of them assume that input–output data are available at every sampling instant. For multi-rate sampled data systems, these algorithms cannot be applied directly. Multi-rate systems are abundant in industry; for example, many soft-sensor design problems are related to modeling, parameter identification, or state estimation involving multi-rate systems. The research on multi-rate systems can be traced back to the 1950s. The first important work was the switch decomposition technology (i.e., the later lifting technology) proposed by Kranc in 1957. A multi-rate sampling data system is always a time-varying system, even if the corresponding continuous system is time-invariant. Later, the introduction of lifting technology by Friedland and Khargoneckar, etc., can transform a periodic discrete time-varying system into a time-invariant system. Lifting technology has now become a standard tool for handling multi-rate systems. Due to the system upgrade, the upgraded system is a multivariable system with more causal constraints and more variables. Different from the conventional single-rate system identification method: to study the identification of multi-rate systems, this causal constraint needs to be considered; at the same time, how to extract single-rate models from the elevated multi-rate model needs to be studied.

Time-delay systems are very common in practical situations in nature, for instance, transmission problems, communications, population growth models, and other biological systems. Many physical control processes can be best modeled by systems involving time delay in the state or in the control. Effects of time delay on the stability and performance of control systems have drawn the attention of many investigators in different engineering disciplines, including structural systems, chemical processes, remotely controlled undersea and aerospace robots and structures, and manufacturing processes.

The convergence analysis of the identification algorithm is an important method to prove the effectiveness of the algorithm in the field of control. Earlier convergence analysis assumed that the input and output signals of the system under consideration had finite nonzero power, and the noise was an independent and uniformly distributed random sequence with finite fourth-order moments. Such ideal assumptions are difficult to satisfy in practice. The convergence rate of the standard recursive least square parameter estimation is obtained by assuming that the process noise $v(t)$ has finite second-order and higher-order moments. Since then, most of the convergent results of the least square algorithm have made so-called "weak" assumptions.

It focuses on the modeling and identification problem for dual-rate state-space systems with time delay. The key idea is to use the Kalman filter and the recursive least squares algorithm to directly identify the parameters of the lifted models of dual-rate systems. The main contributions of this paper are as follows.

1. Identifiability of the dual-rate state-space systems with time delays is discussed under mild condition.
2. The decomposed subsystems lead to less amount of calculation and storage.
3. The convergence of the proposed algorithm is analyzed under the weak continuous excitation condition.

### 2.4.1 Problem formulation

In this chapter, we consider a general dual-rate system with additive disturbance described in Fig. 2.1, and a zero-order hold $G_{\mathrm{T}_1}$ with period $\mathrm{T}_1$ generates input $u(t)$ to process a discrete-time signal $u(k\mathrm{T}_1)$; to yield a discrete-time signal $y(k\mathrm{T}_2)$, the output $y(t)$ of a continuous-time process $H(s)$ is sampled by a sampler $R_{\mathrm{T}_2}$. For such a dual-rate state-space system with time delay, the known input–output data are $\{u(k\mathrm{T}_1), y(k\mathrm{T}_2)\}$. Since the zero-order hold is used, we have $u(t) = u(k\mathrm{T}_1)$, $k\mathrm{T}_1 \leq t \leq (k+1)\mathrm{T}_1$.

Throughout the paper, we assume $H(s)$ is a linear time-invariant (LTI) continuous-time process with the following state-space representation:

$$\boldsymbol{x}(t+1) = \boldsymbol{A}_1\boldsymbol{x}(t) + \boldsymbol{A}_d\boldsymbol{x}(t-d) + \boldsymbol{b}_1u(t) + \boldsymbol{f}_1w(t), \tag{2.78}$$

$$\boldsymbol{y}(t) = \boldsymbol{x}(t) + \boldsymbol{v}(t), \tag{2.79}$$

where $x(t)\in R^n$ is the state vector, $u(t)\in R$ is the system input, $y(t)\in R^n$ is the system output, $v(t)\in R^n$ is a random noise with zero mean, $w(t)\in R$ is a process noise, and $A_1 \in R^{n\times n}$, $A_d\in R^{n\times n}$, $b_1 \in R^n$, and $f_1 \in R^n$ are the system parameter matrices/vectors.

Although $H(s)$ is LTI, the system from $u(k\mathrm{T}_1)$ to $y(k\mathrm{T}_2)$ in Fig. 2.14 is linear periodically time-varying due to different updating and sampling periods. For such a time-varying dual-rate system, our objective is twofold.

**Remark 5**: This chapter is to establish time-invariant models of the system from the available input to the available output data by using the lifting technique and to develop the Kalman filter and recursive algorithm for combined parameter and state estimation based on the given dual-rate measurement data $\{u(k\mathrm{T}_1), y(k\mathrm{T}_2)\}$, taking causality constraints into consideration.

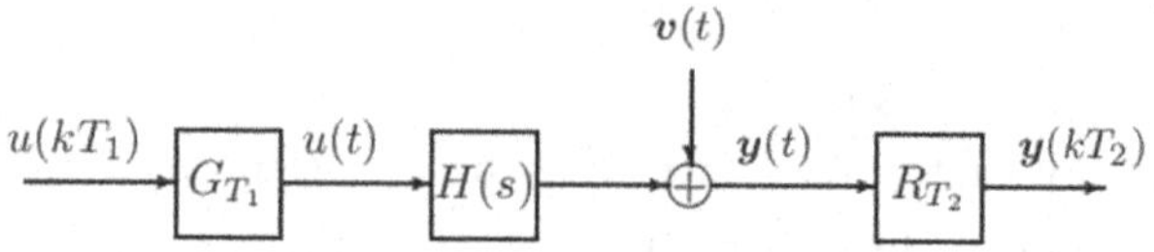


**Figure 2.14** The diagram of the dual-rate system.

### 2.4.2 Augmented state estimation

For the dual-rate system from input to output in Fig. 2.14, we intend to lift the input and output to get a lifted state-space model with time delay,

$$\begin{aligned} \boldsymbol{x}((k+1)T) &= \boldsymbol{A}_1\boldsymbol{x}(kT) + \boldsymbol{A}_d\boldsymbol{x}(kT-d) + \boldsymbol{b}_1 u(kT) + \boldsymbol{f}_1 w(kT), \\ \boldsymbol{y}(kT) &= \boldsymbol{x}(kT) + \boldsymbol{v}(kT). \end{aligned}$$

For convenience, consider the following state-space system with time delay,

$$\boldsymbol{x}(k+1) = \boldsymbol{A}_1\boldsymbol{x}(k) + \boldsymbol{A}_d\boldsymbol{x}(k-d) + \boldsymbol{b}_1 u(k) + \boldsymbol{f}_1 w(k), \tag{2.80}$$

$$\boldsymbol{y}(k) = \boldsymbol{x}(k) + \boldsymbol{v}(k), \tag{2.81}$$

The identifiability of a system depends on its controllability and observability. Therefore, it is very important whether the lifted state-space models with time delay in the above are controllable and observable. We assume that the state-space model with time delay is a minimal implementation. The controllability and observability can be achieved under mild condition.

**Remark 6**: In real industry, not all states can be measured by sensors. In the face of this problem, one approach is to consider the corresponding input–output representation by eliminating the available state vector. But such an approach cannot address the identification and state estimation of the system under consideration. In this chapter, a recursive state estimation algorithm is presented to update the state estimation by constructing a state observer.

Define an expanded state vector and some matrix/vectors:

$$\boldsymbol{X}(k) := \begin{bmatrix} \boldsymbol{x}(k) \\ \boldsymbol{x}(k-d) \end{bmatrix} \in \mathbb{R}^{2n},$$

$$\boldsymbol{A} := \begin{bmatrix} \boldsymbol{A}_1 & \boldsymbol{A}_d \\ \boldsymbol{I} & 0 \end{bmatrix} \in \mathbb{R}^{(2n)\times(2n)}, \qquad \boldsymbol{b} := \begin{bmatrix} \boldsymbol{b}_1 \\ 0 \end{bmatrix} \in \mathbb{R}^{2n}, \qquad \boldsymbol{f} := \begin{bmatrix} \boldsymbol{f}_1 \\ 0 \end{bmatrix} \in \mathbb{R}^{2n}.$$

Combining the above definitions, Eqs. (2.80) and (2.81) can be equivalently rewritten as

$$\begin{aligned}\boldsymbol{X}(k+1) &= \begin{bmatrix} \boldsymbol{A}_1 & \boldsymbol{A}_d \\ \boldsymbol{I} & 0 \end{bmatrix} \boldsymbol{X}(k) + \begin{bmatrix} \boldsymbol{b}_1 \\ 0 \end{bmatrix} u(k) + \begin{bmatrix} \boldsymbol{f}_1 \\ 0 \end{bmatrix} w(k) \\ &= \boldsymbol{AX}(k) + \boldsymbol{b}u(k) + \boldsymbol{f}w(k),\end{aligned} \tag{2.82}$$

$$\begin{aligned} y(k) &= \boldsymbol{x}(k) + \boldsymbol{v}(k) \\ &= [1, \ \ 0]\boldsymbol{X}(k) + \boldsymbol{v}(k) \\ &=: \boldsymbol{cX}(k) + \boldsymbol{v}(k). \end{aligned} \tag{2.83}$$

where $v(t)$ is the observation noise which is assumed to be zero mean Gaussian white noise with covariance $R(t)$, $R(t) = \text{cov}[v(t)]$.

The state of the filter is represented by the posteriori state estimate $\hat{X}(k|k)$ and the posteriori error covariance matrix $P(k|k)$.

Let $\hat{X}(k){:=}\hat{X}(k|k)$, $P(k){:=}P(k|k)$. $k+i|k$ stands for the estimation data on time $k+i$ which is based on the data of time $k$.

Assume that the state filter adopts the form,

$$\hat{\boldsymbol{X}}(k+1|k) = A\hat{\boldsymbol{X}}(k) + \boldsymbol{b}u(k) + \boldsymbol{f}w(k) + \boldsymbol{L}(k)\left[y(k) - c\hat{\boldsymbol{X}}(k)\right]. \tag{2.84}$$

The objective is to determine a gain vector $L(k)$ such that the state estimation error vector $\tilde{X}(k){:=}X(k) - \hat{X}(k)$ is minimal. Construct the following state estimation error

$$\begin{aligned} \tilde{\boldsymbol{X}}(k+1|k) &= \boldsymbol{X}(k+1|k) - \hat{\boldsymbol{X}}(k+1|k) \\ = \boldsymbol{AX}(k) + \boldsymbol{b}u(k) + \boldsymbol{f}w(k) &- \left\{A\hat{\boldsymbol{X}}(k) + \boldsymbol{b}u(k) + \boldsymbol{f}w(k) + \boldsymbol{L}(k)\left[y(k) - c\hat{\boldsymbol{X}}(k)\right]\right\} \\ &= \boldsymbol{AX}(k) - A\hat{\boldsymbol{X}}(k) - \boldsymbol{L}(k)y(k) + \boldsymbol{L}(k)\boldsymbol{cX}(k) - \boldsymbol{L}(k)c\tilde{\boldsymbol{X}}(k) \\ &= A\hat{\boldsymbol{X}}(k) - \boldsymbol{L}(t)\left[y(k) - \boldsymbol{cX}(k) + c\tilde{\boldsymbol{X}}(k)\right] \\ &= [\boldsymbol{A} - \boldsymbol{L}(k)\boldsymbol{c}]\tilde{\boldsymbol{X}}(k) - \boldsymbol{L}(k)\boldsymbol{v}(k). \end{aligned} \tag{2.85}$$

Define the state estimation error covariance matrix:

$$P(k){:=}\mathrm{E}\left\{\left(\tilde{\boldsymbol{X}}(k) - \mathrm{E}\left[\tilde{\boldsymbol{X}}(k)\right]\right)\left(\hat{\boldsymbol{X}}(k) - \mathrm{E}\left[\tilde{\boldsymbol{X}}(k)\right]\right)^{\mathrm{T}}\right\}.$$

Taking the expectation of both sides of (2.85) gives the mean value of $\tilde{X}$:

$$\mathrm{E}\left[\tilde{\boldsymbol{X}}(k+1|k)\right] = \boldsymbol{A}\mathrm{E}\left[\tilde{\boldsymbol{X}}(k)\right].$$

Because of $\mathrm{E}[X(0)] = X_0$, if $\mathrm{E}[\hat{X}(0)] = X_0$, the mean value of the state estimation error is zero and is independent of $L(k)$.

$$\begin{aligned} \boldsymbol{P}(k+1) &= \mathrm{E}\left[\tilde{\boldsymbol{X}}(k+1|k)\tilde{\boldsymbol{X}}^{\mathrm{T}}(k+1|k)\right] \\ &= [\boldsymbol{A}-\boldsymbol{L}(k)\boldsymbol{c}]\boldsymbol{P}(k+1|k)[\boldsymbol{A}-\boldsymbol{L}(k)\boldsymbol{c}]^{\mathrm{T}} + \boldsymbol{L}(k)\sigma^2\boldsymbol{L}^{\mathrm{T}}(k). \end{aligned} \tag{2.86}$$

because $\tilde{X}(k)$ and $v(k)$ are independent. Let $P(0)=p_0 I$. From (2.86), we know if $P(k)$ is the nonnegative definite matrix, $P(k+1)$ is also the nonnegative definite matrix, making up $P(k+1)$ with the following form:

$$\begin{aligned} \boldsymbol{P}(k+1) &= \boldsymbol{AP}(k+1|k)\boldsymbol{A}^{\mathrm{T}} - \boldsymbol{L}(k)\boldsymbol{cP}(k+1|k)\boldsymbol{A}^{\mathrm{T}} - \boldsymbol{AP}(k+1|k)\boldsymbol{c}^{T}\boldsymbol{L}^{T}(k) \\ &\quad + \boldsymbol{L}(k)\left[\sigma^2 + \boldsymbol{cP}(k+1|k)\boldsymbol{c}^{\mathrm{T}}\right]\boldsymbol{L}^{\mathrm{T}}(k) \\ &= \boldsymbol{AP}(k+1|k)\boldsymbol{A}^{\mathrm{T}} - \boldsymbol{AP}(k+1|k)\boldsymbol{c}^{\mathrm{T}}\left[\sigma^2+\boldsymbol{cP}(k+1|k)\boldsymbol{c}^{\mathrm{T}}\right]^{-1}\boldsymbol{cP}(k+1|k)\boldsymbol{A}^{\mathrm{T}} \\ &\quad + \left\{\boldsymbol{L}(k) - \boldsymbol{AP}(k+1|k)\boldsymbol{c}^{\mathrm{T}}\left[\sigma^2+\boldsymbol{cP}(k+1|k)\boldsymbol{c}^{\mathrm{T}}\right]^{-1}\right\}\left[\sigma^2 + \boldsymbol{cP}(k+1|k)\boldsymbol{c}^{\mathrm{T}}\right] \\ &\quad \times \left\{\boldsymbol{L}(k)-\boldsymbol{AP}(k+1|k)\boldsymbol{c}^{\mathrm{T}}\left[\sigma^2+\boldsymbol{cP}(k+1|k)\boldsymbol{c}^{\mathrm{T}}\right]^{-1}\right\}^{\mathrm{T}}. \end{aligned} \tag{2.87}$$

By minimizing the estimate error covariance matrix $P(k+1)$, from (2.87), we can conclude the optimal gain vector $L(k)$. The above covariance matrix $P(k+1)$ contains the sum of four items: the top three items on the right side has nothing to do with $L(k)$, because $\sigma^2 + cP(k+1|k)c^{\mathrm{T}}$ is positive. Choosing the gain $L(k)$ to make the right last item of (2.87) to be zero, thus we have

$$\boldsymbol{L}(k) = \boldsymbol{AP}(k+1|k)\boldsymbol{c}^{\mathrm{T}}\left[\sigma^2+\boldsymbol{cP}(k+1|k)\boldsymbol{c}^{\mathrm{T}}\right]^{-1}, \tag{2.88}$$

$$\boldsymbol{P}(k+1) = \boldsymbol{AP}(k+1|k)\boldsymbol{A}^{\mathrm{T}} - \boldsymbol{AP}(k+1|k)\boldsymbol{c}^{\mathrm{T}}\left(\sigma^2+\boldsymbol{cP}(k+1|k)\boldsymbol{c}^{\mathrm{T}}\right)^{-1}\boldsymbol{cP}(k+1|k)\boldsymbol{A}^{\mathrm{T}}. \tag{2.89}$$

Eqs. (2.84), (2.88), and (2.89) are called Kalman Filter, also called one-step ahead Kalman state estimation algorithm, and rewritten as follows,

$$\hat{\boldsymbol{X}}(k+1) = A\hat{\boldsymbol{X}}(k) + \boldsymbol{b}u(k) + \boldsymbol{f}w(k) + \boldsymbol{L}(k)\left[y(k) - c\hat{\boldsymbol{X}}(k)\right], \tag{2.90}$$

$$\boldsymbol{L}(k) = \boldsymbol{AP}(k+1|k)\boldsymbol{c}^{\mathrm{T}}\left[\sigma^2+\boldsymbol{cP}(k+1|k)\boldsymbol{c}^{\mathrm{T}}\right]^{-1}, \tag{2.91}$$

$$\boldsymbol{P}(k+1) = \boldsymbol{AP}(k+1|k)\boldsymbol{A}^{\mathrm{T}} - \boldsymbol{AP}(k+1|k)\boldsymbol{c}^{T}\left[\sigma^2+\boldsymbol{cP}(k+1|k)\boldsymbol{c}^{\mathrm{T}}\right]^{-1}\boldsymbol{cP}(k+1|k)\boldsymbol{A}^{\mathrm{T}}. \tag{2.92}$$

The Kalman filter algorithm of Eqs. (2.80)–(2.81) is deduced below.

Define

$$\boldsymbol{L}(k) = \begin{bmatrix} \boldsymbol{L}_1(k) \\ \boldsymbol{L}_2(k) \end{bmatrix} \in \mathbb{R}^{2n},$$
$$\boldsymbol{P}(k+1|k) = \begin{bmatrix} \boldsymbol{P}_1(k+1|k) & \boldsymbol{P}_{12}(k+1|k) \\ \boldsymbol{P}_{12}^T(k+1|k) & \boldsymbol{P}_2(k+1|k) \end{bmatrix} \in \mathbb{R}^{(2n)\times(2n)}.$$
$$\hat{\boldsymbol{X}}(k|k) := \begin{bmatrix} \hat{\boldsymbol{x}}(k|k) \\ \hat{\boldsymbol{x}}(k-1|k) \end{bmatrix} \in \mathbb{R}^{2n}.$$

From Eq. (2.90), we have

$$\hat{\boldsymbol{x}}(k+1) = \boldsymbol{A}_1\hat{\boldsymbol{x}}(k) + \boldsymbol{A}_d\hat{\boldsymbol{x}}(k-1|k) + \boldsymbol{b}_1 u(k) + \boldsymbol{f}_1 w(k) + \boldsymbol{L}_1(k)[y(k) - \hat{\boldsymbol{x}}(k)],$$
$$\hat{\boldsymbol{x}}(k|k+1) = \hat{\boldsymbol{x}}(k) + \boldsymbol{L}_2(k)[y(k) - \hat{\boldsymbol{x}}(k)].$$

Decomposing Eq. (2.91) gives

$$\boldsymbol{L}_1(k) = \left[\boldsymbol{A}_1\boldsymbol{P}_1(k+1|k) + \boldsymbol{A}_d\boldsymbol{P}_{12}^{\mathrm{T}}(k+1|k)\right]\varsigma^{-1}(k),$$
$$\boldsymbol{L}_2(k) = \boldsymbol{P}_1(k+1|k)\varsigma^{-1}(k).$$

Similarly, from Eq. (2.92), we have

$$\begin{aligned}\boldsymbol{P}_1(k+1) = {} & \left[\boldsymbol{A}_1\boldsymbol{P}_1(k+1|k) + \boldsymbol{A}_d\boldsymbol{P}_{12}^{\mathrm{T}}(k+1|k)\right]\boldsymbol{A}_1^{\mathrm{T}} + [\boldsymbol{A}_1\boldsymbol{P}_{12}(k+1|k) + \boldsymbol{A}_d\boldsymbol{P}_2(k+1|k)]\boldsymbol{A}_d^{\mathrm{T}} \\ & - \left[\boldsymbol{A}_1\boldsymbol{P}_1(k+1|k) + \boldsymbol{A}_d\boldsymbol{P}_{12}^{\mathrm{T}}(k+1|k)\right]\left[\boldsymbol{P}_1(k+1|k)\boldsymbol{A}_1^{\mathrm{T}} + \boldsymbol{P}_{12}(k+1|k)\boldsymbol{A}_d^{\mathrm{T}}\right]\varsigma^{-1}(k),\end{aligned}$$
$$\begin{aligned}\boldsymbol{P}_{12}(k+1) = {} & \boldsymbol{A}_1\boldsymbol{P}_1(k+1|k) + \boldsymbol{A}_d\boldsymbol{P}_{12}^{\mathrm{T}}(k+1|k) - \left[\boldsymbol{A}_1\boldsymbol{P}_1(k+1|k) + \boldsymbol{A}_d\boldsymbol{P}_{12}^{\mathrm{T}}(k+1|k)\right] \\ & \times \boldsymbol{P}_1(k+1|k)\varsigma^{-1}(k),\end{aligned}$$
$$\boldsymbol{P}_2(k+1) = \boldsymbol{P}_1(k+1|k) - \boldsymbol{P}_1^2(k+1|k)\varsigma^{-1}(k).$$

If the parameter matrices/vectors $A_1$, $A_d$, $b_1$, and $f_1$ are known, then apply the following Kalman filter to generate the estimate $\hat{x}(k)$ of the state vector $x(k)$:

$$\hat{\boldsymbol{x}}(k+1) = \boldsymbol{A}_1\hat{\boldsymbol{x}}(k) + \boldsymbol{A}_d\hat{\boldsymbol{x}}(k-1|k) + \boldsymbol{b}_1 u(k) + \boldsymbol{f}_1 w(k) + \boldsymbol{L}_1(k)[y(k) - \hat{\boldsymbol{x}}(k)],$$
$$\hat{\boldsymbol{x}}(k|k+1) = \hat{\boldsymbol{x}}(k) + \boldsymbol{L}_2(k)[y(k) - \hat{\boldsymbol{x}}(k)],$$
$$\boldsymbol{L}_1(k) = \left[\boldsymbol{A}_1\boldsymbol{P}_1(k+1|k) + \boldsymbol{A}_d\boldsymbol{P}_{12}^{\mathrm{T}}(k+1|k)\right]\varsigma^{-1}(k),$$
$$\boldsymbol{L}_2(k) = \boldsymbol{P}_1(k+1|k)\varsigma^{-1}(k),$$
$$\begin{aligned}\boldsymbol{P}_1(k+1) = {} & \left[\boldsymbol{A}_1\boldsymbol{P}_1(k+1|k) + \boldsymbol{A}_d\boldsymbol{P}_{12}^{\mathrm{T}}(k+1|k)\right]\boldsymbol{A}_1^{\mathrm{T}} + [\boldsymbol{A}_1\boldsymbol{P}_{12}(k+1|k) + \boldsymbol{A}_d\boldsymbol{P}_2(k+1|k)]\boldsymbol{A}_d^{\mathrm{T}} \\ & - \left[\boldsymbol{A}_1\boldsymbol{P}_1(k+1|k) + \boldsymbol{A}_d\boldsymbol{P}_{12}^{\mathrm{T}}(k+1|k)\right]\left[\boldsymbol{P}_1(k+1|k)\boldsymbol{A}_1^{\mathrm{T}} + \boldsymbol{P}_{12}(k+1|k)\boldsymbol{A}_d^{\mathrm{T}}\right]\varsigma^{-1}(k),\end{aligned}$$
$$\begin{aligned}\boldsymbol{P}_{12}(k+1) = {} & \boldsymbol{A}_1\boldsymbol{P}_1(k+1|k) + \boldsymbol{A}_d\boldsymbol{P}_{12}^{\mathrm{T}}(k+1|k) \\ & - \left[\boldsymbol{A}_1\boldsymbol{P}_1(k+1|k) + \boldsymbol{A}_d\boldsymbol{P}_{12}^{\mathrm{T}}(k+1|k)\right]\boldsymbol{P}_1(k+1|k)\varsigma^{-1}(k),\end{aligned}$$
$$\boldsymbol{P}_2(k+1) = \boldsymbol{P}_1(k+1|k) - \boldsymbol{P}_1^2(k+1|k)\varsigma^{-1}(k),$$

where $\varsigma(k) = \sigma^2 + P_1(k+1|k)$.

When the parameter matrices/vectors $A_1$, $A_d$, $b_1$, and $f_1$ are unknown, then we use the estimated parameter vector $\hat{\theta}(k)=[\hat{A}_1(k),\hat{A}_d(k),\hat{b}_1(k),\hat{f}_1(k)]^{\mathrm{T}}$ to construct the estimates $\hat{A}_1(k)$, $\hat{A}_d(k)$, $\hat{b}_1(k)$, and $\hat{f}_1(k)$ of $A_1$, $A_d$, $b_1$, and $f_1$ and use the estimates to compute the estimate $\hat{x}(k)$ of the state vector $x(k)$:

$$\hat{\boldsymbol{x}}(k+1)=\hat{\boldsymbol{A}}_1(k)\hat{\boldsymbol{x}}(k)+\hat{\boldsymbol{A}}_2(k)\hat{\boldsymbol{x}}(k-1|k)+\hat{\boldsymbol{b}}_1(k)u(k)+\hat{\boldsymbol{f}}_1(k)w(k)+\boldsymbol{L}_1(k)[y(k)-\hat{\boldsymbol{x}}(k)],$$

$$\hat{\boldsymbol{x}}(k|k+1)=\hat{\boldsymbol{x}}(k)+\boldsymbol{L}_2(k)[y(k)-\hat{\boldsymbol{x}}(k)],$$

$$\boldsymbol{L}_1(k)=\left[\hat{\boldsymbol{A}}_1(k)\boldsymbol{P}_1(k+1|k)+\hat{\boldsymbol{A}}_2(k)\boldsymbol{P}_{12}^{\mathrm{T}}(k+1|k)\right]\varsigma^{-1}(k),$$

$$\boldsymbol{L}_2(k)=\boldsymbol{P}_1(k+1|k)\varsigma^{-1}(k),$$

$$\begin{aligned}
\boldsymbol{P}_1(k+1)=&\left[\hat{\boldsymbol{A}}_1(k)\boldsymbol{P}_1(k+1|k)+\hat{\boldsymbol{A}}_2(k)\boldsymbol{P}_{12}^{\mathrm{T}}(k+1|k)\right]\hat{\boldsymbol{A}}_1^{\mathrm{T}}(k)+[\hat{\boldsymbol{A}}_1(k)\boldsymbol{P}_{12}(k+1|k)\\
&+\hat{\boldsymbol{A}}_2(k)\boldsymbol{P}_2(k+1|k)]\hat{\boldsymbol{A}}_2^{\mathrm{T}}(k)-\left[\hat{\boldsymbol{A}}_1(k)\boldsymbol{P}_1(k+1|k)+\hat{\boldsymbol{A}}_2(k)\boldsymbol{P}_{12}^{\mathrm{T}}(k+1|k)\right]\\
&\times\left[\boldsymbol{P}_1(k+1|k)\hat{\boldsymbol{A}}_1^{\mathrm{T}}(k)+\boldsymbol{P}_{12}(k+1|k)\hat{\boldsymbol{A}}_2^{\mathrm{T}}(k)\right]\varsigma^{-1}(k),
\end{aligned}$$

$$\begin{aligned}
\boldsymbol{P}_{12}(k+1)=&\hat{\boldsymbol{A}}_1(k)\boldsymbol{P}_1(k+1|k)+\hat{\boldsymbol{A}}_2(k)\boldsymbol{P}_{12}^{\mathrm{T}}(k+1|k)\\
&-\left[\hat{\boldsymbol{A}}_1(k)\boldsymbol{P}_1(k+1|k)+\hat{\boldsymbol{A}}_2(k)\boldsymbol{P}_{12}^{\mathrm{T}}(k+1|k)\right]\boldsymbol{P}_1(k+1|k)\varsigma^{-1}(k),
\end{aligned}$$

$$\boldsymbol{P}_2(k+1)=\boldsymbol{P}_1(k+1|k)-\boldsymbol{P}_1^2(k+1|k)\varsigma^{-1}(k),$$

where $\varsigma(k)=\sigma^2+P_1(k+1|k)$.

**Remark 7**: According to the state equation of different time, the state vector is described by the measurable input–output variables, and the identification model of the system is derived. Then, the model algorithm is generalized, and the corresponding residual-based enhanced least squares algorithm is derived. The estimated parameters are used to compute the system state. The proposed algorithm has low computational cost and high accuracy.

### 2.4.3 Recursive parameter estimation

The following gives the state estimation-based recursive least squares (SE-RLS) parameter identification algorithm:

$$\hat{\boldsymbol{\theta}}'(k)=\hat{\boldsymbol{\theta}}'(k-1)+\boldsymbol{P}(k)\hat{\phi}(k)\left[y(k)-\hat{\phi}^{\mathrm{T}}(k)\hat{\boldsymbol{\theta}}'(k-1)\right], \tag{2.93}$$

$$\boldsymbol{P}^{-1}(k)=\boldsymbol{P}^{-1}(k-1)+\hat{\phi}(k)\hat{\phi}^{\mathrm{T}}(k),\quad \boldsymbol{P}(0)=p_0\boldsymbol{I}, \tag{2.94}$$

$$\hat{\phi}(k)=\left[\hat{\boldsymbol{x}}^{\mathrm{T}}(k-2),\ \hat{\boldsymbol{x}}^{\mathrm{T}}(k-3),\cdots,u(k-1),\ u(k-2),\cdots,u(k-n)\right]^{\mathrm{T}}. \tag{2.95}$$

Define the parameter matrix $\theta'$ and the information vector $\varphi(k)$ as

$$\boldsymbol{\theta}' := \left[\boldsymbol{A}_1,\ \ \boldsymbol{A}_d,\ \ \boldsymbol{b}_1,\ \ \boldsymbol{f}_1\right] \in \mathbb{R}^{n \times (2n+2)}, \tag{2.96}$$

$$\phi(k) := \left[\boldsymbol{x}^{\mathrm{T}}(k),\ \ \boldsymbol{x}^{\mathrm{T}}(k-d),\ \ u(k),\ \ w(k)\right]^{\mathrm{T}} \in \mathbb{R}^{2n+2}. \tag{2.97}$$

Eq. (2.80) is equivalently written as

$$\boldsymbol{x}(k+1) = \boldsymbol{\theta}'\phi(k) \tag{2.98}$$

Eq. (2.98) is the identification model of the system, where the estimate of the parameter matrix $\theta'$ is obtained by the following least squares equation,

$$\hat{\boldsymbol{\theta}}'(k) = \left[\sum_{j=1}^{t} \boldsymbol{x}(j+1)\phi^{\mathrm{T}}(j)\right]\left[\sum_{j=1}^{t} \phi(j)\phi^{\mathrm{T}}(j)\right]^{-1},$$

which is computed from the state information and input information $\{x(k), u(k), w(k)\}$.

Eq. (2.96) shows that the parameter matrix $A$ in $\theta'$ contains a great number of zero elements, this will increase computation, and however, these zero elements do not need to estimate in practice. Decomposing the original identification problem into two subproblems with small sizes to reduce computation.

Substituting the parameters $A_1$, $A_d$, $b_1$, and $f_1$ into (2.80), we have

$$\begin{aligned}\boldsymbol{x}(k+1) = {} & \left[x_2(k),\ \ x_3(k), \ldots, x_n(k),\ \ \boldsymbol{a}^{\mathrm{T}}\boldsymbol{x}^{\mathrm{T}}(k)\right]^{\mathrm{T}} + \left[\boldsymbol{a}_{d1}^{\mathrm{T}},\ \ \boldsymbol{a}_{d2}^{\mathrm{T}},\ \ \ldots \boldsymbol{a}_{d,n-1}^{\mathrm{T}},\ \ \boldsymbol{a}_{dn}^{\mathrm{T}}\right]^{\mathrm{T}}\boldsymbol{x}(k-d)\\ & + [b_1,\ \ b_2, \ldots, b_{n-1},\ \ b_n]^{\mathrm{T}}u(k) + \left[f_1,\ \ f_2, \ldots, f_n\right]^{\mathrm{T}}w(k),\\ \boldsymbol{a} := {} & [a_n,\ \ a_{n-1}, \ldots, a_1].\end{aligned}$$

Decompose the above equation into two subsystems:

$$\begin{aligned}S_1{:}\boldsymbol{x}_{n-1}(k+1) - \boldsymbol{\xi}(k) = {} & \left[\boldsymbol{a}_{d1}^{\mathrm{T}}, \ldots, \boldsymbol{a}_{d,n-1}^{\mathrm{T}}\right]^{\mathrm{T}}\boldsymbol{x}(k-d) + [b_1,\ \ b_2, \ldots, b_{n-1}]^{\mathrm{T}}u(k)\\ & + \left[f_1, f_2,\ \ \ldots, f_{n-1}\right]^{\mathrm{T}}w(k),\end{aligned} \tag{2.99}$$

$$S_2{:}\boldsymbol{x}_n(k+1) = \boldsymbol{a}\boldsymbol{x}(k) + \boldsymbol{a}_{dn}\boldsymbol{x}(k-d) + b_n u(k) + f_n w(k), \tag{2.100}$$

where

$$\begin{aligned}&\boldsymbol{x}_{n-1}(k) := [x_1(k),\ \ x_2(k), \cdots, x_{n-1}(k)]^{\mathrm{T}} \in \mathbb{R}^{n-1},\\ &\boldsymbol{\xi}(k) := [x_2(k),\ \ x_3(k), \cdots, x_n(k)]^{\mathrm{T}} \in \mathbb{R}^{n-1},\\ &\boldsymbol{x}(k) := \begin{bmatrix}\boldsymbol{x}_{n-1}(k)\\ x_n(k)\end{bmatrix} \in \mathbb{R}^n,\end{aligned}$$

Define the parameter matrix $\theta$ and the information vector $\phi_1(k)$ of subsystem $S_1$ as

$$\boldsymbol{\theta}^{\mathrm{T}} := \begin{bmatrix} \boldsymbol{a}_{d1} & b_1 & f_1 \\ \boldsymbol{a}_{d2} & b_2 & f_2 \\ \vdots & \vdots & \vdots \\ \boldsymbol{a}_{d,\ n-1} & b_{n-1} & f_{n-1} \end{bmatrix} \in \mathbb{R}^{(n-1)\times(n+2)},$$

$$\varphi_1(k) := \begin{bmatrix} \boldsymbol{x}(k-d) \\ u(k) \\ w(k) \end{bmatrix} \in \mathbb{R}^{n+2}, \tag{2.101}$$

and the parameter vector $\vartheta$ and the information vector $\phi_2(k)$ of subsystem $S_2$ as

$$\vartheta := \left[\boldsymbol{a},\ \boldsymbol{a}_m,\ b_n,\ f_n\right]^{\mathrm{T}} \in \mathbb{R}^{2n+2},$$

$$\varphi_2(k) := \left[\boldsymbol{x}^{\mathrm{T}}(k),\ \boldsymbol{x}^{\mathrm{T}}(k-d),\ u(k),\ w(k)\right]^{\mathrm{T}} \in \mathbb{R}^{2n+2}.$$

The identification models of subsystems $S_1$ and $S_2$ are as follows

$$S_1:\ \boldsymbol{x}_{n-1}(k+1) - \boldsymbol{\xi}(k) = \boldsymbol{\theta}^{\mathrm{T}}\varphi_1(k), \tag{2.102}$$

$$S_2:\ x_n(k+1) = \varphi_2^{\mathrm{T}}(k)\vartheta. \tag{2.103}$$

**Remark 8**: The identification model in (2.98) contains $n(2n+2)$ parameters, subsystem (2.102) contains $(n-1)\times(n+2)$ parameters in $\theta$, and subsystem (2.103) contains $2n+2$ parameters in $\vartheta$. Such a decomposition leads to less amount of calculation and storage.

Because the multiple output $y(k)$ in (2.80) involves only the white noise $v(k)$, we will derive an algorithm for estimating $\theta$ and $\vartheta$ using the input–output data $u(k)$ and $y(k)$.

Based on the identification model in (2.102)–(2.103), defining and minimizing a quadratic criterion function, the state estimation-based recursive least squares (SE-RLS) parameter identification algorithm is as follows:

$$\hat{\boldsymbol{\theta}}(k) = \hat{\boldsymbol{\theta}}(k-1) + \boldsymbol{P}_1(k)\hat{\varphi}_1(k)\left[\boldsymbol{y}_{n-1}(k) - \boldsymbol{\xi}(k-1) - \hat{\boldsymbol{\theta}}^{\mathrm{T}}\hat{\varphi}_1(k-1)\right], \tag{2.104}$$

$$\hat{\vartheta}(k) = \hat{\vartheta}(k-1) + \boldsymbol{P}_2(k)\hat{\varphi}_2(k)\left[y_n(k) - \hat{\varphi}_2^{\mathrm{T}}(k-1)\hat{\vartheta}\right], \tag{2.105}$$

$$\boldsymbol{P}_1^{-1}(k) = \boldsymbol{P}_1^{-1}(k-1) + \hat{\varphi}_1(k)\hat{\varphi}_1^{\mathrm{T}}(k), \tag{2.106}$$

$$\boldsymbol{P}_2^{-1}(k) = \boldsymbol{P}_2^{-1}(k-1) + \hat{\varphi}_2(k)\hat{\varphi}_2^{\mathrm{T}}(k), \tag{2.107}$$

$$\varphi_1(k) = \left[\boldsymbol{x}^{\mathrm{T}}(k-d),\ \ u(k), w(k)\right]^{\mathrm{T}}, \tag{2.108}$$

$$\varphi_2(k) = \left[\boldsymbol{x}^{\mathrm{T}}(k),\ \ \boldsymbol{x}^{\mathrm{T}}(k-d),\ \ u(k), w(k)\right]^{\mathrm{T}}, \tag{2.109}$$

Equations (2.104)–(2.109) form the recursive least squares algorithm of dual-rate system whose initial values $\hat{\theta}(0)$ and $\hat{\vartheta}(0)$ are taken as zero matrix or zero vector of appropriate sizes, $\hat{y}(i)$, $\hat{\phi}_1(i)$, $\hat{\phi}_2(i)$, $\hat{x}(i)$, $\hat{w}(i)$, $\hat{v}(i)$, $u(i)$, and $y(i)$ as zero vectors or zero matrices of appropriate sizes for $i \leq 0$, and $P_1(0) = p_0 I$, $P_2(0) = p_0 I$, $p_0 = 10^6$, $I$ is an identity matrix of appropriate dimensions.

**Theorem 2**: For the system in (2.80)–(2.81) and the RLS algorithm, suppose that $\{v(k)\}$ is a white noise sequence with zero mean, $\mathrm{E}[v(k)] = 0$, $\mathrm{E}[v(k)v^{\mathrm{T}}(i)] = 0$, $t \neq i$,$\mathrm{E}[||v(k)||^2] = \sigma_v^2(k) \leq \sigma_v^2 < \infty$, and that there exist constant $0 < \alpha_1 \leq \beta_1 < \infty$, $0 < \alpha_2 \leq \beta_2 < \infty$ and $k_0 > 0$ for $k \geq k_0$, the persistent excitation condition holds:

$$\alpha_1 \boldsymbol{I} \leq \frac{1}{k}\sum_{i=1}^{k} \varphi_1(i)\varphi_1^{\mathrm{T}}(i) \leq \beta_1 \boldsymbol{I},\ \ a.s.,$$

$$\alpha_2 \boldsymbol{I} \leq \frac{1}{k}\sum_{i=1}^{k} \varphi_2(i)\varphi_2^{\mathrm{T}}(i) \leq \beta_2 \boldsymbol{I},\ \ a.s.,$$

assume that $\mathrm{E}[||\hat{\theta}(0) - \theta||^2] = \delta_1 < \infty$, $\mathrm{E}[||\hat{\vartheta}(0) - \vartheta||^2] = \delta_2 < \infty$, $\hat{\theta}(0)$ and $v_{n-1}(k)$ are uncorrelated, $\hat{\vartheta}(0)$ and $v_n(k)$ are uncorrelated. Then, the parameter estimates $\hat{\theta}(k)$ and $\hat{\vartheta}(k)$ given by the RLS algorithm converge uniformly to the true parameters $\theta$ and $\vartheta$ at the speed of $(1/\sqrt{k})$,

$$\begin{aligned}\mathrm{E}\left[||\hat{\boldsymbol{\theta}}(t) - \boldsymbol{\theta}||^2\right] &\leq \frac{2\lambda_{\max}^2\left[\boldsymbol{P}_1^{-1}(0)\right]\delta_1}{\alpha_1^2 k^2} + \frac{2n\sigma_v^2}{\alpha_1 k} \\ &= \frac{2\delta_1}{\alpha_1^2 p_1^2 k^2} + \frac{2n\sigma_v^2}{\alpha_1 k} =: f_1(k),\ k \geq k_0,\end{aligned}$$

or

$$\lim_{t\to\infty} \hat{\boldsymbol{\theta}}(k) = \boldsymbol{\theta},\ \ m.s.,$$

(C2)

$$\begin{aligned}
\mathrm{E}\left[\|\hat{\vartheta}(k)-\vartheta\|^2\right] &\leq \frac{2\lambda_{\max}^2\left[\boldsymbol{P}_2^{-1}(0)\right]\delta_2}{\alpha_2^2k^2}+\frac{2n\sigma_v^2}{\alpha_2 k}\\
&=\frac{2\delta_2}{\alpha_2^2p_2^2k^2}+\frac{2n\sigma_v^2}{\alpha_2 k}=:f_2(k),\quad k\geq k_0,
\end{aligned}$$

or

$$\lim_{k\to\infty}\hat{\vartheta}(k)=\vartheta,\quad m.s.,$$

where $\lambda_{\max}[X]$ is the maximum eigenvalue of matrix $X$.

**Proof** Define the parameter estimation error matrix/vector: $\tilde{\theta}(k):=\hat{\theta}(k)-\theta$, $\tilde{\vartheta}(k):=\hat{\vartheta}(k)-\vartheta$.

Minusing $\theta$ for both sides of Eq. (2.104) to get

$$\begin{aligned}
\tilde{\boldsymbol{\theta}}(k)&:=\hat{\boldsymbol{\theta}}(k-1)-\boldsymbol{\theta}+\boldsymbol{P}_1(k)\boldsymbol{\varphi}_1(k)\left[\boldsymbol{\varphi}_1^{\mathrm{T}}(k)\boldsymbol{\theta}+\boldsymbol{v}_{n-1}^{\mathrm{T}}(k)-\boldsymbol{\varphi}_1^{\mathrm{T}}(k)\hat{\boldsymbol{\theta}}(k-1)\right]\\
&=\tilde{\boldsymbol{\theta}}(k-1)+\boldsymbol{P}_1(k)\boldsymbol{\varphi}_1(k)\left[-\boldsymbol{\varphi}_1^{\mathrm{T}}(k)\tilde{\boldsymbol{\theta}}(k-1)+\boldsymbol{v}_{n-1}^{\mathrm{T}}(k)\right]\\
&=\left[\boldsymbol{I}-\boldsymbol{P}_1(k)\boldsymbol{\varphi}_1(k)\boldsymbol{\varphi}_1^{\mathrm{T}}(k)\right]\left[\tilde{\boldsymbol{\theta}}(k-1)+\boldsymbol{P}_1(k)\boldsymbol{\varphi}_1(k)\boldsymbol{v}_{n-1}^{\mathrm{T}}(k)\right],
\end{aligned}$$

$$\begin{aligned}
\tilde{\vartheta}(k)&:=\tilde{\vartheta}(k-1)-\vartheta+\boldsymbol{P}_2(k)\boldsymbol{\varphi}_2(k)\left[\boldsymbol{\varphi}_2^{\mathrm{T}}(k)\vartheta+v_n(k)-\boldsymbol{\varphi}_2^{\mathrm{T}}(k)\tilde{\vartheta}(k-1)\right]\\
&=\tilde{\vartheta}(k-1)+\boldsymbol{P}_2(k)\boldsymbol{\varphi}_2(k)\left[-\boldsymbol{\varphi}_2^{\mathrm{T}}(k)\tilde{\vartheta}(k-1)+v_n(k)\right]\\
&=\left[\boldsymbol{I}-\boldsymbol{P}_2(k)\boldsymbol{\varphi}_2(k)\boldsymbol{\varphi}_2^{\mathrm{T}}(k)\right]\left[\tilde{\vartheta}(k-1)+\boldsymbol{P}_2(k)\boldsymbol{\varphi}_2(k)\boldsymbol{v}_n(k)\right].
\end{aligned}$$

Pre-multiplying both sides of Eq. (2.106) by $P_1(k)$ and bringing in the above equation to get

$$\begin{aligned}
\tilde{\boldsymbol{\theta}}(k)&=\boldsymbol{P}_1(k)\boldsymbol{P}_1^{-1}(k-1)\tilde{\boldsymbol{\theta}}(k-1)+\boldsymbol{P}_1(k)\boldsymbol{\varphi}_1(k)\boldsymbol{v}_{n-1}^{\mathrm{T}}(k)\\
&=\boldsymbol{P}_1(k)\boldsymbol{P}_1^{-1}(k-1)\left[\boldsymbol{P}_1(k-1)\boldsymbol{P}_1^{-1}(k-2)\tilde{\boldsymbol{\theta}}(k-2)\right.\\
&\quad\left.+\boldsymbol{P}_1(k-1)\boldsymbol{\varphi}_1(k-1)\boldsymbol{v}_{n-1}^{\mathrm{T}}(k-1)\right]+\boldsymbol{P}_1(k)\boldsymbol{\varphi}_1(k)\boldsymbol{v}_{n-1}^{\mathrm{T}}(k)\\
&=\boldsymbol{P}_1(k)\boldsymbol{P}_1^{-1}(k-2)\tilde{\boldsymbol{\theta}}(k-2)+\boldsymbol{P}_1(k)\boldsymbol{\varphi}_1(k-1)\boldsymbol{v}_{n-1}^{\mathrm{T}}(k-1)+\boldsymbol{P}_1(k)\boldsymbol{\varphi}_1(k)\boldsymbol{v}_{n-1}^{\mathrm{T}}(k)\\
&=\boldsymbol{P}_1(k)\boldsymbol{P}_1^{-1}(0)\tilde{\boldsymbol{\theta}}(0)+\boldsymbol{P}_1(k)\sum_{i-1}^{k}\boldsymbol{\varphi}_1(i)\boldsymbol{v}_{n-1}^{\mathrm{T}}(i)=:\gamma_1(k)+\gamma_2(k),
\end{aligned}\tag{2.110}$$

$$\begin{aligned}\tilde{\vartheta}(k) &= \boldsymbol{P}_2(k)\boldsymbol{P}_2^{-1}(k-1)\tilde{\vartheta}(k-1) + \boldsymbol{P}_2(k)\boldsymbol{\varphi}_2(k)\boldsymbol{v}_n(k) \\ &= \boldsymbol{P}_2(k)\boldsymbol{P}_2^{-1}(k-1)\left[\boldsymbol{P}_2(k-1)\boldsymbol{P}_2^{-1}(k-2)\tilde{\vartheta}(k-2)\right. \\ &\quad \left.+ \boldsymbol{P}_2(k-1)\boldsymbol{\varphi}_2(k-1)\boldsymbol{v}_n(k-1)\right] + \boldsymbol{P}_2(k)\boldsymbol{\varphi}_2(k)\boldsymbol{v}_n(k) \\ &= \boldsymbol{P}_2(k)\boldsymbol{P}_2^{-1}(k-2)\tilde{\vartheta}(k-2) + \boldsymbol{P}_2(k)\boldsymbol{\varphi}_2(k-1)\boldsymbol{v}_n(k-1) + \boldsymbol{P}_2(k)\boldsymbol{\varphi}_2(k)\boldsymbol{v}_n(k) \\ &= \boldsymbol{P}_2(k)\boldsymbol{P}_2^{-1}(0)\tilde{\vartheta}(0) + \boldsymbol{P}_2(k)\sum_{i-1}^{k}\boldsymbol{\varphi}_2(i)\boldsymbol{v}_n(i) \\ &=: \eta_1(k) + \eta_2(k),\end{aligned} \tag{2.111}$$

where

$$\gamma_1(k) = \boldsymbol{P}_1(k)\boldsymbol{P}_1^{-1}(0)\tilde{\boldsymbol{\theta}}(0),$$

$$\gamma_2(k) = \boldsymbol{P}_1(k)\sum_{i=1}^{k}\boldsymbol{\varphi}_1(i)\boldsymbol{v}_{n-1}^{\mathrm{T}}(i),$$

$$\eta_1(k) = \boldsymbol{P}_2(k)\boldsymbol{P}_2^{-1}(0)\tilde{\vartheta}(0),$$

$$n_2(k) = \boldsymbol{P}_2(k)\sum_{i=1}^{k}\boldsymbol{\varphi}_2(i)\boldsymbol{v}_n(i).$$

Let

$$\boldsymbol{\Phi}_{k1} = \left[\boldsymbol{\varphi}_1(1), \boldsymbol{\varphi}_1(2), \ldots, \boldsymbol{\varphi}_1(k)\right]^{\mathrm{T}},$$

$$\boldsymbol{V}_{k1} = [\boldsymbol{v}_{n-1}(1), \boldsymbol{v}_{n-1}(2), \ldots, \boldsymbol{v}_{n-1}(k)]^{\mathrm{T}},$$

$$\boldsymbol{\Phi}_{k2} = \left[\boldsymbol{\varphi}_2(1), \boldsymbol{\varphi}_2(2), \ldots, \boldsymbol{\varphi}_2(k)\right]^{\mathrm{T}},$$

$$\boldsymbol{V}_{k2} = [v_n(1), v_n(2), \ldots, v_n(k)]^{\mathrm{T}}.$$

Combining the above definitions and Eq. (2.106) gives

$$\begin{aligned}\boldsymbol{P}_1^{-1}(k) &= \boldsymbol{\Phi}_{k1}^{\mathrm{T}}\boldsymbol{\Phi}_{k1} + \boldsymbol{P}_1^{-1}(0) \\ &= \sum_{i=1}^{k}\boldsymbol{\varphi}_1(i)\boldsymbol{\varphi}_1^{\mathrm{T}}(i) + \boldsymbol{P}_1^{-1}(0),\end{aligned}$$

$$\begin{aligned}\boldsymbol{P}_1^{-1}(k) &= \boldsymbol{\Phi}_{k2}^{\mathrm{T}}\boldsymbol{\Phi}_{k2} + \boldsymbol{P}_2^{-1}(0) \\ &= \sum_{i=1}^{k}\boldsymbol{\varphi}_2(i)\boldsymbol{\varphi}_2^{\mathrm{T}}(i) + \boldsymbol{P}_2^{-1}(0).\end{aligned}$$

According to the persistent excitation condition, we have

$$\begin{aligned}
\alpha_1 k\boldsymbol{I} &\leq \boldsymbol{\Phi}_{k1}^{\mathrm{T}}\boldsymbol{\Phi}_{k1} \leq \beta_1 k\boldsymbol{I}, \ a.s., \\
\alpha_1 k\boldsymbol{I} + \boldsymbol{P}_1^{-1}(0) &\leq \boldsymbol{P}_1^{-1}(k) \leq \beta_1 k\boldsymbol{I} + \boldsymbol{P}_1^{-1}(0), \quad a.s., \\
\frac{\boldsymbol{I}}{\beta_1 k + 1/p_1} &\leq \boldsymbol{P}_1(k) \leq \frac{\boldsymbol{I}}{\alpha_1 k}, \quad a.s., \\
\alpha_2 k\boldsymbol{I} &\leq \boldsymbol{\Phi}_{k2}^{\mathrm{T}}\boldsymbol{\Phi}_{k2} \leq \beta_2 k\boldsymbol{I}, \quad a.s., \\
\alpha_2 k\boldsymbol{I} + \boldsymbol{P}_2^{-1}(0) &\leq \boldsymbol{P}_2^{-1}(k) \leq \beta_2 k\boldsymbol{I} + \boldsymbol{P}_2^{-1}(0), \quad a.s., \\
\frac{\boldsymbol{I}}{\beta_2 k + 1/p_2} &\leq \boldsymbol{P}_2(k) \leq \frac{\boldsymbol{I}}{\alpha_2 k}, \quad a.s.
\end{aligned}$$

Thus, we have

$$\begin{aligned}
\mathrm{tr}\left[\boldsymbol{P}_1(k)\boldsymbol{\Phi}_{k1}^{\mathrm{T}}\boldsymbol{\Phi}_{k1}\right] &= \mathrm{tr}\left[\boldsymbol{I} - \boldsymbol{P}_1(k)\boldsymbol{P}_1^{-1}(0)\right] \\
&= \mathrm{tr}\left[\boldsymbol{I} - \boldsymbol{P}_1(k)/p_1\right] \leq n, \\
\mathrm{E}\left[||\gamma_1(k)||^2\right] &= \mathrm{E}\left\{\mathrm{tr}\left[\tilde{\boldsymbol{\theta}}^{\mathrm{T}}(0)\boldsymbol{P}_1^{-1}(0)\boldsymbol{P}_1^2(k)\boldsymbol{P}_1^{-1}(0)\tilde{\boldsymbol{\theta}}(0)\right]\right\} \\
&\leq \lambda_{\max}^2[\boldsymbol{P}_1(k)]\mathrm{E}\left[\tilde{\boldsymbol{\theta}}^{\mathrm{T}}(0)\boldsymbol{P}_1^{-2}(0)\tilde{\boldsymbol{\theta}}(0)\right] \\
&= \frac{\lambda_{\max}^2\left[\boldsymbol{P}_1^{-1}(0)\right]\delta_1}{\alpha_1^2 k^2}, k \geqslant k_0, \\
\mathrm{E}\left[||\gamma_2(k)||^2\right] &= \mathrm{E}\left[||\boldsymbol{P}_1(k)\sum_{i=1}^{k}\varphi_1(i)\boldsymbol{v}_{n-1}^{\mathrm{T}}(i)||^2\right] \\
&= E\left[||\boldsymbol{P}_1(k)\boldsymbol{\Phi}_{k1}^{\mathrm{T}}\boldsymbol{V}_{k1}||^2\right] \\
&= \mathrm{E}\left\{\mathrm{tr}\left[\boldsymbol{P}_1(k)\boldsymbol{\Phi}_{k1}^{\mathrm{T}}\boldsymbol{V}_{k1}\boldsymbol{V}_{k1}^{\mathrm{T}}\boldsymbol{\Phi}_{k1}\boldsymbol{P}_1(k)\right]\right\} \\
&\leq \mathrm{tr}\left\{\mathrm{E}\left[\boldsymbol{\Phi}_{k1}\boldsymbol{P}_1^2(k)\boldsymbol{\Phi}_{k1}^{\mathrm{T}}\right]\right\}\sigma_v^2 \\
&\leq \frac{n\sigma_v^2}{\alpha_1 k}, k \geqslant k_0, \\
\mathrm{tr}\left[\boldsymbol{P}_2(k)\boldsymbol{\Phi}_{k2}^{\mathrm{T}}\boldsymbol{\Phi}_{k2}\right] &= \mathrm{tr}\left[\boldsymbol{I} - \boldsymbol{P}_2(k)/p_2\right] \leq n, \\
\mathrm{E}\left[||\eta_1(k)||^2\right] &= \frac{\lambda_{\max}^2\left[\boldsymbol{P}_2^{-1}(0)\right]\delta_2}{\alpha_2^2 k^2}, k \geqslant k_0, \\
\mathrm{E}\left[||\eta_2(k)||^2\right] &= \mathrm{E}\left[||\boldsymbol{P}_2(k)\sum_{i=1}^{k}\varphi_2(i)v_2(i)||^2\right] \\
&\leq \frac{n\sigma_v^2}{\alpha_2 k}, k \geqslant k_0.
\end{aligned}$$

Taking the norms and expectations for both sides of Eq. (2.111) to get

$$\begin{aligned}0 \leq \mathrm{E}\left[||\tilde{\boldsymbol{\theta}}(k)||^2\right] &= \mathrm{E}\left[||\gamma_1(k)+\gamma_2(k)||^2\right]\\ &= \mathrm{E}\left[||\gamma_1(k)||^2\right]+\mathrm{E}\left[||\gamma_2(k)||^2\right]+\mathrm{tr}\left\{\mathrm{E}\left[\gamma_1^{\mathrm{T}}(k)\gamma_2(k)\right]\right\}\\ &\leq 2\mathrm{E}\left[||\gamma_1(k)||^2\right]+2\mathrm{E}\left[||\gamma_2(k)||^2\right]\\ &\leq \frac{2\lambda_{\max}^2\left[\boldsymbol{P}_1^{-1}(0)\right]\delta_1}{\alpha_1^2k^2}+\frac{2n\sigma_v^2}{\alpha_1 k}, k \geqslant k_0,\end{aligned}$$

$$0 \leq \mathrm{E}\left[||\tilde{\vartheta}(k)||^2\right] \leq \frac{2\lambda_{\max}^2\left[\boldsymbol{P}_2^{-1}(0)\right]\delta_2}{\alpha_2^2k^2}+\frac{2n\sigma_v^2}{\alpha_2 k}, k \geqslant k_0.$$

If $\mathrm{tr}\left\{\mathrm{E}[\gamma_1^{\mathrm{T}}(k)\gamma_2(k)]\right\}=0$, $\mathrm{tr}\left\{\mathrm{E}[\eta_1^{\mathrm{T}}(k)\eta_2(k)]\right\}=0$, the following equations about the upper bound of the estimation error hold:

$$\begin{aligned}\mathrm{E}\left[||\tilde{\boldsymbol{\theta}}(k)||^2\right] &= \mathrm{E}\left[||\gamma_1(k)+\gamma_2(k)||^2\right]\\ &= \mathrm{E}\left[||\gamma_1(k)||^2\right]+\mathrm{E}\left[||\gamma_2(k)||^2\right]+\mathrm{tr}\left\{\mathrm{E}\left[\gamma_1^{\mathrm{T}}(k)\gamma_2(k)\right]\right\}\\ &\leq \mathrm{E}\left[||\gamma_1(k)||^2\right]+\mathrm{E}\left[||\gamma_2(k)||^2\right]\\ &\leq \frac{\lambda_{\max}^2\left[\boldsymbol{P}_1^{-1}(0)\right]\delta_1}{\alpha_1^2k^2}+\frac{n\sigma_v^2}{\alpha_1 k}, k \geqslant k_0,\end{aligned}$$

$$\mathrm{E}\left[||\tilde{\vartheta}(k)||^2\right] \leq \frac{\lambda_{\max}^2\left[\boldsymbol{P}_2^{-1}(0)\right]\delta_2}{\alpha_2^2k^2}+\frac{n\sigma_v^2}{\alpha_2 k}, k \geqslant k_0.$$

## 2.4.4 Case study

Consider the following state-space system with time delay:

$$\begin{aligned}\boldsymbol{x}(k+1) = &\begin{bmatrix}0 & 1\\ -0.25 & -0.95\end{bmatrix}\boldsymbol{x}(k)+\begin{bmatrix}0.35 & -0.38\\ 0.18 & -0.05\end{bmatrix}x(k-1)\\ &+\begin{bmatrix}1.00\\ -1.00\end{bmatrix}u(k)+\begin{bmatrix}1.00\\ 0.97\end{bmatrix}w(k),\\ &\boldsymbol{y}(k)=\boldsymbol{x}(k)+\boldsymbol{v}(k).\end{aligned}$$

In simulation, the input $\{u(k)\}$ is taken as an uncorrelated persistent excitation signal sequence with zero mean and unit variance and $\{v(k)\}$ as a white noise sequence with zero mean and variances $\sigma^2=0.10^2$ and $\sigma^2=0.50^2$. Apply the proposed algorithm to estimate the parameters of this example system, and the parameter estimates and their estimation errors are shown in Tables 2.8–2.9 and Figs. 2.15–2.19.

**Table 2.8** The parameter estimates and errors with $\sigma^2 = 0.50^2$.

| $K$ | 100 | 200 | 500 | 1000 | 2000 | 3000 |
|---|---|---|---|---|---|---|
| $a_1 = -0.25$ | −0.33525 | −0.28095 | −0.25048 | −0.24896 | −0.25789 | −0.25759 |
| $a_2 = -0.95$ | −0.98993 | −0.91664 | −0.91787 | −0.92302 | −0.95503 | −0.95990 |
| $b_{21} = 0.18$ | 0.17029 | 0.16224 | 0.16971 | 0.17816 | 0.18697 | 0.18745 |
| $b_{22} = -0.05$ | −0.10814 | −0.04455 | −0.02085 | −0.02857 | −0.05779 | −0.05399 |
| $f_2 = -1.00$ | −0.99300 | −1.00160 | −1.00451 | −0.98968 | −0.99836 | −0.99821 |
| $g_2 = 0.97$ | 1.01024 | 0.94950 | 0.98804 | 0.96431 | 0.97949 | 0.97858 |
| $b_{11} = 0.35$ | 0.33398 | 0.34289 | 0.34241 | 0.35291 | 0.35606 | 0.34878 |
| $b_{12} = -0.38$ | −0.34817 | −0.33635 | −0.35828 | −0.35574 | −0.36590 | −0.36977 |
| $f_1 = 1.00$ | 1.02751 | 1.03195 | 1.01957 | 1.00827 | 1.00688 | 1.00797 |
| $g_1 = 1.00$ | 0.75219 | 0.94818 | 0.95514 | 0.99924 | 0.98418 | 0.98300 |
| $\delta(\%)$ | 12.19427 | 4.04025 | 3.17803 | 1.95805 | 1.25696 | 1.21002 |

**Table 2.9** The parameter estimates and errors with $\sigma^2 = 0.10^2$.

| $K$ | 100 | 200 | 500 | 1000 | 2000 | 3000 |
|---|---|---|---|---|---|---|
| $a_1 = -0.25$ | −0.26399 | −0.25426 | −0.24912 | −0.24401 | −0.25495 | −025148 |
| $a_2 = -0.95$ | −0.96078 | −0.94307 | −0.94232 | −0.93860 | −0.95422 | −0.95184 |
| $b_{21} = 0.18$ | 0.17273 | 0.17103 | 0.17933 | 0.18115 | 0.18146 | 0.18047 |
| $b_{22} = -0.05$ | −0.05890 | −0.04254 | −0.04315 | −0.04081 | −0.05417 | −0.05022 |
| $f_2 = -1.00$ | −0.98371 | −0.99405 | −0.99760 | −0.99615 | −0.99878 | −0.99905 |
| $g_2 = 0.97$ | 0.72608 | 0.82876 | 0.93568 | 0.94028 | 0.96826 | 0.97125 |
| $b_{11} = 0.35$ | 0.33903 | 0.34153 | 0.35004 | 0.35390 | 0.35013 | 0.34874 |
| $b_{12} = -0.38$ | −0.38443 | −0.38131 | −0.37458 | −0.37553 | −0.37624 | −0.37910 |
| $f_1 = 1.00$ | 1.00367 | 1.00671 | 1.00388 | 1.00166 | 1.00130 | 1.00160 |
| $g_1 = 1.00$ | 0.84029 | 0.98227 | 0.96844 | 1.00587 | 0.98853 | 0.98538 |
| $\delta(\%)$ | 12.84049 | 6.29245 | 2.11585 | 1.53185 | 0.64056 | 0.66005 |

From Tables 2.8–2.9 and Figs. 2.15–2.19, we can draw the following conclusions.

1. The proposed algorithm is effective for estimating the parameters of the dual-rate state-space model. With the data length increasing, the parameter estimation errors become smaller and converge to zero.
2. A low noise variance leads to higher accuracy of parameter estimates.
3. As the data length $k$ increases, the parameter estimates approach their true values.
4. It is clear that the proposed state observer can generate accurate state estimates because the state estimates are close to their true values as $k$ increases.
5. The predicted outputs match well with the actual outputs.

## 2.5 Conclusion

This chapter studies parameter and state estimation methods for a dual-rate state-space system with time delay. The identification model is transformed into two parts: one with the parameter matrix and the other with the parameter vector. The unknown states of the system are obtained under the framework of the state observer. The unknown parameters are

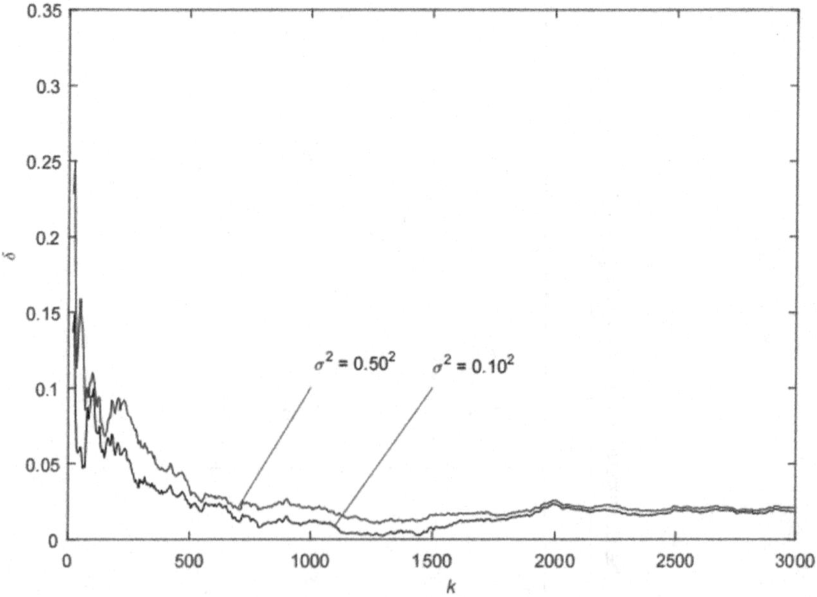


**Figure 2.15** The parameter estimation error $\delta$ versus $k$.

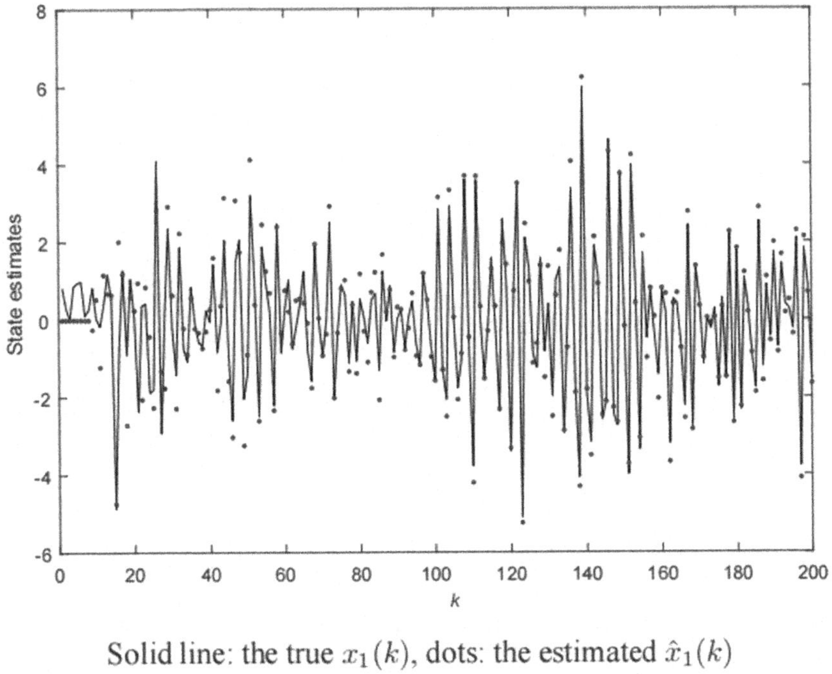


**Figure 2.16** The state estimate of $x_1(k)$.

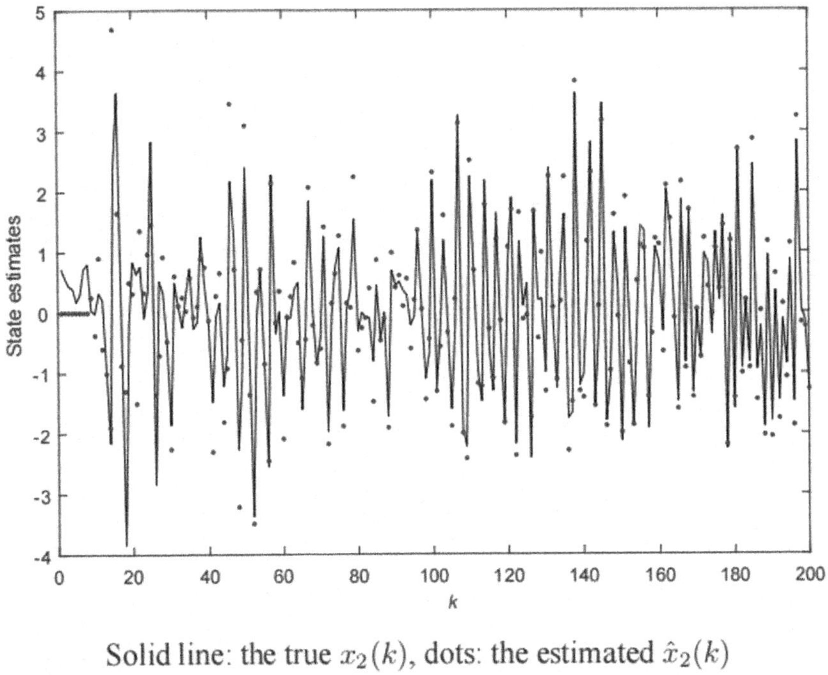


**Figure 2.17** The state estimate of $x_2(k)$.

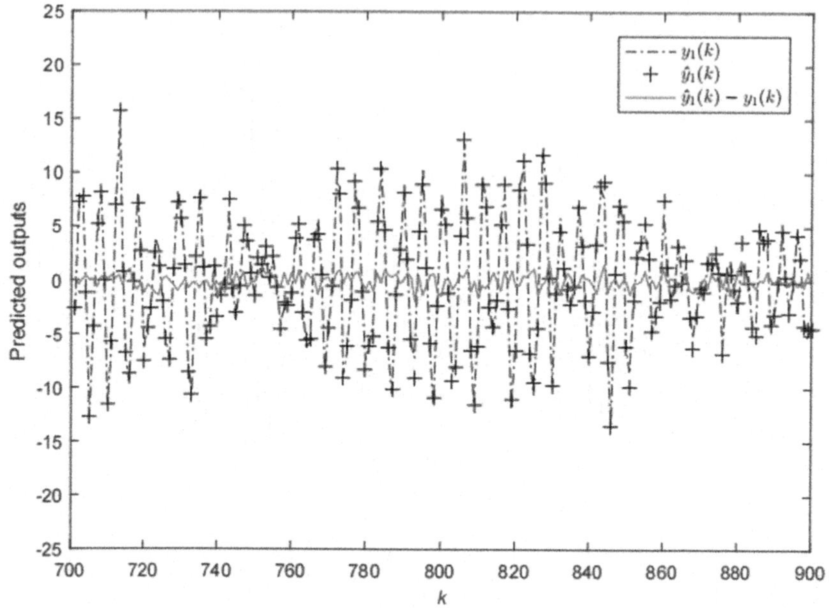


**Figure 2.18** The actual output $y_1(k)$ and the estimated output $\hat{y}_1(k)$ versus $k$.

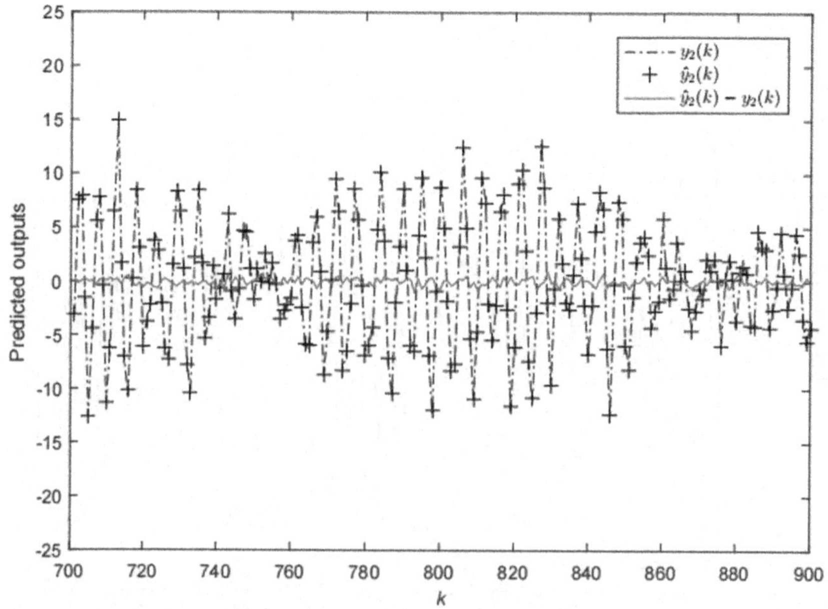


**Figure 2.19** The actual output $y_2(k)$ and the estimated output $\hat{y}_2(k)$ versus $k$.

estimated by the least squares from sampled data. The numerical example shows that the parameter estimates converge to their true values, and the state observer based on the estimated parameters makes the estimated state curves match the actual state curves. The proposed recursive identification algorithm for state-space systems with time delay can be applied to the fields of information processing, communication and network control system, etc.

## References

Ding, F., 2014. Combined state and least squares parameter estimation algorithms for dynamic systems. Applied Mathematics Modelling 38, 403–412.

Fattah, S.A., Zhu, W.P., Ahmad, M.O., 2011. Identification of autoregressive moving average systems based on noise compensation in the correlation domain. IET Signal Processing 5, 292–305.

Hmida, F.B., Khemiri, K., Ragot, J., Gossa, M., 2012. Three-stage Kalman filter for state and fault estimation of linear stochastic systems with unknown inputs. Journal of Franklin Institute 349, 2369–2388.

Ko, C.N., 2012. Identification of non-linear systems using radial basis function neural networks with time-varying learning algorithm. IET Signal Processing 6, 91–98.

# CHAPTER 3

# Multiple state-delay identification

## 3.1 Parameter estimation and convergence for state-space model with time delay

System identification deals with the problem of estimating the unknown parameters of systems by using the measured input—output data; in other words, identification is to build mathematical models of dynamical systems based on observed data from the systems. System identification has been widely used in many areas, for example, aerospace engineering, mechanical systems, biological systems, and environmental systems. A large number of works have been published in this research field.

The least squares method has been the dominant algorithm for parameter estimation due to its simplicity in concept and convenience in implementation. The convergence analysis of identification algorithms has always been one of the important projects in the field of control.

It focuses on the convergence and parameter estimation problems of time-delay systems. In the recognition of the impact on system function, mathematical models incorporating time delays have been developed for an ever widening range of mode control and stabilization of systems (Qin et al., 2011). The objective of this paper is to present a parameter identification algorithm based on the input—output representation of the state-space model with time delay, and further to analyze the convergence properties of the proposed algorithm.

### 3.1.1 The system description and identification model

Let us introduce some notation. $A := X$ or $X := A$ stands for "$A$ is defined as $X$"; the symbol $I(I_n)$ stands for an identity matrix of appropriate size $(n \times n)$; the superscript $T$ denotes the matrix transpose; $|X| := \det[X]$ represents the determinant of a square matrix $\chi$; $z$ represents a unit forward shift operator: $zx(t) = x(t+1)$ and $z^{-1}x(t) = x(t-1)$; the norm of a matrix $\chi$ is defined by $||X||^2 := \mathrm{tr}[XX^{\mathrm{T}}]$; $\lambda_{\min}[X]$ represents the minimum eigenvalues of $\chi$; for $g(t) \geq 0$, we write $f(t) = O(g(t))$ if there exists a positive constant $\delta_1$ such that $|f(t)| \leq \delta_1 g(t)$; $\hat{\theta}(t)$ denotes the estimate of $\theta$ at time $t$.

*State Space Systems With Time-Delays Analysis, Identification, and Applications*
DOI: https://doi.org/10.1016/B978-0-323-91768-1.00011-3

Consider the following state-space system with time delay,

$$\boldsymbol{x}(t+1) = \boldsymbol{A}\boldsymbol{x}(t) + \boldsymbol{B}_1\boldsymbol{x}(t-1) + \boldsymbol{B}_2\boldsymbol{x}(t-2) + \cdots + \boldsymbol{B}_n\boldsymbol{x}(t-n) + \boldsymbol{f}u(t), \tag{3.1}$$

$$y(t) = \boldsymbol{c}\boldsymbol{x}(t) + v(t),$$
$$\boldsymbol{A} := \begin{bmatrix} 0 & 1 & 0 & \cdots & 0 \\ 0 & 0 & 1 & \ddots & \vdots \\ \vdots & \vdots & & \ddots & 0 \\ 0 & 0 & \ldots & 0 & 1 \\ a_n & a_{n-1} & a_{n-2} & \cdots & a_1 \end{bmatrix} \in \mathbb{R}^{n \times n}, \ \boldsymbol{f} := \begin{bmatrix} f_1 \\ f_2 \\ \vdots \\ f_n \end{bmatrix} \in \mathbb{R}^n, \tag{3.2}$$

$$\boldsymbol{B}_i := \begin{bmatrix} \boldsymbol{b}_{i1} \\ \boldsymbol{b}_{i2} \\ \vdots \\ \boldsymbol{b}_{in} \end{bmatrix} \in \mathbb{R}^{n \times n}, \ \ \boldsymbol{b}_{ij} \in \mathbb{R}^{1 \times n}, i,j = 1,2,\cdots,n, \ \ \boldsymbol{c} := [1,0,0,\cdots,0] \in \mathbb{R}^{1 \times n},$$

where $x(t) = [x_1(t), x_2(t), \ldots, x_n(t)]^{\mathrm{T}} \in R^n$ is the state vector, $u(t) \in R$ is the system input, $y(t) \in R$ is the system output, and $v(t) \in R$ is a random noise with zero mean, $A \in R^{n \times n}, B_i \in R^{n \times n}, f \in R^n$, and $c \in R^{1 \times n}$ are the system parameter matrices/vectors, where the system matrices $A$, $B_i$, and the system vector $f$ are the unknown parameters to be estimated from input–output data $\{u(t), y(t)\}$. The following transforms the time-delay state-space model in (3.1) and (3.2) into an input–output representation and gives its identification model.

According to the definition of $A$, $B_i$, and $f$, expanding (1) gives

$$x_j(t+1) = x_{j+1}(t) + \boldsymbol{b}_{1j}\boldsymbol{x}(t-1) + \boldsymbol{b}_{2j}\boldsymbol{x}(t-2) + \cdots + \boldsymbol{b}_{nj}\boldsymbol{x}(t-n) + f_j u(t),$$
$$j = 1,2,\ldots,n-1, \tag{3.3}$$

$$x_n(t+1) = \boldsymbol{a}\boldsymbol{x}(t) + \boldsymbol{b}_{1n}\boldsymbol{x}(t-1) + \boldsymbol{b}_{2n}\boldsymbol{x}(t-2) + \cdots + \boldsymbol{b}_{nn}\boldsymbol{x}(t-n) + f_n u(t),$$
$$\boldsymbol{a} = [a_n, a_{n-1}, \cdots, a_1]. \tag{3.4}$$

Multiplying Eq. (3.3) by $z^{-j}$ and (3.4) by $z^{-n}$, adding all expressions give

$$x_1(t) = \boldsymbol{b}_{11}\boldsymbol{x}(t-2) + (\boldsymbol{b}_{12} + \boldsymbol{b}_{21})\boldsymbol{x}(t-3) + \cdots + (\boldsymbol{a} + \boldsymbol{b}_{1,n-1} + \boldsymbol{b}_{2,n-2} + \cdots + \boldsymbol{b}_{n-1,1})\boldsymbol{x}(t-n)$$
$$+ \cdots + \boldsymbol{b}_{nn}\boldsymbol{x}(t-2n) + f_1 u(t-1) + f_2 u(t-2) + \cdots + f_n u(t-n). \tag{3.5}$$

Define the information vector $\varphi(t)$ and the parameter vector $\theta$ as

$$\varphi(t) := \left[\boldsymbol{x}^{\mathrm{T}}(t-2), \boldsymbol{x}^{\mathrm{T}}(t-3), \cdots, \boldsymbol{x}^{\mathrm{T}}(t-n), \cdots, \boldsymbol{x}^{\mathrm{T}}(t-2n), u(t-1),\right.$$
$$\left. u(t-2, \cdots, u(t-n)\right] \in \mathbb{R}^{n+n\times(2n-1)},$$
$$\boldsymbol{\theta} := \left[\boldsymbol{b}_{11}, \boldsymbol{b}_{12}+\boldsymbol{b}_{21}, \cdots, \boldsymbol{a}+\boldsymbol{b}_{1,n-1}+\boldsymbol{b}_{2,n-2}+\cdots+\boldsymbol{b}_{n-1,1}, \cdots, \boldsymbol{b}_{nn}, \boldsymbol{f}^{\mathrm{T}}\right] \in \mathbb{R}^{n+n\times(2n-1)}.$$

Substituting (3.5) into (3.2) gives the following identification model of the state-space system with time delay,

$$\begin{aligned} y(t) &= \boldsymbol{b}_{11}\boldsymbol{x}(t-2) + (\boldsymbol{b}_{12}+\boldsymbol{b}_{21})\boldsymbol{x}(t-3) \\ &\quad + \cdots + (\boldsymbol{a}+\boldsymbol{b}_{1,n-1}+\boldsymbol{b}_{2,n-2}+\cdots+\boldsymbol{b}_{n-1,1})\boldsymbol{x}(t-n) \\ &\quad + \cdots + \boldsymbol{b}_{nn}\boldsymbol{x}(t-2n) + f_1 u(t-1) + f_2 u(t-2) + \cdots + f_n u(t-n) + v(t) \\ &= \varphi^{\mathrm{T}}(t)\boldsymbol{\theta} + v(t). \end{aligned} \tag{3.6}$$

### 3.1.2 The parameter estimation algorithm

This section discusses the residual-based parameter identification algorithm to directly estimate the parameters $A, B_i$, and $f$. We know that $y(t)$ and $u(t)$ are available, but $\varphi(t)$ is unknown because it contains the unavailable state vector $x(t)$. If the unavailable $x(t)$ in the information vector $\varphi(t)$ is replaced with its estimate $\hat{x}(t)$, then the identification problem of the parameter matrix $\theta$ can be solved. Based on the identification model in (3.6), we can obtain the following residual-based least squares algorithm for estimating $\theta$:

$$\hat{\boldsymbol{\theta}}(t) = \hat{\boldsymbol{\theta}}(t-1) + \boldsymbol{P}(t)\hat{\varphi}(t)\left[y(t) - \hat{\varphi}^{\mathrm{T}}(t)\hat{\boldsymbol{\theta}}(t-1)\right], \tag{3.7}$$

$$\boldsymbol{P}^{-1}(t) = \boldsymbol{P}^{-1}(t-1) + \hat{\varphi}(t)\hat{\varphi}^{\mathrm{T}}(t), \boldsymbol{P}(0) = p_0\boldsymbol{I}, \tag{3.8}$$

$$\hat{v}(t) = y(t) - \hat{\varphi}^{\mathrm{T}}(t)\hat{\boldsymbol{\theta}}(t), \tag{3.9}$$

$$\hat{\varphi}(t) = \left[\hat{\boldsymbol{x}}^{\mathrm{T}}(t-2), \hat{\boldsymbol{x}}^{\mathrm{T}}(t-3), \cdots \hat{\boldsymbol{x}}^{\mathrm{T}}(t-n), \cdots, \hat{\boldsymbol{x}}^{\mathrm{T}}(t-2n), u(t-1), u(t-2), \cdots, u(t-n)\right]^{\mathrm{T}}. \tag{3.10}$$

This algorithm is commonly used for convergence analysis. To avoid computing the matrix inversion, this algorithm is equivalently expressed as

$$\hat{\boldsymbol{\theta}}(t) = \hat{\boldsymbol{\theta}}(t-1) + \boldsymbol{L}(t)\left[y(t) - \varphi^{\mathrm{T}}(t)\hat{\boldsymbol{\theta}}(t-1)\right], \tag{3.11}$$

$$\boldsymbol{L}(t) = \boldsymbol{P}(t)\hat{\boldsymbol{\varphi}}(t) = \frac{\boldsymbol{P}(t-1)\hat{\boldsymbol{\varphi}}(t)}{1 + \hat{\boldsymbol{\varphi}}^{\mathrm{T}}(t)\boldsymbol{P}(t-1)\hat{\boldsymbol{\varphi}}(t)}, \tag{3.12}$$

$$\boldsymbol{P}(t) = \left[\boldsymbol{I} - \boldsymbol{L}(t)\hat{\boldsymbol{\varphi}}^{\mathrm{T}}(t)\right]\boldsymbol{P}(t-1), \boldsymbol{P}(0) = p_0\boldsymbol{I}, \tag{3.13}$$

$$\hat{v}(t) = y(t) - \hat{\boldsymbol{\varphi}}^{\mathrm{T}}(t)\hat{\boldsymbol{\varphi}}(t), \tag{3.14}$$

$$\hat{\boldsymbol{\varphi}}(t) = \left[\hat{\boldsymbol{x}}^{\mathrm{T}}(t-2), \hat{\boldsymbol{x}}^{\mathrm{T}}(t-3), \cdots, \hat{\boldsymbol{x}}^{\mathrm{T}}(t-n), \cdots, \hat{\boldsymbol{x}}^{\mathrm{T}}(t-2n), u(t-1), u(t-2), \cdots, u(t-n)\right]^{\mathrm{T}}, \tag{3.15}$$

where $L(t) \in R^{n+n\times(2n-1)}$ is the gain vector.

### 3.1.3 Main convergence results

Eqs. (3.1) and (3.2) can be rewritten as

$$\boldsymbol{X}(t+1) = \boldsymbol{A}_0\boldsymbol{X}(t) + \boldsymbol{F}_0 u(t), \tag{3.16}$$

$$\begin{aligned}
&y(t) = \boldsymbol{C}_0\boldsymbol{X}(t) + v(t),\\
&\boldsymbol{A}_0 = \begin{bmatrix} \boldsymbol{A} & \boldsymbol{B}_1 & \cdots & \boldsymbol{B}_{n-1} & \boldsymbol{B}_n \\ \boldsymbol{I} & \boldsymbol{0} & \cdots & \boldsymbol{0} & \boldsymbol{0} \\ \boldsymbol{0} & \boldsymbol{I} & \cdots & \boldsymbol{0} & \boldsymbol{0} \\ \vdots & \vdots & \ddots & \vdots & \vdots \\ \boldsymbol{0} & \boldsymbol{0} & \cdots & \boldsymbol{I} & \boldsymbol{0} \end{bmatrix} \in \mathbb{R}^{n^2\times n^2}, \boldsymbol{F}_0 = \begin{bmatrix} \boldsymbol{f} \\ \boldsymbol{0} \\ \vdots \\ \boldsymbol{0} \end{bmatrix} \in \mathbb{R}^{n^2\times 1},\\
&\boldsymbol{C}_0 = [\boldsymbol{c}, \boldsymbol{0}, \boldsymbol{0}, \cdots, \boldsymbol{0}] \in \mathbb{R}^{1\times n^2},
\end{aligned} \tag{3.17}$$

where $X(t) := [x(t), x(t-1), \ldots, x(t-n)]^{\mathrm{T}} \in R^{n^2}$ is the state vector, $A_0$ is the above-row companion matrix, $F_0$ is an matrix whose the first $n \times 1$ block is an arbitrary vector and the rest is a zero matrix, and $C_0$ is an matrix whose the first $1 \times n$ block is an identity vector and the rest is a zero matrix.

Define the matrices,

$$\boldsymbol{H} := z\boldsymbol{I} - \boldsymbol{A}_0 = \begin{bmatrix} z\boldsymbol{I} - \boldsymbol{A} & -\boldsymbol{B}_1 & \cdots & -\boldsymbol{B}_{n-1} & -\boldsymbol{B}_n \\ -\boldsymbol{I} & z\boldsymbol{I} & \cdots & \boldsymbol{0} & \boldsymbol{0} \\ \boldsymbol{0} & -\boldsymbol{I} & \cdots & \boldsymbol{0} & \boldsymbol{0} \\ \vdots & \vdots & \ddots & \vdots & \vdots \\ \boldsymbol{0} & \boldsymbol{0} & \cdots & -\boldsymbol{I} & z\boldsymbol{I} \end{bmatrix}, \boldsymbol{H}^{-1} := \begin{bmatrix} \boldsymbol{h}_{11} & * & \cdots & * \\ \boldsymbol{h}_{21} & * & \cdots & * \\ \boldsymbol{h}_{31} & * & \cdots & * \\ \vdots & \vdots & & \vdots \\ \boldsymbol{h}_{n+1,1} & * & \cdots & * \end{bmatrix}.$$

According to $HH^{-1} = I$, we have

$$\begin{bmatrix} z\boldsymbol{I}-\boldsymbol{A} & -\boldsymbol{B}_1 & \cdots & -\boldsymbol{B}_{n-1} & -\boldsymbol{B}_n \\ -\boldsymbol{I} & z\boldsymbol{I} & \cdots & 0 & 0 \\ 0 & -\boldsymbol{I} & \cdots & 0 & 0 \\ \vdots & \vdots & \ddots & \vdots & \vdots \\ 0 & 0 & \cdots & -\boldsymbol{I} & z\boldsymbol{I} \end{bmatrix} \begin{bmatrix} \boldsymbol{h}_{11} & * & \cdots & * \\ \boldsymbol{h}_{21} & * & \cdots & * \\ \boldsymbol{h}_{31} & * & \cdots & * \\ \vdots & \vdots & & \vdots \\ \boldsymbol{h}_{n+1,1} & * & \cdots & * \end{bmatrix} = \begin{bmatrix} \boldsymbol{I} & 0 & \cdots & 0 \\ 0 & \boldsymbol{I} & \cdots & 0 \\ \vdots & \vdots & \ddots & \vdots \\ 0 & 0 & \cdots & \boldsymbol{I} \end{bmatrix}.$$

Expanding this matrix equation, its first row gives the following equations:

$$\begin{cases} (z\boldsymbol{I}-\boldsymbol{A})\boldsymbol{h}_{11} - \boldsymbol{B}_1\boldsymbol{h}_{21} - \boldsymbol{A}_3\boldsymbol{h}_{31} - \cdots - \boldsymbol{B}_n\boldsymbol{h}_{n+1,1} = \boldsymbol{I}, \\ -\boldsymbol{h}_{11} + z\boldsymbol{h}_{21} = \boldsymbol{0}, \\ -\boldsymbol{h}_{21} + z\boldsymbol{h}_{31} = \boldsymbol{0}, \\ \vdots \\ -\boldsymbol{h}_{n1} + z\boldsymbol{h}_{n+1,1} = \boldsymbol{0}. \end{cases}$$

From the above equations, we can get the solution of $h_{11}$,

$$\boldsymbol{h}_{11} = \left(z^n\boldsymbol{I} - \boldsymbol{A}z^{n-1} - \boldsymbol{B}_1 z^{n-2} - \cdots - \boldsymbol{B}_n\right)^{-1} z^{n-1}.$$

Thus, we have the transfer matrix from the input $u(t)$ to the output $y(t)$:

$$\begin{aligned} G(z) &= \boldsymbol{C}_0(z\boldsymbol{I}-\boldsymbol{A}_0)^{-1}\boldsymbol{F}_0 \\ &= [\boldsymbol{c}, \boldsymbol{0}, \boldsymbol{0}, \cdots, \boldsymbol{0}] \begin{bmatrix} z\boldsymbol{I}-\boldsymbol{A} & -\boldsymbol{B}_1 & \cdots & -\boldsymbol{B}_{n-1} & -\boldsymbol{B}_n \\ -\boldsymbol{I} & z\boldsymbol{I} & \cdots & 0 & 0 \\ 0 & -\boldsymbol{I} & \cdots & 0 & 0 \\ \vdots & \vdots & \ddots & \vdots & \vdots \\ 0 & 0 & \cdots & -\boldsymbol{I} & z\boldsymbol{I} \end{bmatrix}^{-1} \begin{bmatrix} \boldsymbol{f} \\ 0 \\ \vdots \\ 0 \end{bmatrix} \\ &= [\boldsymbol{c}, \boldsymbol{0}, \boldsymbol{0}, \cdots, \boldsymbol{0}] \begin{bmatrix} \boldsymbol{h}_{11} & * & \cdots & * \\ \boldsymbol{h}_{21} & * & \cdots & * \\ \boldsymbol{h}_{31} & * & \cdots & * \\ \vdots & \vdots & & \vdots \\ \boldsymbol{h}_{n+1,1} & * & \cdots & * \end{bmatrix} \begin{bmatrix} \boldsymbol{f} \\ 0 \\ \vdots \\ 0 \end{bmatrix} \\ &= \boldsymbol{c}\left(z^n\boldsymbol{I} - \boldsymbol{A}z^{n-1} - \boldsymbol{B}_1 z^{n-2} - \cdots - \boldsymbol{B}_n\right)^{-1}\boldsymbol{f}z^{n-1} \\ &= \frac{\boldsymbol{c}\,\mathrm{adj}\left[z^n\boldsymbol{I} - \boldsymbol{A}z^{n-1} - \boldsymbol{B}_1 z^{n-2} - \cdots - \boldsymbol{B}_n\right]\boldsymbol{f}z^{n-1}}{\det[z^n\boldsymbol{I} - \boldsymbol{A}z^{n-1} - \boldsymbol{B}_1 z^{n-2} - \cdots - \boldsymbol{B}_n]} \\ &=: \frac{B(z)}{A(z)}. \end{aligned}$$

Hence, we have the input–output representation:

$$y(t) = \frac{B(z)}{A(z)}u(t) + v(t),$$

where $A(z)$ and $B(z)$ are polynomials in a unit backward shift operator $z^{-1}[z^{-1}y(t) = y(t-1)]$, and

$$A(z) := z^{n^2}\det\left[z^n\boldsymbol{I} - \boldsymbol{A}z^{n-1} - \boldsymbol{B}_1 z^{n-2} - \cdots - \boldsymbol{B}_n\right],$$
$$B(z) := z^{n^2+n-1}\text{cadj}\left[z^n\boldsymbol{I} - \boldsymbol{A}z^{n-1} - \boldsymbol{B}_1 z^{n-2} - \cdots - \boldsymbol{B}_n\right]\boldsymbol{f}.$$

Let $r(t) := \frac{B(z)}{A(z)}u(t)$, then $y(t) = r(t) + v(t)$; according to (3.14), the estimate of $r(t)$ is computed by

$$\hat{r}(t) = y(t) - \hat{v}(t) = \hat{\varphi}^{\mathrm{T}}(t)\hat{\boldsymbol{\theta}}(t). \tag{3.18}$$

**Theorem 3.1.**: For the system in (3.1–3.2) and algorithm in (3.7–3.10), assume that $\{v(t), \mathscr{F}_t\}$ is a martingale difference vector sequence defined on a probability space $\{\Omega, \mathscr{F}, \mathrm{P}\}$, where $\{\mathscr{F}_t\}$ is the $\sigma$ algebra sequence generated by the observations up to and including time $t$. The noise sequence $\{v(t)\}$ satisfies the following assumptions:

(A1) $\mathrm{E}[v(t)|\mathscr{F}_{t-1}] = 0, \quad \text{a.s.},$

(A2) $\mathrm{E}\left[v^2(t)|\mathscr{F}_{t-1}\right] \leqslant \sigma_v^2 < \infty, \quad \text{a.s.},$

(A3) $H(z) = A^{-1}(z)\frac{1}{2}$ is strictly positive real.

Then, the following inequality holds,

$$E[T(t) + S(t)|\mathscr{F}_{t-1}] \leqslant T(t-1) + S(t-1) + 2\hat{\varphi}^{\mathrm{T}}(t)\boldsymbol{P}(t)\hat{\varphi}^{\mathrm{T}}\sigma_v^2,$$

where

$$T(t) := \tilde{\boldsymbol{\theta}}^{\mathrm{T}}(t)\boldsymbol{P}^{-1}(t)\tilde{\boldsymbol{\theta}}(t), \tag{3.19}$$

$$S(t) := 2\sum_{i=1}^{t}\tilde{u}(i)\tilde{y}(i) \geqslant 0,$$

$$\tilde{y}(t) := \frac{1}{2}\hat{\varphi}^{\mathrm{T}}\tilde{\boldsymbol{\theta}}(t) + \left[y(t) - \hat{\varphi}^{\mathrm{T}}(t)\hat{\boldsymbol{\theta}}(t) - v(t)\right], \tag{3.20}$$

$$\tilde{u}(t) := -\hat{\varphi}^{\mathrm{T}}(t)\tilde{\boldsymbol{\theta}}(t), \tag{3.21}$$

$$\tilde{\boldsymbol{\theta}}(t) := \hat{\boldsymbol{\theta}}(t) - \boldsymbol{\theta}. \tag{3.22}$$

**Proof** Define the innovation vector, $\hat{\boldsymbol{\theta}}$

$$e(t) := y(t) - \hat{\boldsymbol{\varphi}}^{\mathrm{T}}(t)\hat{\boldsymbol{\theta}}(t-1). \tag{3.23}$$

Using (3.9), it follows that

$$\hat{v}(t) = \left[1 - \hat{\boldsymbol{\varphi}}^{\mathrm{T}}(t)\boldsymbol{P}(t)\hat{\boldsymbol{\varphi}}(t)\right]e(t) = \frac{e(t)}{1 + \hat{\boldsymbol{\varphi}}^{\mathrm{T}}(t)\boldsymbol{P}(t-1)\hat{\boldsymbol{\varphi}}(t)}. \tag{3.24}$$

Substituting (3.7) into (3.22) and using (3.23) and (3.24), it is not difficult to get

$$\begin{aligned} \tilde{\boldsymbol{\theta}}(t) &= \hat{\boldsymbol{\theta}}(t) - \boldsymbol{\theta} = \tilde{\boldsymbol{\theta}}(t-1) + \boldsymbol{P}(t)\hat{\boldsymbol{\varphi}}(t)e(t) \\ &= \tilde{\boldsymbol{\theta}}(t-1) + \boldsymbol{P}(t-1)\hat{\boldsymbol{\varphi}}(t)\hat{v}(t), \end{aligned} \tag{3.25}$$

or

$$\boldsymbol{P}^{-1}(t-1)\tilde{\boldsymbol{\theta}}(t) = \boldsymbol{P}^{-1}(t-1)\tilde{\boldsymbol{\theta}}(t-1) + \hat{\boldsymbol{\varphi}}(t)\hat{v}(t). \tag{3.26}$$

Premultiplying (3.26) by $\tilde{\theta}^{\mathrm{T}}(t)$ and using (3.25) yield

$$\begin{aligned} \tilde{\boldsymbol{\theta}}^{\mathrm{T}}(t)\boldsymbol{P}^{-1}(t-1)\tilde{\boldsymbol{\theta}}(t) &= \left[\tilde{\boldsymbol{\theta}}(t-1)+\boldsymbol{P}(t-1)\hat{\boldsymbol{\varphi}}(t)\hat{v}(t)\right]^{\mathrm{T}}\boldsymbol{P}^{-1}(t-1)\tilde{\boldsymbol{\theta}}(t-1) \\ &\quad + \tilde{\boldsymbol{\theta}}^{\mathrm{T}}(t)\hat{\boldsymbol{\varphi}}(t)\hat{v}(t) \\ &= \tilde{\boldsymbol{\theta}}^{\mathrm{T}}(t-1)\boldsymbol{P}^{-1}(t-1)\tilde{\boldsymbol{\theta}}(t-1) + \hat{v}(t)\hat{\boldsymbol{\varphi}}^{\mathrm{T}}(t)\tilde{\boldsymbol{\theta}}(t-1) \\ &\quad + \tilde{\boldsymbol{\theta}}^{\mathrm{T}}(t)\hat{\boldsymbol{\varphi}}(t)(\hat{v}(t). \end{aligned}$$

Using (3.8) and (3.23–3.25), from (3.19), we obtain

$$\begin{aligned} T(t) &= T(t-1) + \tilde{\boldsymbol{\theta}}^{\mathrm{T}}(t)\hat{\boldsymbol{\varphi}}(t)\hat{\boldsymbol{\varphi}}^{\mathrm{T}}(t)\tilde{\boldsymbol{\theta}}(t) + \hat{v}(t)\hat{\boldsymbol{\varphi}}^{\mathrm{T}}(t)\tilde{\boldsymbol{\theta}}(t-1) + \tilde{\boldsymbol{\theta}}^{\mathrm{T}}(t)\hat{\boldsymbol{\varphi}}(t)\hat{v}(t) \\ &= T(t-1) + \tilde{\boldsymbol{\theta}}^{\mathrm{T}}(t)\hat{\boldsymbol{\varphi}}(t)\hat{\boldsymbol{\varphi}}^{\mathrm{T}}(t)\tilde{\boldsymbol{\theta}}(t) + \hat{v}(t)\hat{\boldsymbol{\varphi}}^{\mathrm{T}}(t)\left[\tilde{\boldsymbol{\theta}}(t) - \boldsymbol{P}(t)\hat{\boldsymbol{\varphi}}(t)e(t)\right] \\ &\quad + \tilde{\boldsymbol{\theta}}^{\mathrm{T}}(t)\hat{\boldsymbol{\varphi}}(t)\hat{v}(t) \\ &= T(t-1) + \tilde{\boldsymbol{\theta}}^{\mathrm{T}}(t)\hat{\boldsymbol{\varphi}}(t)\hat{\boldsymbol{\varphi}}^{\mathrm{T}}(t)\tilde{\boldsymbol{\theta}}(t) + 2\hat{\boldsymbol{\varphi}}^{\mathrm{T}}(t)\tilde{\boldsymbol{\theta}}(t)\hat{v}(t) \\ &\quad - \hat{\boldsymbol{\varphi}}^{\mathrm{T}}(t)\boldsymbol{P}(t)\hat{\boldsymbol{\varphi}}(t)\hat{v}(t)e(t) \\ &= T(t-1) + \tilde{\boldsymbol{\theta}}^{\mathrm{T}}(t)\hat{\boldsymbol{\varphi}}(t)\hat{\boldsymbol{\varphi}}^{\mathrm{T}}(t)\tilde{\boldsymbol{\theta}}(t) + 2\hat{\boldsymbol{\varphi}}^{\mathrm{T}}(t)\tilde{\boldsymbol{\theta}}(t)\hat{v}(t) \\ &\quad - \hat{\boldsymbol{\varphi}}^{\mathrm{T}}(t)\boldsymbol{P}(t)\hat{\boldsymbol{\varphi}}(t)\left[1 - \hat{\boldsymbol{\varphi}}^{\mathrm{T}}(t)\boldsymbol{P}(t)\hat{\boldsymbol{\varphi}}(t)\right]e^2(t) \\ &\leqslant T(t-1) + \tilde{\boldsymbol{\theta}}^{\mathrm{T}}(t)\hat{\boldsymbol{\varphi}}(t)\hat{\boldsymbol{\varphi}}^{\mathrm{T}}(t)\tilde{\boldsymbol{\theta}}(t) + 2\hat{\boldsymbol{\varphi}}^{\mathrm{T}}(t)\tilde{\boldsymbol{\theta}}(t)\hat{v}(t) \\ &= T(t-1) + 2\hat{\boldsymbol{\varphi}}^{\mathrm{T}}(t)\tilde{\boldsymbol{\theta}}(t)\left[\frac{1}{2}\tilde{\boldsymbol{\theta}}^{\mathrm{T}}(t)\hat{\boldsymbol{\varphi}}(t) + (\hat{v}(t) - v(t))\right] + 2\hat{\boldsymbol{\varphi}}^{\mathrm{T}}(t)\tilde{\boldsymbol{\theta}}(t)\hat{v}(t). \end{aligned}$$

Using (3.20), (3.21), (3.24), and (3.25), and $0 \leq \hat{\varphi}^{\mathrm{T}}(t)P(t)\hat{\varphi}(t) \leq 1$, we have

$$\begin{aligned}\mathrm{T}(t) &\leqslant T(t-1) - 2\overline{u}(t)\overline{y}(t) + 2\hat{\boldsymbol{\varphi}}^{\mathrm{T}}(t)\left[\tilde{\boldsymbol{\theta}}(t-1) + \boldsymbol{P}(t)\hat{\varphi}(t)e(t)\right]v(t)\\ &= T(t-1) - 2\overline{u}(t)\overline{y}(t) + 2\hat{\boldsymbol{\varphi}}^{\mathrm{T}}(t)\tilde{\boldsymbol{\theta}}(t-1)v(t)\\ &\quad + 2\hat{\boldsymbol{\varphi}}^{\mathrm{T}}(t)\boldsymbol{P}(t)\hat{\varphi}(t)[e(t) - v(t)]v(t) + v^2(t).\end{aligned}$$

Since $\hat{\varphi}^{\mathrm{T}}(t)\tilde{\theta}(t-1)$, $e(t) - v(t)$, $\hat{\varphi}^{\mathrm{T}}(t)P(t)\hat{\varphi}(t)$ are uncorrelated with $v(t)$ and are $\mathscr{F}_{t-1}$-measurable, taking the conditional expectation of both sides of the above equation with respect to $\mathscr{F}_{t-1}$ and using (A1–A2) give

$$\mathrm{E}[T(t)|F_{t-1}] \leqslant T(t-1) - 2\mathrm{E}[\tilde{u}(t)\tilde{y}(t)|\mathscr{F}_{t-1}] + 2\hat{\boldsymbol{\varphi}}^{\mathrm{T}}(t)\boldsymbol{P}(t)\hat{\varphi}(t)\sigma_v^2, \text{ a.s.} \tag{3.27}$$

According to (3.18), since

$$\begin{aligned}A(z)[\hat{v}(t) - v(t)] &= A(z)\hat{v}(t) - A(z)y(t) + B(z)u(t)\\ &= A(z)\left[-\hat{\boldsymbol{\varphi}}^{\mathrm{T}}(t)\hat{\boldsymbol{\theta}}(t)\right] + B(z)u(t)\\ &= -A(z)\hat{r}(t) + B(z)u(t)\\ &= -\hat{r}(t) + \hat{\boldsymbol{\varphi}}^{\mathrm{T}}(t)\boldsymbol{\theta}(t)\\ &= -\hat{\boldsymbol{\varphi}}^{\mathrm{T}}(t)\hat{\boldsymbol{\theta}}(t) + \hat{\boldsymbol{\varphi}}^{\mathrm{T}}(t)\boldsymbol{\theta}(t)\\ &= -\hat{\boldsymbol{\varphi}}^{\mathrm{T}}(t)\hat{\boldsymbol{\theta}}(t) = \tilde{u}(t).\end{aligned} \tag{3.28}$$

Using (3.9), (3.21), and (3.28), from (3.20), we get

$$\begin{aligned}\tilde{y}(t) &= \frac{1}{2}\hat{\boldsymbol{\varphi}}^{\mathrm{T}}(t)\hat{\boldsymbol{\theta}}(t) + [\hat{v}(t) - v(t)]\\ &= -\frac{1}{2}\tilde{u}(t) + A^{-1}(z)\tilde{u}(t)\\ &= \left[A^{-1}(z) - \frac{1+\rho}{2}\right]\tilde{u}(t) + \frac{\rho}{2}\tilde{u}(t)\\ &=: \tilde{y}_1(t) + \frac{\rho}{2}\tilde{u}(t),\end{aligned}$$

where

$$\tilde{y}_1(t) := H_1(z)\tilde{u}(t), \quad H_1(z) := A^{-1}(z) - \frac{1+\rho}{2}.$$

Since $H(z)$ is a strictly positive real function, there exists a constant $\rho > 0$ such that $H(z)$ is also strictly positive real. The following inequalities holds

$$2\sum_{i=1}^{t}\tilde{u}(i)\tilde{y}_1(i) \geqslant 0, \text{ a.s.},$$

$$S(t) = 2\sum_{i=1}^{t}\tilde{u}(i)\tilde{y}_1(i) + \rho\sum_{i=1}^{t}\tilde{u}^2(i) \geqslant 0, \text{ a.s.}$$

Adding both sides of (3.27) by $S(t)$ gives the conclusion of Theorem 3.1.

**Theorem 3.2.**: For the system in (3.1–3.2) and the algorithm in (3.7–3.10), assume that (A1–A3) hold, $A(z)$ is stable; that is, all zeros of $A(z)$ are inside the unit circle, and then, the parameter estimation error satisfies

$$\|\hat{\boldsymbol{\theta}}(t)-\boldsymbol{\theta}\|^2 = O\left(\frac{[\ln r(t)]^c}{\lambda_{\min}[\boldsymbol{P}^{-1}(t)]}\right), \text{ a.s., for any } c>1.$$

**Proof** From the definition of $T(t)$, we have

$$\|\tilde{\boldsymbol{\theta}}(t)\|^2 \leqslant \frac{\tilde{\boldsymbol{\theta}}^{\mathrm{T}}(t)\boldsymbol{P}^{-1}(t)\tilde{\boldsymbol{\theta}}(t)}{\lambda_{\min}[\boldsymbol{P}^{-1}(t)]} = \frac{T(t)}{\lambda_{\min}[\boldsymbol{P}^{-1}(t)]}. \tag{3.29}$$

Let

$$W(t) := \frac{T(t)+S(t)}{[\ln|\boldsymbol{P}^{-1}(t)|]^c}, c>1.$$

Since $\ln|P^{-1}(t)|$ is nondecreasing, using Theorem 3.1 yields

$$\begin{aligned} E[W(t)|\mathscr{F}_{t-1}] &\leqslant \frac{T(t-1)+S(t-1)}{[\ln|\boldsymbol{P}^{-1}(t)|]^c} + \frac{2\hat{\boldsymbol{\varphi}}^{\mathrm{T}}(t)\boldsymbol{P}(t)\hat{\boldsymbol{\varphi}}(t)}{[\ln|\boldsymbol{P}^{-1}(t)|]^c}\sigma_v^2 \\ &\leqslant V(t-1) + \frac{2\hat{\boldsymbol{\varphi}}^{\mathrm{T}}(t)\boldsymbol{P}(t)\hat{\boldsymbol{\varphi}}(t)}{[\ln|\boldsymbol{P}^{-1}(t)|]^c}\sigma_v^2, \text{ a.s.} \end{aligned} \tag{3.30}$$

Using Theorem 3.1, it is clear that for $c>1$, the sum for $t$ from 1 to $\infty$ of the last term on the right-hand side of (3.30) is finite. Now applying the martingale convergence theorem to (3.30), we conclude that $W(t)$ converges almost surely to a finite random variable, say $W_0$; that is,

$$W(t) = \frac{T(t)+S(t)}{[\ln|\boldsymbol{P}^{-1}(t)|]^c} \to W_0 < \infty, \text{ a.s.},$$

or

$$T(t) = O\left([\ln|\boldsymbol{P}^{-1}(t)|]^c\right), \text{ a.s.}, \tag{3.31}$$

$$S(t)=O\left(\left[\ln\left|\boldsymbol{P}^{-1}(t)\right|\right]^c\right),\ \text{a.s.} \tag{3.32}$$

Since $H(z)$ is a strictly positive real function, it follows that

$$\sum_{i=1}^{t}\tilde{u}^2(i)=O\left(\left[\ln\left|\boldsymbol{P}^{-1}(t)\right|\right]^c\right). \tag{3.33}$$

Using (3.30–3.32), we have

$$\left\|\hat{\boldsymbol{\theta}}(t)-\boldsymbol{\theta}\right\|^2=O\left(\frac{\left[\ln\left|\boldsymbol{P}^{-1}(t)\right|\right]^c}{\lambda_{\min}\left[\boldsymbol{P}^{-1}(t)\right]}\right)=O\left(\frac{[\ln r(t)]^c}{\lambda_{\min}\left[\boldsymbol{P}^{-1}(t)\right]}\right),\ \text{a.s., for any } c>1.$$

Assume that there exist positive constants $r, c_1, c_2$, and $t_0$ such that the following generalized persistent excitation condition (unbounded condition number) holds:

$$c_1\boldsymbol{I}\leqslant\frac{1}{t}\sum_{j=1}^{t}\varphi(i)=\varphi^{\mathrm{T}}(j)\leqslant c_2t^r\boldsymbol{I},\ \text{a.s., for } t\geqslant t_0.$$

Then for any $c>1$, we have

$$\left\|\hat{\boldsymbol{\theta}}(t)-\boldsymbol{\theta}\right\|^2=O\left(\frac{[\ln t]^c}{t}\right)\to 0,\ \text{a.s., for any } c>1.$$

### 3.1.4 Example

Consider the following state-space system with two-step time delay:

$$\begin{aligned}\boldsymbol{x}(t+1)=&\begin{bmatrix}0 & 1\\ -0.45 & -0.80\end{bmatrix}\boldsymbol{x}(t)+\begin{bmatrix}0.20 & -0.15\\ 0.15 & -0.20\end{bmatrix}\\ &\boldsymbol{x}(t-1)+\begin{bmatrix}0.20 & -0.18\\ 0.18 & -0.05\end{bmatrix}\boldsymbol{x}(t-2)+\begin{bmatrix}1\\ -1\end{bmatrix}u(t),\\ y(t)=&\left[1,0\right]\boldsymbol{x}(t)+v(t).\end{aligned}$$

In simulation, the input $\{u(t)\}$ is taken as an uncorrelated persistent excitation signal sequence with zero mean and unit variance and $\{v(t)\}$ as a white noise sequence with zero mean and variance $\sigma^2=0.10^2$ and $\sigma^2=0.50^2$, respectively. Apply the least squares parameter estimation algorithm to estimate the parameters. The parameter estimates and their estimation errors are shown in Table 3.1 with $\sigma^2=0.10^2$ and $\sigma^2=0.50^2$, and the parameter estimation errors $\delta$ versus $t$ are shown in Fig. 3.1.

**Table 3.1** The parameter estimates and errors with $\sigma^2 = 0.10^2$ and $\sigma^2 = 0.50^2$.

| $\sigma^2$ | $t$ | $\theta_1$ | $\theta_2$ | $\theta_3$ | $\theta_4$ | $\theta_5$ | $\theta_6$ | $\theta_7$ | $\theta_8$ | $\delta$ (%) |
|---|---|---|---|---|---|---|---|---|---|---|
| $0.10^2$ | 100 | −0.28299 | −0.97586 | 0.32661 | −0.38171 | 0.16798 | −0.06587 | 1.00546 | −0.98879 | 2.95817 |
| | 200 | −0.26318 | −0.95017 | 0.33684 | −0.37992 | 0.16909 | −0.04759 | 1.00712 | −0.99403 | 1.30710 |
| | 500 | −0.25526 | −0.94766 | 0.34848 | −0.37166 | 0.17979 | −0.04759 | 1.00461 | −0.99816 | 0.64431 |
| | 1000 | −0.24648 | −0.94091 | 0.35348 | −0.37390 | 0.18190 | −0.0424 | 1.001 3 | −0.99664 | 0.81912 |
| | 2000 | −0.25437 | −0.95348 | 0.34938 | −0.37713 | 0.18073 | −0.05303 | 1.00133 | −0.99911 | 0.39905 |
| | 3000 | −0.24761 | −0.94788 | 0.34864 | −0.38137 | 0.17933 | −0.04617 | 1.00150 | −0.99939 | 0.31130 |
| $0.50^2$ | 100 | −0.39755 | −1.05834 | 0.23756 | −0.39829 | 0.11678 | −0.11493 | 1.02158 | −1.02473 | 13.06648 |
| | 200 | −0.31071 | −0.94434 | 0.28621 | −0.38066 | 0.12506 | −0.03269 | 1.0331 | −1.01146 | 6.14516 |
| | 500 | −0.27513 | −0.93693 | 0.34237 | −0.33959 | 0.17795 | −0.03720 | 1.02235 | −1.00702 | 3.13488 |
| | 1000 | −0.23201 | −0.90412 | 0.36723 | −0.35002 | 0.18910 | −0.01198 | 1.00886 | −0.99151 | 4.02927 |
| | 2000 | −0.27173 | −0.96726 | 0.34678 | −0.36580 | 0.18346 | −0.06491 | 1.00647 | −0.99977 | 1.96240 |
| | 3000 | −0.23804 | −0.93936 | 0.34319 | −0.38689 | 0.17660 | −0.03080 | 1.00741 | −0.99976 | 1.55023 |
| True values | | −0.25000 | −0.95000 | 0.35000 | −0.38000 | 0.18000 | −0.05000 | 1.00000 | −1.00000 | |

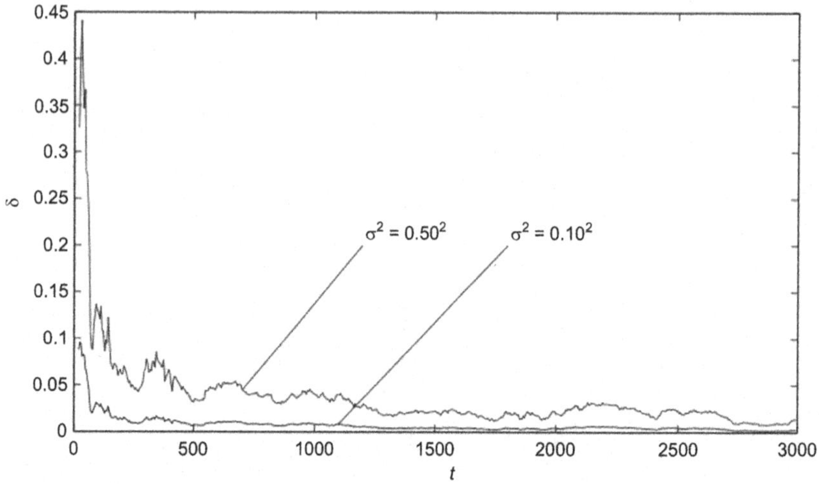


**Figure 3.1** The parameter estimation errors $\delta$ versus $t$ with $\sigma^2 = 0.10^2$ and $\sigma^2 = 0.50^2$.

From Table 3.1 and Fig. 3.1, we can draw the following conclusions: (1) the parameter estimation errors $\delta$ become smaller (in general) with the increasing of $t$ – see Fig. 3.1; (2) a lower noise level leads to a faster rate of convergence of the parameter estimates to the true parameters – see Fig. 3.1; (3) the parameter estimation accuracy becomes higher as the data length $t$ increases – see Table 3.1; (4) the parameter estimation errors under the same data lengths become higher for low noise levels – see Table 3.1.

### 3.1.5 Conclusions

In this chapter, the identification problems for linear systems based on the state-space models with unknown parameters are studied. A new parameter estimation algorithm has been presented directly from input–output data. The analysis using the martingale convergence theorem indicates that the proposed algorithms can give consistent parameter estimation. The simulation results show that the proposed algorithms are effective.

## 3.2 Iterative parameter estimation for state-space model with multistate delays based on decomposition

It is well known that time delays are frequently encountered in various practical systems such as chemical processes, communication networks, and digital signal analysis (Kim, 2011). Such delay is usually a source of instability and performance deterioration. Consequently, a rigorous study on time-delay systems has been made, and the analysis and synthesis of time-delay systems have received growing interest in recent years (Shi and Yu, 2011). Several approaches have been proposed; in particular, the linear matrix inequality (LMI) approach has been extensively used because of its computational efficiency, and the free weighting matrices approach method is less conservative and can be extended to deal with many problems such as output feedback stabilization control.

Parameter estimation is the eternal theme of system identification, and many identification methods were developed for various systems, for example, the iterative parameter estimation algorithm for multivariable systems and for generalized coupled Sylvester matrix equations, the hierarchical gradient-based identification algorithm and the hierarchical least squares identification method for multivariable systems, the auxiliary model-based least squares algorithms for dual-rate sampled-data nonlinear systems and for state-space model with a unit time delay.

In the field of system modeling and parameter estimation, the iterative methods are used for offline identification to estimate the parameters of linear and nonlinear systems. Iterative identification methods are very effective in identifying systems with unknown variables in the information vector and have been successfully applied to many different models. The proposed algorithm requires less computation compared with the least squares-based iterative algorithm.

The significance and difficulty of estimating state-space systems are widely admitted. Consequently, there is extremely large and active research effort directed toward the problem. A primary aspect is that the states in the identification process should be identified first or simultaneously with the system parameters since the states are usually unknown.

It is to develop highly accurate system identification algorithms using the iterative techniques and hierarchical identification principle. We frame our study in the identification of state-space systems with multi-state delays for which the information vector in the corresponding identification representation contains unknown state variables. The basic idea of the proposed methods is to use the iterative techniques to deal with the identification problem by adopting the interactive estimation theory which is a hierarchical computation process. Based on this idea, we present a hierarchical gradient-based iterative and a hierarchical least squares-based iterative identification algorithms for state-space model with multistate delays.

### 3.2.1 System description and identification model

Let us first introduce the necessary notation. "$A =: X$" or "$X := A$" stands for "$A$ is defined as $X$"; the symbol $I(I_n)$ stands for an identity matrix of appropriate size ($n \times n$); $z$ represents a unit forward shift operator: $zx(t) = x(t+1)$ and $z^{-1}x(t) = x(t-1)$; the superscript T denotes the matrix/vector transpose; the norm of a matrix $X$ is defined as $||X||^2 := \mathrm{tr}[XX^{\mathrm{T}}]$; $\lambda_{\max}[X]$ represents the maximum eigenvalue of the nonnegative definite matrix $X$; $\hat{\vartheta}(t)$ denotes the estimate of $\vartheta$ at time $t$; $1_n$ represents an $n \times 1$ vector whose elements are all unity.

Based on the hierarchical identification principle, this paper studies the gradient-based and least squares-based iterative identification algorithms for the time-delay systems described by the following state-space model,

$$\boldsymbol{x}(t+1) = \boldsymbol{A}\boldsymbol{x}(t) + \boldsymbol{B}_1\boldsymbol{x}(t-1) + \boldsymbol{B}_2\boldsymbol{x}(t-2) + \cdots + \boldsymbol{B}_r\boldsymbol{x}(t-r) + \boldsymbol{g}u(t), \tag{3.34}$$

where $x(t) = [x_1(t), x_2(t), \ldots, x_n(t)]^{\mathrm{T}} \in R^n$ is the state vector, and $u(t)$ denotes the system input. Assuming that the order $n$ is known and $u(t) = 0$ for $t \leq 0$. The system matrices $A, B_i (i = 1, 2, \ldots, r)$ and vector $g$ are the unknown parameters to be estimated from the measured state data $x(t-l)(l = 0, 1, \ldots, r)$ and input data $u(t)$.

Define the parameter matrices/vector $A$, $B_i$, and $g$ as

$$\boldsymbol{A} := \begin{bmatrix} 0 & 1 & 0 & \cdots & 0 \\ 0 & 0 & 1 & \ddots & \vdots \\ \vdots & \vdots & \ddots & \ddots & 0 \\ 0 & 0 & \cdots & 0 & 1 \\ a_n & a_{n-1} & a_{n-2} & \cdots & a_1 \end{bmatrix} \in \mathbb{R}^{n \times n}, \quad \boldsymbol{g} := \begin{bmatrix} g_1 \\ g_2 \\ \vdots \\ g_n \end{bmatrix} \in \mathbb{R}^n,$$

$$\boldsymbol{B}_i := \begin{bmatrix} \boldsymbol{b}_{i1} \\ \boldsymbol{b}_{i2} \\ \vdots \\ \boldsymbol{b}_{in} \end{bmatrix} \in \mathbb{R}^{n \times n}, \quad \boldsymbol{b}_{ij} \in \mathbb{R}^{1 \times n}, \quad i = 1, 2, \cdots, r, \quad j = 1, 2, \cdots, n.$$

In order to improve the parameter estimation accuracy, we apply the iterative identification method to identify this model.

Define the parameter matrix $\vartheta$ and the information vector $\phi(t)$ as

$$\vartheta^{\mathrm{T}} := [\boldsymbol{A}, \boldsymbol{B}_1, \boldsymbol{B}_2, \cdots, \boldsymbol{B}_r, \boldsymbol{g}] \in \mathbb{R}^{n \times (n+nr+1)}, \tag{3.35}$$

$$\phi(t) := [\boldsymbol{x}^{\mathrm{T}}(t), \boldsymbol{x}^{\mathrm{T}}(t-1), \boldsymbol{x}^{\mathrm{T}}(t-2), \cdots, \boldsymbol{x}^{\mathrm{T}}(t-r), u(t)]^{\mathrm{T}} \in \mathbb{R}^{n+nr+1}. \tag{3.36}$$

Eq. (3.34) can be equivalently written as

$$x(t+1) = \vartheta^{\mathrm{T}}\phi(t). \tag{3.37}$$

Eq. (3.37) is the identification model for the state-space systems with multistate delays in (3.34). The goal is to obtain the estimates of the parameter matrix $\vartheta$ from the state information and input information $\{x(t), u(t)\}$, while from (3.35) we can see that the parameter matrix $A$ in $\vartheta^{\mathrm{T}}$ contains a great number of zero elements which will increase the computation; however, these zero elements do not need to estimate in practice. Thus, this section applies the decomposition technique to avoid estimating zero elements and to develop a new iterative identification algorithm for estimating the system parameters $A, B_i$, and $g$, and transforms the original identification problem into two subproblems with small sizes. Substituting the coefficients into Eq. (3.34), we have

$$x(t+1) = \begin{bmatrix} 0 & 1 & 0 & \cdots & 0 \\ 0 & 0 & 1 & \ddots & \vdots \\ \vdots & \vdots & \ddots & \ddots & 0 \\ 0 & 0 & \cdots & 0 & 1 \\ a_n & a_{n-1} & a_{n-2} & \cdots & a_1 \end{bmatrix} x(t) + \begin{bmatrix} \boldsymbol{b}_{11} \\ \boldsymbol{b}_{12} \\ \vdots \\ \boldsymbol{b}_{1n} \end{bmatrix}$$

$$x(t-1) + \cdots + \begin{bmatrix} \boldsymbol{b}_{r1} \\ \boldsymbol{b}_{r2} \\ \vdots \\ \boldsymbol{b}_{rn} \end{bmatrix} x(t-r) + \begin{bmatrix} g_1 \\ g_2 \\ \vdots \\ g_n \end{bmatrix} u(t).$$

The above equation can be rewritten as

$$x(t+1) = \begin{bmatrix} x_2(t) \\ x_3(t) \\ \vdots \\ x_n(t) \\ \boldsymbol{a}x(t) \end{bmatrix} + \begin{bmatrix} \boldsymbol{b}_{11} \\ \boldsymbol{b}_{12} \\ \vdots \\ \boldsymbol{b}_{1,n-1} \\ \boldsymbol{b}_{1n} \end{bmatrix} x(t-1) + \cdots + \begin{bmatrix} \boldsymbol{b}_{r1} \\ \boldsymbol{b}_{r2} \\ \vdots \\ \boldsymbol{b}_{r,n-1} \\ \boldsymbol{b}_{rn} \end{bmatrix}$$

$$x(t-r) + \begin{bmatrix} g_1 \\ g_2 \\ \vdots \\ g_{n-1} \\ g_n \end{bmatrix} u(t), \quad \boldsymbol{a} = [a_n, a_{n-1}, \cdots, a_1].$$

Hence, in order to make the calculation more simple, we can decompose the above equation into two subsystems,

$$S_1{:}\boldsymbol{x}'(t+1)-\boldsymbol{x}'(t)=\begin{bmatrix}\boldsymbol{b}_{11}\\ \boldsymbol{b}_{12}\\ \vdots\\ \boldsymbol{b}_{1,n-1}\end{bmatrix}\boldsymbol{x}(t-1)+\cdots+\begin{bmatrix}\boldsymbol{b}_{r1}\\ \boldsymbol{b}_{r2}\\ \vdots\\ \boldsymbol{b}_{r,n-1}\end{bmatrix}\boldsymbol{x}(t-r)+\begin{bmatrix}g_1\\ g_2\\ \vdots\\ g_{n-1}\end{bmatrix}u(t), \tag{3.38}$$

$$S_2{:}x(t+1)-\boldsymbol{a}\boldsymbol{x}(t)+\boldsymbol{b}_{1n}\boldsymbol{x}(t-1)+\cdots+\boldsymbol{b}_{rn}\boldsymbol{x}(t-r)+g_nu(t), \tag{3.39}$$

where

$$\boldsymbol{x}'(t)=[x_1(t),x_2(t),\cdots,x_{n-1}(t)]^{\mathrm{T}}\in\mathbb{R}^{n-1},\boldsymbol{x}(t+1)=\begin{bmatrix}\boldsymbol{x}'(t+1)\\ x(t+1)\end{bmatrix}\in\mathbb{R}^n.$$

Define the parameter matrix $\theta_1$ and the information vector $\varphi_1(t)$ of subsystem 1 as

$$\boldsymbol{\theta}_1^{\mathrm{T}}:=\begin{bmatrix}\boldsymbol{b}_{11} & \boldsymbol{b}_{21} & \cdots & \boldsymbol{b}_{r1} & g_1\\ \boldsymbol{b}_{12} & \boldsymbol{b}_{22} & \cdots & \boldsymbol{b}_{r2} & g_2\\ \vdots & \vdots & & \vdots & \vdots\\ \boldsymbol{b}_{1,n-1} & \boldsymbol{b}_{2,n-1} & \cdots & \boldsymbol{b}_{r,n-1} & g_{n-1}\end{bmatrix}\in\mathbb{R}^{(n-1)\times(nr+1)},$$

$$\boldsymbol{\varphi}_1(t):=\begin{bmatrix}\boldsymbol{x}(t-1)\\ \boldsymbol{x}(t-2)\\ \vdots\\ \boldsymbol{x}(t-r)\\ u(t)\end{bmatrix}\in\mathbb{R}^{nr+1},$$

and the parameter vector $\theta_2$ and the information vector $\varphi_2(t)$ of subsystem 2 as

$$\boldsymbol{\theta}_2:=\left[\boldsymbol{a},\boldsymbol{b}_{1n},\boldsymbol{b}_{2n},\cdots,\boldsymbol{b}_{rn},g_n\right]^{\mathrm{T}}\in\mathbb{R}^{n+nr+1},$$

$$\boldsymbol{\varphi}_2(t):=\left[\boldsymbol{x}^{\mathrm{T}}(t),\boldsymbol{x}^{\mathrm{T}}(t-1),\boldsymbol{x}^{\mathrm{T}}(t-2),\cdots,\boldsymbol{x}^{\mathrm{T}}(t-r),u(t)\right]^{\mathrm{T}}\in\mathbb{R}^{n+nr+1}.$$

The identification models of subsystem 1 and subsystem 2 can be written as

$$S_1{:}\boldsymbol{x}'(t+1)-\boldsymbol{x}'(t)=\boldsymbol{\theta}_2^{\mathrm{T}}\boldsymbol{\varphi}_1(t), \tag{3.40}$$

$$S_2{:}x(t+1)=\boldsymbol{\varphi}_2^{\mathrm{T}}(t)\boldsymbol{\theta}_2. \tag{3.41}$$

The identification model in (3.37) contains $n \times (n + nr + 1)$ parameters, subsystem (3.40) contains $(n - 1) \times (nr + 1)$ parameters in $\theta_1$, and subsystem (3.41) contains $n + nr + 1$ parameters in $\theta_2$. Such a decomposition leads to less amount of calculation and storage.

## 3.2.2 The hierarchical gradient-based iterative algorithm

From two submodels in (3.40) and (3.41), we can see that directly estimating the parameter matrix $\theta_1$ in $S_1$ and the parameter vector $\theta_2$ in $S_2$ is impossible in that $S_1$ and $S_2$ contain the unknown state vector $x(t)$. This motivates us to study the hierarchical identification algorithms to estimate $\theta_1$ and $\theta_2$.

Define the stack information matrices $\Phi_1$ and $\Phi_2$, the stack state matrices $X$, $X_1$ and vector $X_2$ as

$$\boldsymbol{\Phi}_1 := \left[\varphi_1(1), \varphi_1(2), \ldots, \varphi_1(L)\right] \in \mathbb{R}^{(nr+1) \times L}, \tag{3.42}$$

$$\boldsymbol{\Phi}_2 := \left[\varphi_2(1), \varphi_2(2), \ldots, \varphi_2(L)\right] \in \mathbb{R}^{(n+nr+1) \times L}, \tag{3.43}$$

$$\boldsymbol{X} := [\boldsymbol{x}'(2), \boldsymbol{x}'(3), \ldots, \boldsymbol{x}'(L+1)] \in \mathbb{R}^{(n-1) \times L}, \tag{3.44}$$

$$\boldsymbol{X}_1 := [\boldsymbol{x}'(1), \boldsymbol{x}'(2), \ldots, \boldsymbol{x}'(L)] \in \mathbb{R}^{(n-1) \times L}, \tag{3.45}$$

$$\boldsymbol{X}_2 := [x(2), x(3), \ldots, x(L+1)]^{\mathrm{T}} \in \mathbb{R}^{L}. \tag{3.46}$$

From (3.40) and (3.41), we have

$$S_1 : \boldsymbol{X} - \boldsymbol{X}_1 = \boldsymbol{\theta}_1^{\mathrm{T}} \boldsymbol{\Phi}_1,$$
$$S_2 : \boldsymbol{X}_2 = \boldsymbol{\Phi}_2^{\mathrm{T}} \boldsymbol{\theta}_2.$$

Define two cost functions,

$$J_1(\boldsymbol{\theta}_1) := \left\| \boldsymbol{X} - \boldsymbol{X}_1 = \boldsymbol{\theta}_1^{\mathrm{T}} \boldsymbol{\Phi}_1 \right\|^2, \tag{3.47}$$

$$J_2(\boldsymbol{\theta}_2) := \left\| \boldsymbol{X}_2 - \boldsymbol{\Phi}_2^{\mathrm{T}} \boldsymbol{\theta}_2 \right\|^2. \tag{3.48}$$

Let $k = 1, 2, 3, \ldots$ be an iterative variable, $\hat{\theta}_{1,k}$ and $\hat{\theta}_{2,k}$ denote the estimates of $\theta_1$ and $\theta_2$ at iteration $k$, respectively, and $\mu_k \geq 0$ be a convergence factor. Provided that $\varphi_1(t)$ and $\varphi_2(t)$ are persistently exciting, minimizing $J_1(\theta_1)$ and $J_2(\theta_2)$ by using the negative gradient search leads to the iterative algorithm of estimating $\theta_1$ and $\theta_2$ as follows:

$$\begin{aligned}\hat{\boldsymbol{\theta}}_{1,k} &= \hat{\boldsymbol{\theta}}_{1,k-1} - \frac{\mu_k}{2}\text{grad}\left[J_1\left(\hat{\boldsymbol{\theta}}_{1,k-1}\right)\right] \\ &= \hat{\boldsymbol{\theta}}_{1,k-1} + \mu_k\boldsymbol{\Phi}_1\left[\boldsymbol{X}-\boldsymbol{X}_1-\hat{\boldsymbol{\theta}}^{\mathrm{T}}_{1,k-1}\boldsymbol{\Phi}_1\right]^{\mathrm{T}},\end{aligned} \tag{3.49}$$

$$\begin{aligned}\hat{\boldsymbol{\theta}}_{2,k} &= \hat{\boldsymbol{\theta}}_{2,k-1} - \frac{\mu_k}{2}\text{grad}\left[J_2\left(\hat{\boldsymbol{\theta}}_{2,k-1}\right)\right] \\ &= \hat{\boldsymbol{\theta}}_{2,k-1} + \mu_k\boldsymbol{\Phi}_2\left[\boldsymbol{X}_2 - \boldsymbol{\Phi}_2^{\mathrm{T}}\hat{\boldsymbol{\theta}}_{2,k-1}\right].\end{aligned} \tag{3.50}$$

For the above two equations, we adopt the hierarchical identification principle to solve the difficulty of identification problems by replacing unknown $\theta_1$ and $\theta_2$ with their estimates $\hat{\theta}_{1,k-1}$ and $\hat{\theta}_{2,k-1}$, and unknown state $x(t-i)$ in $\varphi_1(t)$ with its estimate $\hat{x}_k(t-i)(i=1,2,\ldots,r)$, and $x(t-l)$ in $\varphi_2(t)$ with its estimate $\hat{x}_k(t-l)(l=0,1,\ldots,r)$.

Construct the estimate $\hat{\varphi}_{1,k}(t)$ of $\varphi_1(t)$ and $\hat{\varphi}_{2,k}(t)$ of $\varphi_2(t)$ as

$$\hat{\varphi}_{1,k}(t) = \begin{bmatrix}\hat{\boldsymbol{x}}_{k-1}(t-1) \\ \hat{\boldsymbol{x}}_{k-1}(t-2) \\ \vdots \\ \hat{\boldsymbol{x}}_{k-1}(t-r)u(t)\end{bmatrix} \in \mathbb{R}^{nr+1}, \quad \hat{\varphi}_{2,k}(t) = \begin{bmatrix}\hat{\boldsymbol{x}}_{k-1}(t) \\ \hat{\boldsymbol{x}}_{k-1}(t-1) \\ \vdots \\ \hat{\boldsymbol{x}}_{k-1}(t-r)u(t)\end{bmatrix} \in \mathbb{R}^{n+nr+1}.$$

Define

$$\begin{aligned}\hat{\boldsymbol{\Phi}}_{1,k} &:= \left[\hat{\varphi}_{1,k}(1), \hat{\varphi}_{1,k}(2), \ldots, \hat{\varphi}_{1,k}(L)\right] \in \mathbb{R}^{(nr+1)\times L}, \\ \hat{\boldsymbol{\Phi}}_{2,k} &:= \left[\hat{\varphi}_{2,k}(1), \hat{\varphi}_{2,k}(2), \ldots, \hat{\varphi}_{2,k}(L)\right] \in \mathbb{R}^{(n+nr+1)\times L}.\end{aligned}$$

Replacing the unknown $\Phi_1$ and $\Phi_2$ in (16) and (17) with $\hat{\Phi}_{1,k}$ and $\hat{\Phi}_{2,k}$, we have

$$\begin{aligned}\hat{\boldsymbol{\theta}}_{1,k} &= \hat{\boldsymbol{\theta}}_{1,k-1} + \mu_k\hat{\boldsymbol{\Phi}}_{1,k}\left[\boldsymbol{X}-\boldsymbol{X}_1-\hat{\boldsymbol{\theta}}^{\mathrm{T}}_{1,k-1}\hat{\boldsymbol{\Phi}}_{1,k}\right]^{\mathrm{T}}, \\ \hat{\boldsymbol{\theta}}_{2,k} &= \hat{\boldsymbol{\theta}}_{2,k-1} + \mu_k\hat{\boldsymbol{\Phi}}_{2,k}\left[\boldsymbol{X}_2 - \hat{\boldsymbol{\Phi}}^{\mathrm{T}}_{2,k}\hat{\boldsymbol{\theta}}_{2,k-1}\right].\end{aligned}$$

Or

$$\begin{aligned}\hat{\boldsymbol{\theta}}_{1,k} &= \left[\boldsymbol{I} - \mu_k\hat{\boldsymbol{\Phi}}_{1,k}\hat{\boldsymbol{\Phi}}^{\mathrm{T}}_{1,k}\right]\hat{\boldsymbol{\theta}}_{1,k-1} + \mu_k\hat{\boldsymbol{\Phi}}_{1,k}[\boldsymbol{X}-\boldsymbol{X}_1]^{\mathrm{T}}, \\ \hat{\boldsymbol{\theta}}_{2,k} &= \left[\boldsymbol{I} - \mu_k\hat{\boldsymbol{\Phi}}_{2,k}\hat{\boldsymbol{\Phi}}^{\mathrm{T}}_{2,k}\right]\hat{\boldsymbol{\theta}}_{2,k-1} + \mu_k\hat{\boldsymbol{\Phi}}_{2,k}\boldsymbol{X}_2,\end{aligned}$$

where $I$ is an identity matrix of appropriate size. The necessary convergence condition for the parameter estimation $\hat{\theta}_{1,k}$ and $\hat{\theta}_{2,k}$ is that all eigenvalues of $[I - \mu_k \hat{\Phi}_{1,k} \hat{\Phi}^{\mathrm{T}}_{1,k}]$ and $[I - \mu_k \hat{\Phi}_{2,k} \hat{\Phi}^{\mathrm{T}}_{2,k}]$ have to be inside the unit circle. So the convergence factor $\mu_k$ should satisfy

$$0 \leqslant \mu_k \leqslant 2\left\{\lambda_{\max}\left[\hat{\mathbf{\Phi}}_{1,k}\hat{\mathbf{\Phi}}^{\mathrm{T}}_{1,k}\right]\right\}^{-1},$$
$$0 \leqslant \mu_k \leqslant 2\left\{\lambda_{\max}\left[\hat{\mathbf{\Phi}}_{2,k}\hat{\mathbf{\Phi}}^{\mathrm{T}}_{2,k}\right]\right\}^{-1}.$$

One conservative choice of $\mu_k$ is

$$0 \leqslant \mu_k \leqslant 2\left(\|\hat{\mathbf{\Phi}}_{1,k}\|^2 + \|\hat{\mathbf{\Phi}}_{2,k}\|^2\right)^{-1}.$$

Summarizing the above expressions gives the following hierarchical gradient-based iterative identification algorithm for state-space system with multistate delays (the HGI algorithm for short):

$$\hat{\boldsymbol{\theta}}_{1,k} = \hat{\boldsymbol{\theta}}_{1,k-1} + \mu_k \sum_{t=1}^{L} \hat{\varphi}_{1,k}(t)\left[\hat{\boldsymbol{x}}'(t+1)_k - \hat{\boldsymbol{x}}'(t)_k - \hat{\boldsymbol{\theta}}^{\mathrm{T}}_{1,k-1}\hat{\varphi}_{1,k}(t)\right]^{\mathrm{T}}, \tag{3.51}$$

$$\hat{\boldsymbol{\theta}}_{2,k} = \hat{\boldsymbol{\theta}}_{2,k-1} + \mu_k \sum_{t=1}^{L} \hat{\varphi}_{2,k}(t)\left[\hat{x}(t+1)_k - \hat{\varphi}^{\mathrm{T}}_{2,k}(t)\hat{\boldsymbol{\theta}}_{2,k-1}\right], \tag{3.52}$$

$$\hat{\varphi}_{1,k}(t) = \left[\hat{\boldsymbol{x}}^{\mathrm{T}}_{k-1}(t-1), \hat{\boldsymbol{x}}^{\mathrm{T}}_{k-1}(t-2), \cdots, \hat{\boldsymbol{x}}^{\mathrm{T}}_{k-1}(t-r), u(t)\right]^{\mathrm{T}}, \tag{3.53}$$

$$\hat{\varphi}_{2,k}(t) = \left[\hat{\boldsymbol{x}}^{\mathrm{T}}_{k-1}(t), \hat{\boldsymbol{x}}^{\mathrm{T}}_{k-1}(t-1), \hat{\boldsymbol{x}}^{\mathrm{T}}_{k-1}(t-2), \cdots, \hat{\boldsymbol{x}}^{\mathrm{T}}_{k-1}(t-r), u(t)\right]^{\mathrm{T}}, \tag{3.54}$$

$$\mu_k \leqslant 2\left(\sum_{t=1}^{L}\left[\|\hat{\varphi}_{1,k}(t)\|^2 + \|\hat{\varphi}_{2,k}(t)\|^2\right]\right)^{-1}, \tag{3.55}$$

$$\hat{\boldsymbol{\theta}}^{\mathrm{T}}_{1,k} = \begin{bmatrix} \hat{\boldsymbol{b}}_{11,k} & \hat{\boldsymbol{b}}_{21,k} & \cdots & \hat{\boldsymbol{b}}_{r1,k} & \hat{g}_{1,k} \\ \hat{\boldsymbol{b}}_{12,k} & \hat{\boldsymbol{b}}_{22,k} & \cdots & \hat{\boldsymbol{b}}_{r2,k} & \hat{g}_{2,k} \\ \vdots & \vdots & & \vdots & \vdots \\ \hat{\boldsymbol{b}}_{1,n-1,k} & \hat{\boldsymbol{b}}_{2,n-1,k} & \cdots & \hat{\boldsymbol{b}}_{r,n-1,k} & \hat{g}_{n-1,k} \end{bmatrix}, \tag{3.56}$$

$$\hat{\boldsymbol{\theta}}_{2,k} = \left[\hat{\boldsymbol{a}}_k, \hat{\boldsymbol{b}}_{1n,k}, \hat{\boldsymbol{b}}_{2n,k}, \cdots, \hat{\boldsymbol{b}}_{rn,k}, \hat{g}_{n,k}\right]^{\mathrm{T}}. \tag{3.57}$$

The steps involved in the HGI algorithm to compute $\hat{\theta}_{1,k}$ and $\hat{\theta}_{2,k}$ are listed in the following.

- Collect the input/output data $\{u(t), x(t+1){:}t = 1, 2, \ldots, L\}$, and form $\hat{\varphi}_{1,k}$ and $\hat{\varphi}_{2,k}$ by (3.53) and (3.54).
- To initialize, let $k = 1$, $\theta_1^{\mathrm{T}}(0) = 1_{(n-1)\times(nr+1)}/p_0, \theta_2(0) = 1_{n+nr+1}/p_0, p_0 = 10^6$.
- Choose a suitable $\mu_k$ satisfying (3.55) and update $\hat{\theta}_{1,k}$ and $\hat{\theta}_{2,k}$ by (3.51) and (3.52), respectively.
- If $||\hat{\theta}_{1,k} - \hat{\theta}_{1,k-1}|| + ||\hat{\theta}_{2,k} - \hat{\theta}_{2,k-1}|| \leq \varepsilon$, then terminate the procedure and obtain the iterative times $k$ and estimates $\hat{\theta}_{1,k}$ and $\hat{\theta}_{2,k}$; otherwise, increase $k$ by 1 and go to Step 3.

### 3.2.3 The hierarchical least squares-based iterative algorithm

The hierarchical gradient-based iterative algorithm converges slowly. To improve the convergence rate, the solution here is to derive a hierarchical least squares-based iterative algorithm.

By minimizing Eqs. (3.47) and (3.48), we can obtain the least squares-based iterative solutions $\hat{\theta}_1$ and $\hat{\theta}_2$ of $\theta_1$ and $\theta_2$, respectively,

$$\hat{\boldsymbol{\theta}}_{1,k} = \hat{\boldsymbol{\theta}}_{1,k-1} + \mu_k\left[\boldsymbol{\Phi}_1\boldsymbol{\Phi}_1^{\mathrm{T}}\right]^{-1}\boldsymbol{\Phi}_1\left[\mathbf{X} - \mathbf{X}_1 - \hat{\boldsymbol{\theta}}_{1,k-1}^{\mathrm{T}}\boldsymbol{\Phi}_1\right]^{\mathrm{T}}, \tag{3.58}$$

$$\hat{\boldsymbol{\theta}}_{2,k} = \hat{\boldsymbol{\theta}}_{2,k-1} + \mu_k\left[\boldsymbol{\Phi}_2\boldsymbol{\Phi}_2^{\mathrm{T}}\right]^{-1}\boldsymbol{\Phi}_2\left[\mathbf{X}_2 - \boldsymbol{\Phi}_2^{\mathrm{T}}\hat{\boldsymbol{\theta}}_{2,k-1}\right], \tag{3.59}$$

where $\mu_k \geq 0$ is the time-varying step size or time-varying convergence factor, which will be given later. Substituting (3.42–3.46) into (3.58) and (3.59) gives

$$\begin{aligned}\hat{\boldsymbol{\theta}}_{1,k} &= \hat{\boldsymbol{\theta}}_{1,k-1} + \mu_k\left[\boldsymbol{\Phi}_1\boldsymbol{\Phi}_1^{\mathrm{T}}\right]^{-1}\sum_{t=1}^{L}\varphi_1(t)\left[\mathbf{X} - \mathbf{X}_1 - \hat{\boldsymbol{\theta}}_{1,k-1}^{\mathrm{T}}\varphi_1(t)\right]^{\mathrm{T}} \\ &= \hat{\boldsymbol{\theta}}_{1,k-1} + \mu_k\left[\sum_{t=1}^{L}\varphi_1(t)\varphi_1^{\mathrm{T}}(t)\right]^{-1}\sum_{t=1}^{L}\varphi_1(t)\left[\boldsymbol{x}'(t+1) - \boldsymbol{x}'(t) - \hat{\boldsymbol{\theta}}_{1,k-1}^{\mathrm{T}}\varphi_1(t)\right]^{\mathrm{T}},\end{aligned} \tag{3.60}$$

$$\begin{aligned}\hat{\boldsymbol{\theta}}_{2,k} &= \hat{\boldsymbol{\theta}}_{2,k-1} + \mu_k\left[\boldsymbol{\Phi}_2\boldsymbol{\Phi}_2^{\mathrm{T}}\right]^{-1}\sum_{t=1}^{L}\varphi_2(t)\left[\mathbf{X}_2 - \varphi_2^{\mathrm{T}}(t)\hat{\boldsymbol{\theta}}_{2,k-1}\right] \\ &= \hat{\boldsymbol{\theta}}_{2,k-1} + \mu_k\left[\sum_{t=1}^{L}\varphi_2(t)\varphi_2^{\mathrm{T}}(t)\right]^{-1}\sum_{t=1}^{L}\varphi_2(t)\left[x(t+1) - \varphi_2^{\mathrm{T}}(t)\hat{\boldsymbol{\theta}}_{2,k-1}\right].\end{aligned} \tag{3.61}$$

Estimating the states in (3.38) and (3.39), we have

$$\boldsymbol{x}'(t+1)_k - \boldsymbol{x}'(t)_k = \begin{bmatrix} \hat{\boldsymbol{b}}_{11,k} \\ \hat{\boldsymbol{b}}_{12,k} \\ \vdots \\ \hat{\boldsymbol{b}}_{1,n-1,k} \end{bmatrix} \hat{\boldsymbol{x}}(t-1)_{k-1} + \cdots + \begin{bmatrix} \hat{\boldsymbol{b}}_{r1,k} \\ \hat{\boldsymbol{b}}_{r2,k} \\ \vdots \\ \hat{\boldsymbol{b}}_{r,n-1,k} \end{bmatrix} \hat{\boldsymbol{x}}(t-r)_{k-1} + \begin{bmatrix} \hat{g}_{1,k} \\ \hat{g}_{2,k} \\ \vdots \\ \hat{g}_{n-1,k} \end{bmatrix} u(t), \tag{3.62}$$

$$\hat{x}(t+1)_k = \hat{\boldsymbol{a}}_k \hat{\boldsymbol{x}}(t)_{k-1} + \hat{\boldsymbol{b}}_{1n,k} \hat{\boldsymbol{x}}(t-1)_{k-1} + \cdots + \hat{\boldsymbol{b}}_{rn,k} \hat{\boldsymbol{x}}(t-r)_{k-1} + \hat{g}_{n,k} u(t). \tag{3.63}$$

Replacing $\varphi_1(t)$ and $\varphi_2(t)$ in (3.60–3.61) with their estimates $\hat{\varphi}_{1,k}(t)$ and $\hat{\varphi}_{2,k}(t)$, we can obtain the hierarchical least squares-based iterative algorithm (HLSI) for estimating $\theta_1$ and $\theta_2$ for state-space models with multistate delays as follows:

$$\begin{aligned} \hat{\boldsymbol{\theta}}_{1,k} &= \hat{\boldsymbol{\theta}}_{1,k-1} + \mu_k \left[ \sum_{t=1}^{L} \hat{\boldsymbol{\varphi}}_{1,k}(t) \hat{\boldsymbol{\varphi}}_{1,k}^{\mathrm{T}}(t) \right]^{-1} \\ &\quad \times \sum_{t=1}^{L} \hat{\boldsymbol{\varphi}}_{1,k}(t) \left[ \hat{\boldsymbol{x}}'(t+1)_k - \hat{\boldsymbol{x}}'(t)_k - \hat{\boldsymbol{\theta}}_{1,k-1}^{\mathrm{T}} \hat{\boldsymbol{\varphi}}_{1,k}(t) \right]^{\mathrm{T}}, \end{aligned} \tag{3.64}$$

$$\hat{\boldsymbol{\varphi}}_{1,k}(t) = \left[ \hat{\boldsymbol{x}}_{k-1}^{\mathrm{T}}(t-1), \hat{\boldsymbol{x}}_{k-1}^{\mathrm{T}}(t-2), \cdots, \hat{\boldsymbol{x}}_{k-1}^{\mathrm{T}}(t-r), u(t) \right]^{\mathrm{T}}, \tag{3.65}$$

$$\hat{\boldsymbol{\theta}}_{2,k} = \hat{\boldsymbol{\theta}}_{2,k-1} + \mu_k \left[ \sum_{t=1}^{L} \hat{\boldsymbol{\varphi}}_{2,k}(t) \hat{\boldsymbol{\varphi}}_{2,k}^{\mathrm{T}}(t) \right]^{-1} \sum_{t=1}^{L} \hat{\boldsymbol{\varphi}}_{2,k}(t) \left[ \hat{x}(t+1)_k - \hat{\boldsymbol{\varphi}}_{2,k}^{\mathrm{T}}(t) \hat{\boldsymbol{\theta}}_{2,k-1} \right], \tag{3.66}$$

$$\hat{\boldsymbol{\varphi}}_{2,k}(t) = \left[ \hat{\boldsymbol{x}}_{k-1}^{\mathrm{T}}(t), \hat{\boldsymbol{x}}_{k-1}^{\mathrm{T}}(t-1), \hat{\boldsymbol{x}}_{k-1}^{\mathrm{T}}(t-2), \cdots, \hat{\boldsymbol{x}}_{k-1}^{\mathrm{T}}(t-r), u(t) \right]^{\mathrm{T}}, \tag{3.67}$$

$$\mu_k = \frac{1}{\lambda_{\max}\left\{\hat{\mathbf{\Phi}}_{1,k}^{\mathrm{T}}\left[\hat{\mathbf{\Phi}}_{1,k}\hat{\mathbf{\Phi}}_{1,k}^{\mathrm{T}}\right]^{-1}\hat{\mathbf{\Phi}}_{1,k}\right\} + \lambda_{\max}\left\{\hat{\mathbf{\Phi}}_{2,k}^{\mathrm{T}}\left[\hat{\mathbf{\Phi}}_{2,k}\hat{\mathbf{\Phi}}_{2,k}^{\mathrm{T}}\right]^{-1}\hat{\mathbf{\Phi}}_{2,k}\right\}}, \tag{3.68}$$

$$\hat{\mathbf{\Phi}}_{1,k} = \left[\hat{\varphi}_{1,k}(1), \hat{\varphi}_{1,k}(2), \ldots, \hat{\varphi}_{1,k}(L)\right], \tag{3.69}$$

$$\hat{\mathbf{\Phi}}_{2,k} = \left[\hat{\varphi}_{2,k}(1), \hat{\varphi}_{2,k}(2), \ldots, \hat{\varphi}_{2,k}(L)\right]. \tag{3.70}$$

From (3.64–3.70), we can see that the HLSI algorithm performs a hierarchical interactive process: when computing the parameter estimates $\hat{\theta}_{1,k}$ and $\hat{\theta}_{2,k}$, the unknown state terms $x(t-i)(i=1,2,\ldots,r)$ and $x(t-l)(l=0,1,\ldots,r)$ in the information vector are replaced with their corresponding estimates $\hat{x}_{k-1}(t-i)$ and $\hat{x}_{k-1}(t-j)$ at the $k$–1th iteration, respectively, while the state estimates $\hat{x}_k(t-i)$ and $\hat{x}_k(t-l)$ at iteration $k$ are computed from the parameter estimates $\hat{\theta}_{1,k}$ and $\hat{\theta}_{2,k}$.

The identification steps of the HLSI algorithm to compute $\hat{\theta}_{1,k}$ and $\hat{\theta}_{2,k}$ are listed below.

- Collect the input/output data $\{u(t), x(t+1){:}t=1,2,\ldots,L\}$, form $\hat{\varphi}_{1,k}(t)$ by (3.65) and $\hat{\varphi}_{2,k}(t)$ by (3.67), respectively, and give the estimation precision $\varepsilon = 10^{-3}$.
- To initialize, let $k=1$, $\hat{x}_0(t) =$ random number for all $t$, $\theta_1^{\mathrm{T}}(0) = \mathbf{1}_{(n-1)\times(nr+1)}, \theta_2(0) = \mathbf{1}_{n+nr+1}$, with $\mathbf{1}_{(n-1)\times(nr+1)}$ being an $(n-1)\times(nr+1)$-dimensional matrix whose elements are all 1, and $\mathbf{1}_{n+nr+1}$ being an $(n+nr+1)$-dimensional vector whose elements are all 1.
- Compute $\mu_k$ by (3.68).
- Update the estimates $\hat{\theta}_{1,k}$ by (3.64) and $\hat{\theta}_{2,k}$ by (3.66).
- Compare $\hat{\theta}_{1,k}$ with $\hat{\theta}_{1,k-1}$ and $\hat{\theta}_{2,k}$ with $\hat{\theta}_{2,k-1}$, if they are sufficiently close, or for some pre-set small $\varepsilon$, if $||\hat{\theta}_{1,k}-\hat{\theta}_{1,k-1}|| + ||\hat{\theta}_{2,k}-\hat{\theta}_{2,k-1}|| \leq \varepsilon$, then terminate the procedure, and obtain the iterative times $k$ and estimates $\hat{\theta}_{1,k}$ and $\hat{\theta}_{2,k}$; otherwise, increase $k$ by 1 and go to Step 3.

### 3.2.4 Example

Consider the following state-space system with two-step state delay:

$$\begin{aligned}
\boldsymbol{x}(t+1) &= \begin{bmatrix} 0 & 1 \\ a_2 & a_1 \end{bmatrix}\boldsymbol{x}(t) + \begin{bmatrix} b_{11} & b_{12} \\ b_{13} & b_{14} \end{bmatrix}\boldsymbol{x}(t-1) + \begin{bmatrix} b_{21} & b_{22} \\ b_{23} & b_{24} \end{bmatrix}\boldsymbol{x}(t-2) + \begin{bmatrix} g_1 \\ g_2 \end{bmatrix}u(t) \\
&= \begin{bmatrix} 0 & 1 \\ -0.45 & -0.80 \end{bmatrix}\boldsymbol{x}(t) + \begin{bmatrix} 0.20 & -0.15 \\ 0.15 & -0.20 \end{bmatrix}\boldsymbol{x}(t-1) \\
&\quad + \begin{bmatrix} 0.20 & -0.18 \\ 0.18 & -0.05 \end{bmatrix}\boldsymbol{x}(t-2) + \begin{bmatrix} 1 \\ -1 \end{bmatrix}u(t), \\
\boldsymbol{\theta}_1 &= \left[b_{11}, b_{12}, b_{21}, b_{22}, g_1\right]^{\mathrm{T}} = [-0.20, -0.15, 0.20, -0.18, 1]^{\mathrm{T}}, \\
\boldsymbol{\theta}_2 &= \left[a_2, a_1, b_{13}, b_{14}, b_{23}, b_{24}, g_2\right]^{\mathrm{T}} \\
&= [-0.45, -0.80, 0.15, -0.20, 0.18, -0.05, -1]^{\mathrm{T}}.
\end{aligned}$$

In simulation, the input $\{u(t)\}$ is taken as an uncorrelated persistent excitation signal vector sequence with zero mean and unit variance. We apply the HGI algorithm and the HLSI algorithm to estimate the parameters of this example system, and apply the state estimation algorithm to estimate the state vector $x(t)$. The parameter estimates and their estimation errors are shown in Tables 3.2–3.3, and the parameter estimation errors $\delta$ versus $k$ are shown in Figs. 3.2–3.3, where $\delta := ||\hat{\theta}(t) - \theta||/||\theta||$, and the state estimates $\hat{x}_1(t)$ and $\hat{x}_2(t)$ versus $t$ are shown in Figs. 3.4–3.5.

From the simulation results in Tables 3.2–3.3 and Figs. 3.2–3.5, we can draw the following conclusions: (1) the parameter estimation errors $\delta$ become smaller (in general) with $k$ increasing; (2) the HLSI algorithm converges fast and needs only several iterations to converge to their true values; (3) the parameter estimation accuracy of the HLSI algorithm is higher than that of the HGI algorithm; (4) the state estimates are close to their true values with $k$ increasing.

### 3.2.5 Conclusions

This chapter studies parameter and state estimation problem for a class of linear time-varying dynamic systems whose matrices of state-space model are the scalar parameters. Compared with the recursive least squares methods, the proposed algorithms can not only identify the system unknown parameters, but also identify the system unknown states. The simulation results show that the proposed algorithm are effective. The convergence properties of the proposed algorithm can be analyzed by means of the martingale convergence theorem.

**Table 3.2** The HGI parameter estimates and errors.

| $k$ | 10 | 20 | 50 | 100 | 200 | 500 | 1000 | 1500 | 2000 |
|---|---|---|---|---|---|---|---|---|---|
| $b_{11} = 0.20$ | 0.01788 | 0.02261 | 0.03654 | 0.05639 | 0.08955 | 0.14791 | 0.18298 | 0.19391 | 0.19776 |
| $b_{12} = -0.15$ | −0.14803 | −0.14603 | −0.14047 | −0.13351 | −0.12487 | −0.12317 | −0.13690 | −0.14484 | −0.14806 |
| $b_{21} = 0.20$ | 0.16814 | 0.17036 | 0.17669 | 0.18511 | 0.19722 | 0.20930 | 0.20616 | 0.20254 | 0.20097 |
| $b_{22} = -0.18$ | 0.03593 | 0.03170 | 0.01903 | 0.00038 | −0.03275 | −0.10018 | −0.15089 | −0.16925 | −0.17602 |
| $g_1 = 1.00$ | 0.05614 | 0.10843 | 0.25274 | 0.43458 | 0.67957 | 0.94084 | 0.99599 | 0.99956 | 0.99990 |
| $a_2 = -0.45$ | −0.24479 | −0.25022 | −0.26735 | −0.29421 | −0.34329 | −0.42804 | −0.45834 | −0.45784 | −0.45495 |
| $a_1 = -0.80$ | −0.14997 | −0.18004 | −0.26351 | −0.37014 | −0.51931 | −0.70456 | −0.77196 | −0.78723 | −0.79323 |
| $b_{13} = 0.15$ | 0.30378 | 0.29927 | 0.28520 | 0.26353 | 0.22487 | 0.16159 | 0.14301 | 0.14522 | 0.14763 |
| $b_{14} = -0.20$ | −0.14663 | −0.14183 | −0.12917 | −0.11483 | −0.10028 | −0.10616 | −0.14183 | −0.16686 | −0.18147 |
| $b_{23} = 0.18$ | −0.02254 | −0.00396 | 0.04510 | 0.10164 | 0.16560 | 0.20489 | 0.19576 | 0.18847 | 0.18460 |
| $b_{24} = -0.05$ | 0.25003 | 0.24260 | 0.22114 | 0.19145 | 0.14350 | 0.05938 | 0.00031 | −0.02479 | −0.03699 |
| $g_2 = -1.00$ | −0.04221 | −0.08331 | −0.19978 | −0.35432 | −0.58302 | −0.88733 | −0.98723 | −0.99856 | −0.99984 |
| $\delta$ (%) | 90.37205 | 86.14463 | 74.39763 | 59.37238 | 38.44434 | 13.63933 | 5.28681 | 2.69954 | 1.43764 |

**Table 3.3** The HLSI parameter estimates and errors.

| $k$ | 1 | 2 | 5 | 10 | 20 | 50 |
|---|---|---|---|---|---|---|
| $b_{11} = 0.20$ | 0.01522 | 0.02928 | 0.06536 | 0.10937 | 0.15893 | 0.19647 |
| $b_{12} = -0.15$ | −0.01141 | −0.02196 | −0.04902 | −0.08202 | −0.11920 | −0.14735 |
| $b_{21} = 0.20$ | 0.01522 | 0.02928 | 0.06536 | 0.10937 | 0.15893 | 0.19647 |
| $b_{22} = -0.18$ | −0.01370 | −0.02635 | −0.05883 | −0.09843 | −0.14303 | −0.17682 |
| $g_1 = 1.00$ | 0.07610 | 0.14640 | 0.32682 | 0.54683 | 0.79464 | 0.98234 |
| $a_2 = -0.45$ | −0.03424 | −0.06588 | −0.14707 | −0.24607 | −0.35759 | −0.44205 |
| $a_1 = -0.80$ | −0.06088 | −0.11712 | −0.26146 | −0.43746 | −0.63571 | −0.78587 |
| $b_{13} = 0.15$ | 0.01142 | 0.02196 | 0.04902 | 0.08202 | 0.11920 | 0.14735 |
| $b_{14} = -0.20$ | −0.01522 | −0.02928 | −0.06536 | −0.10937 | −0.15893 | −0.19647 |
| $b_{23} = 0.18$ | 0.01370 | 0.02635 | 0.05883 | 0.09843 | 0.14303 | 0.17682 |
| $b_{24} = -0.05$ | −0.00380 | −0.00732 | −0.01634 | −0.02734 | −0.03973 | −0.04912 |
| $g_2 = -1.00$ | −0.07610 | −0.14640 | −0.32682 | −0.54683 | −0.79464 | −0.98234 |
| $\delta$ (%) | 92.39029 | 85.35964 | 67.31799 | 45.31710 | 20.53639 | 1.76578 |

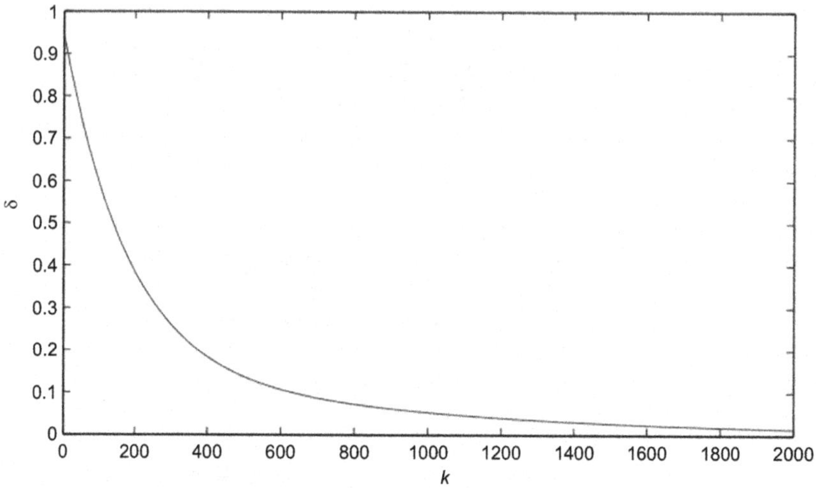


**Figure 3.2** The HGI estimation errors $\delta$ versus $k$.

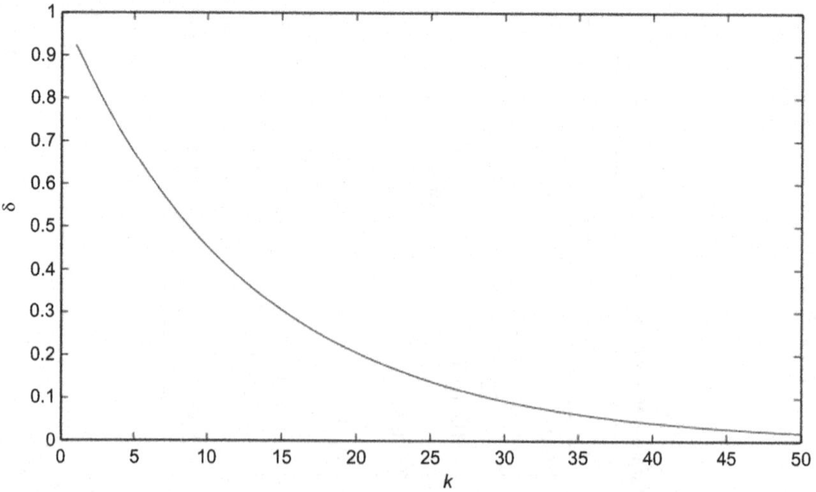


**Figure 3.3** The HLSI estimation errors $\delta$ versus $k$.

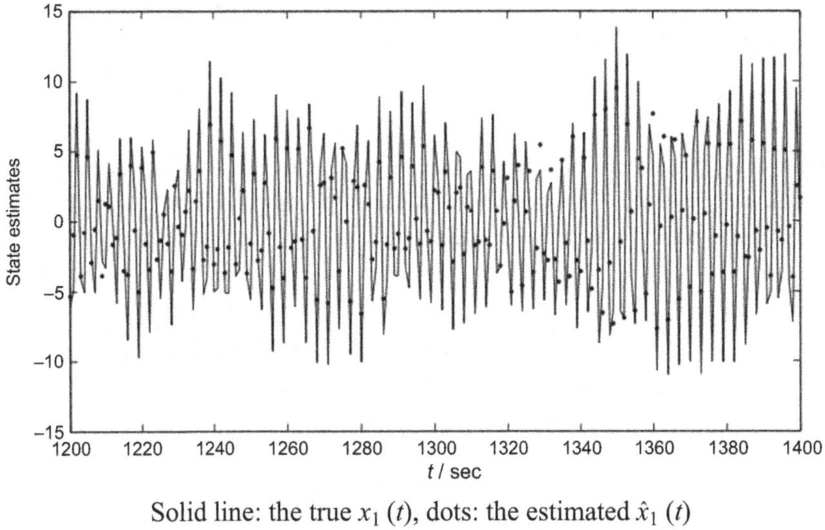


**Figure 3.4** The true state $x_1(t)$ and its predicted $\hat{x}_1(t)$.

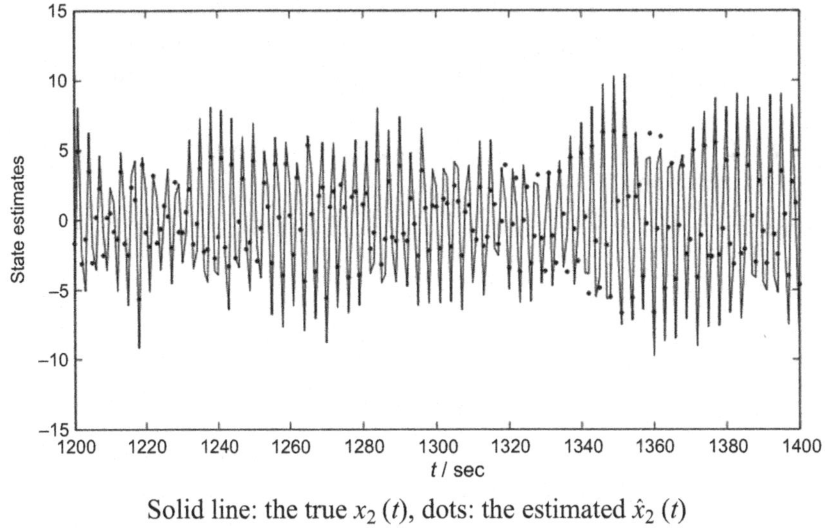


**Figure 3.5** The true state $x_2(t)$ and its predicted $\hat{x}_2(t)$.

## 3.3 Least squares-based iterative parameter estimation algorithm for multiple time delay

The state-space models which can be described by differential/difference equations have wide applications in adaptive control, system analysis, and parameter estimation and thus have received much research attention in parameter and state estimation in several decades. There exist iterative algorithms which are able to make sufficient use of all the measured information so that the accuracy of parameter estimation can be improved. The iterative algorithms can be used for solving not only some matrix equations, but also parameter estimation problems.

An extensive literature has developed for decades on the analysis of time delays in the area of control and identification. Effects of time delay on the function of control systems have received considerable attention of many researchers in industrial processes (Hidayat and Medvedev, 2012).

This paper combines the recursive least squares identification idea and the iterative identification principle to present a least squares-based iterative parameter estimation algorithm for state-space model with time delay. To solve the difficulty of the information matrix including unmeasurable terms, the unknown terms are replaced with their iterative residuals, which are computed through the preceding parameter estimates. Compared with the recursive least squares algorithm, the iterative identification algorithm has less computation burden and is easier to implement.

### 3.3.1 The system description and input–output representation

Consider the following state-space system with time delay,

$$\boldsymbol{x}(t+1) = \boldsymbol{A}_1\boldsymbol{x}(t) + \boldsymbol{A}_2\boldsymbol{x}(t-1) + \boldsymbol{A}_3\boldsymbol{x}(t-2) + \cdots + \boldsymbol{A}_r\boldsymbol{x}(t-r+1) + \boldsymbol{f}u(t), \tag{3.71}$$

$$y(t) = \boldsymbol{c}\boldsymbol{x}(t) + v(t), \tag{3.72}$$

where $x(t) = [x_1(t), x_2(t), \ldots, x_n(t)]^{\mathrm{T}} \in R^n$ is the state vector, $u(t) \in R$ is the system input, $y(t) \in R$ is the system output, $v(t) \in R$ is a random noise with zero mean, $A_i \in R^{n\times n}(i = 1, 2, \ldots, r)$, $f \in R^n$ and $c \in R^{1\times n}$ are the system parameter matrix/vectors.

Let us introduce some notation. "$A =: X$" or "$X := A$" stands for "$A$ is defined as $X$"; the symbol $I(I_n)$ stands for an identity matrix of appropriate size $(n \times n)$; $z$ represents a unit forward shift operator: $zx(t) = x(t+1)$

and $z^{-1}x(t) = x(t-1)$; the superscript T denotes the matrix/vector transpose; $\hat{\theta}(t)$ denotes the estimate of $\theta$ at time $t$; adj$[X]$ denotes the adjoint matrix of the square matrix $X$: adj$[X] = \det[X]X^{-1}$; $\det[X] = |X|$ denotes the determinant of the square matrix $X$.

The goal is to develop least squares-based iterative parameter estimation algorithms to estimate the unknown parameters $(A_i, f)$ by using the available input–output measurement data $\{u(t), y(t)\}$ and to improve the accuracy of parameter estimation. Without loss of generality, assume that $u(t) = 0$, $y(t) = 0$, and $v(t) = 0$ as $t \leq 0$. The following transforms the time-delay state-space model in (3.71) and (3.72) into an input–output representation and gives its identification model.

**Lemma 3.1.**: For the state-space model in (3.71) and (3.72), the transfer function from the input $u(t)$ to the output $y(t)$ is given by

$$G(z) := \frac{c\,\mathrm{adj}\left[z^r\boldsymbol{I} - z^{r-1}\boldsymbol{A}_1 - z^{r-2}\boldsymbol{A}_2 - \cdots - \boldsymbol{A}_r\right]\boldsymbol{f}z^{r-1}}{\det[z^r\boldsymbol{I} - z^{r-1}\boldsymbol{A}_1 - z^{r-2}\boldsymbol{A}_2 - \cdots - \boldsymbol{A}_r]}.$$

**Proof** The first method is to prove this lemma by using the property of the shift operator $z$.

$$z\boldsymbol{x}(t) = \boldsymbol{A}_1\boldsymbol{x}(t) + z^{-1}\boldsymbol{A}_2\boldsymbol{x}(t) + z^{-2}\boldsymbol{A}_3\boldsymbol{x}(t) + \cdots + z^{-r+1}\boldsymbol{A}_r\boldsymbol{x}(t) + \boldsymbol{f}u(t).$$

Eq. (3.71) can be rewritten as

Multiplying both sides by $z^{r-1}$ gives

$$z^r\boldsymbol{x}(t) = z^{r-1}\boldsymbol{A}_1\boldsymbol{x}(t) + z^{r-2}\boldsymbol{A}_2\boldsymbol{x}(t) + z^{r-3}\boldsymbol{A}_3\boldsymbol{x}(t) + \cdots + \boldsymbol{A}_r\boldsymbol{x}(t) + z^{r-1}\boldsymbol{f}u(t),$$

or

$$\boldsymbol{x}(t) = \left(z^r\boldsymbol{I} - z^{r-1}\boldsymbol{A}_1 - z^{r-2}\boldsymbol{A}_2 - \cdots - \boldsymbol{A}_r\right)^{-1}\boldsymbol{f}_z^{\,r-1}u(t).$$

Substituting the above $x(t)$ into Eq. (3.72) gives the output equation:

$$\begin{aligned} y(t) &= c(z^r\boldsymbol{I} - z^{r-1}\boldsymbol{A}_1 - z^{r-2}\boldsymbol{A}_2 - \cdots - \boldsymbol{A}_r)^{-1}\boldsymbol{f}z^{r-1}u(t) + v(t) \\ &=: G(z)u(t) + v(t). \end{aligned}$$

Thus, we have the transfer function of the system from the input $u(t)$ to the output $y(t)$:

$$\begin{aligned} G(z) &= c(z^r\boldsymbol{I} - z^{r-1}\boldsymbol{A}_1 - z^{r-2}\boldsymbol{A}_2 - \cdots - \boldsymbol{A}_r)^{-1}\boldsymbol{f}z^{r-1} \\ &= \frac{c\,\mathrm{adj}\left[z^r\boldsymbol{I} - z^{r-1}\boldsymbol{A}_1 - z^{r-2}\boldsymbol{A}_2 - \cdots - \boldsymbol{A}_r\right]\boldsymbol{f}z^{r-1}}{\det[z^r\boldsymbol{I} - z^{r-1}\boldsymbol{A}_1 - z^{r-2}\boldsymbol{A}_2 - \cdots - \boldsymbol{A}_r]}. \end{aligned}$$

The second method is to define an expanded state vector:

$$\boldsymbol{X}(t) := \begin{bmatrix} \boldsymbol{x}(t) \\ \boldsymbol{x}(t-1) \\ \vdots \\ \boldsymbol{x}(t-r) \end{bmatrix} \in \mathbb{R}^{(r+1)n}.$$

Eqs. (3.71) and (3.72) can be equivalently rewritten as

$$\boldsymbol{X}(t+1) = \begin{bmatrix} \boldsymbol{A}_1 & \boldsymbol{A}_2 & \cdots & \boldsymbol{A}_{r-1} & \boldsymbol{A}_r \\ \boldsymbol{I} & \boldsymbol{0} & \cdots & \boldsymbol{0} & \boldsymbol{0} \\ \boldsymbol{0} & \boldsymbol{I} & \cdots & \boldsymbol{0} & \boldsymbol{0} \\ \vdots & \vdots & \ddots & \vdots & \vdots \\ \boldsymbol{0} & \boldsymbol{0} & \cdots & \boldsymbol{I} & \boldsymbol{0} \end{bmatrix} \boldsymbol{X}(t) + \begin{bmatrix} \boldsymbol{f} \\ \boldsymbol{0} \\ \vdots \\ \boldsymbol{0} \end{bmatrix} u(t),$$

$$y(t) = [\boldsymbol{c}, \boldsymbol{0}, \boldsymbol{0}, \cdots, \boldsymbol{0}]\boldsymbol{X}(t) + v(t).$$

Then, the transfer function from the input $u(t)$ to the output $y(t)$ is given by

$$G(z) = [\boldsymbol{c}, \boldsymbol{0}, \boldsymbol{0}, \cdots, \boldsymbol{0}] \begin{bmatrix} z\boldsymbol{I}-\boldsymbol{A}_1 & -\boldsymbol{A}_2 & \cdots & -\boldsymbol{A}_{r-1} & -\boldsymbol{A}_r \\ -\boldsymbol{I} & z\boldsymbol{I} & \cdots & \boldsymbol{0} & \boldsymbol{0} \\ \boldsymbol{0} & -\boldsymbol{I} & \cdots & \boldsymbol{0} & \boldsymbol{0} \\ \vdots & \vdots & \ddots & \vdots & \vdots \\ \boldsymbol{0} & \boldsymbol{0} & \cdots & -\boldsymbol{I} & z\boldsymbol{I} \end{bmatrix}^{-1} \begin{bmatrix} \boldsymbol{f} \\ \boldsymbol{0} \\ \vdots \\ \boldsymbol{0} \end{bmatrix}. \tag{3.73}$$

Define the matrices:

$$\boldsymbol{H} := \begin{bmatrix} z\boldsymbol{I}-\boldsymbol{A}_1 & -\boldsymbol{A}_2 & \cdots & -\boldsymbol{A}_{r-1} & -\boldsymbol{A}_r \\ -\boldsymbol{I} & z\boldsymbol{I} & \cdots & \boldsymbol{0} & \boldsymbol{0} \\ \boldsymbol{0} & -\boldsymbol{I} & \cdots & \boldsymbol{0} & \boldsymbol{0} \\ \vdots & \vdots & \ddots & \vdots & \vdots \\ \boldsymbol{0} & \boldsymbol{0} & \cdots & -\boldsymbol{I} & z\boldsymbol{I} \end{bmatrix},$$

$$\boldsymbol{H}^{-1} := \begin{bmatrix} h_{11} & * & \cdots & * \\ h_{21} & * & \cdots & * \\ h_{31} & * & \cdots & * \\ \vdots & \vdots & & \vdots \\ h_{r+1,1} & * & \cdots & * \end{bmatrix}.$$

According to $HH^{-1} = I$, we have

$$\begin{bmatrix} z\boldsymbol{I}-\boldsymbol{A}_1 & -\boldsymbol{A}_2 & \cdots & -\boldsymbol{A}_{r-1} & -\boldsymbol{A}_r \\ -\boldsymbol{I} & z\boldsymbol{I} & \cdots & \boldsymbol{0} & \boldsymbol{0} \\ \boldsymbol{0} & -\boldsymbol{I} & \cdots & \boldsymbol{0} & \boldsymbol{0} \\ \vdots & \vdots & \ddots & \vdots & \vdots \\ \boldsymbol{0} & \boldsymbol{0} & \cdots & -\boldsymbol{I} & z\boldsymbol{I} \end{bmatrix} \begin{bmatrix} h_{11} & * & \cdots & * \\ h_{21} & * & \cdots & * \\ h_{31} & * & \cdots & * \\ \vdots & \vdots & & \vdots \\ h_{r+1,1} & * & \cdots & * \end{bmatrix}$$

$$= \begin{bmatrix} \boldsymbol{I} & \boldsymbol{0} & \cdots & \boldsymbol{0} \\ \boldsymbol{0} & & \cdots & \boldsymbol{0} \\ \vdots & \vdots & \ddots & \vdots \\ \boldsymbol{0} & \boldsymbol{0} & \cdots & \boldsymbol{I} \end{bmatrix}.$$

Expanding this matrix equation, its first row gives the following equations:

$$\begin{cases} (z\boldsymbol{I}-\boldsymbol{A}_1)\boldsymbol{h}_{11} - \boldsymbol{A}_2\boldsymbol{h}_{21} - \boldsymbol{A}_3\boldsymbol{h}_{31} - \cdots - \boldsymbol{A}_r\boldsymbol{h}_{r+1,1} = \boldsymbol{I}, \\ -\boldsymbol{h}_{11} + z\boldsymbol{h}_{21} = \boldsymbol{0}, \\ -\boldsymbol{h}_{21} + z\boldsymbol{h}_{31} = \boldsymbol{0}, \\ \vdots \\ -\boldsymbol{h}_{r1} + z\boldsymbol{h}_{r+1,1} = \boldsymbol{0}. \end{cases}$$

From the above equations, we can get the solution of $h_{11}$,

$$\boldsymbol{h}_{11} = \left(z^r\boldsymbol{I} - \boldsymbol{A}_1 z^{r-1} - z^{r-2} - \cdots - \boldsymbol{A}_r\right)^{-1} z^{r-1}.$$

From Eq. (3.73), we have

$$G(z) = [\boldsymbol{c}, \boldsymbol{0}, \boldsymbol{0}, \cdots, \boldsymbol{0}] \begin{bmatrix} z\boldsymbol{I}-\boldsymbol{A}_1 & -\boldsymbol{A}_2 & \cdots & -\boldsymbol{A}_{r-1} & -\boldsymbol{A}_r \\ -\boldsymbol{I} & z\boldsymbol{I} & \cdots & \boldsymbol{0} & \boldsymbol{0} \\ \boldsymbol{0} & -\boldsymbol{I} & \cdots & \boldsymbol{0} & \boldsymbol{0} \\ \vdots & \vdots & \ddots & \vdots & \vdots \\ \boldsymbol{0} & \boldsymbol{0} & \cdots & -\boldsymbol{I} & z\boldsymbol{I} \end{bmatrix}^{-1} \begin{bmatrix} \boldsymbol{f} \\ \boldsymbol{0} \\ \vdots \\ \boldsymbol{0} \end{bmatrix}$$

$$= [\boldsymbol{c}, \boldsymbol{0}, \boldsymbol{0}, \cdots, \boldsymbol{0}] \begin{bmatrix} h_{11} & * & \cdots & * \\ h_{21} & * & \cdots & * \\ h_{31} & * & \cdots & * \\ \vdots & \vdots & & \vdots \\ h_{n+1,1} & * & \cdots & * \end{bmatrix} \begin{bmatrix} \boldsymbol{f} \\ \boldsymbol{0} \\ \vdots \\ \boldsymbol{0} \end{bmatrix}$$

$$= \boldsymbol{c}(z^r\boldsymbol{I} - \boldsymbol{A}_1 z^{r-1} - z^{r-2} - \cdots - \boldsymbol{A}_r)^{-1}\boldsymbol{f} z^{r-1}$$

$$= \frac{\boldsymbol{c}\mathrm{adj}\left[z^r\boldsymbol{I} - \boldsymbol{A}_1 z^{r-1} - z^{r-2} - \cdots - \boldsymbol{A}_r\right]\boldsymbol{f} z^{r-1}}{\det[z^r\boldsymbol{I} - \boldsymbol{A}_1 z^{r-1} - z^{r-2} - \cdots - \boldsymbol{A}_r]}$$

$$=: \frac{\beta(z)}{\alpha(z)}.$$

Hence, we have the input−output representation:

$$y(t) = \frac{\beta(z)}{\alpha(z)} u(t) + v(t),$$

where $\alpha(z)$ and $\beta(z)$ are polynomials in a unit backward shift operator $z^{-1}[z^{-1}y(t) = y(t-1)]$, and

$$\begin{aligned}\alpha(z) &:= z^{-m}\det\left[z^r\boldsymbol{I} - \boldsymbol{A}_1 z^{r-1} - z^{r-2} - \cdots - \boldsymbol{A}_r\right]\\ &= z^{-m}\left(z^m + \alpha_1 z^{m-1} + \alpha_2 z^{m-2} + \cdots + \alpha_m\right)\\ &= 1 + \alpha_1 z^{-1} + \alpha_2 z^{-2} + \cdots + \alpha_m z^{-m},\end{aligned} \tag{3.74}$$

$$\begin{aligned}\beta(z) &:= z^{-m}\mathrm{cadj}\left[z^r\boldsymbol{I} - \boldsymbol{A}_1 z^{r-1} - z^{r-2} - \cdots - \boldsymbol{A}_r\right]\boldsymbol{f} z^{r-1}\\ &= \beta_1 z^{-1} + \beta_2 z^{-2} + \cdots + \beta_m z^{-m}.\end{aligned} \tag{3.75}$$

### 3.3.2 The identification model and parameter estimation algorithm

This section derives the least squares-based iterative parameter estimation algorithm to estimate the parameters of the state-space model with time delay. Compared with the recursive least squares algorithm, the iterative identification algorithm takes full advantage of the system input and output data information and has higher precision of parameter estimation and faster convergence speed.

Define the parameter vector $\theta$ and the information vector $\varphi(t)$ as

$$\boldsymbol{\theta} := \left[\alpha_1, \alpha_2, \cdots, \alpha_m, \beta_1, \beta_2, \cdots, \beta_m\right]^{\mathrm{T}} \in \mathbb{R}^{2m}, \tag{3.76}$$

$$\varphi(t) := [-x(t-1), -x(t-2), \cdots, -x(t-m), u(t-1), u(t-2), \cdots, u(t-m)]^{\mathrm{T}} \in \mathbb{R}^{2m}. \tag{3.77}$$

Define an intermediate variable:

$$x(t) := \frac{\beta(z)}{\alpha(z)} u(t). \tag{3.78}$$

Substituting (3.74) and (3.75) into (3.78) gives

$$\begin{aligned}x(t) &= -\alpha_1 x(t-1) - \alpha_2 x(t-2) - \cdots - \alpha_m x(t-m)\\ &\quad + \beta_1 u(t-1) + \beta_2 u(t-2) + \cdots + \beta_m u(t-m)\\ &= \varphi^{\mathrm{T}}(t)\boldsymbol{\theta}.\end{aligned}$$

Thus, we have the identification model,

$$y(t) = \varphi^{\mathrm{T}}(t)\boldsymbol{\theta} + v(t). \tag{3.79}$$

Define the stacked output vector $Y(L)$, the stacked information matrix $\Phi(L)$ and the stacked white noise vector $V(L)$:

$$\begin{aligned}
\boldsymbol{Y}(L) &:= [y(1), y(2), \cdots, y(L)]^{\mathrm{T}} \in \mathbb{R}^{L},\\
\boldsymbol{\Phi}(L) &:= \left[\varphi(1), \varphi(2), \cdots, \varphi(L)\right]^{\mathrm{T}} \in \mathbb{R}^{L \times 2m},\\
\boldsymbol{V}(L) &:= [v(1), v(2), \cdots, v(L)]^{\mathrm{T}} \in \mathbb{R}^{L},
\end{aligned}$$

where $Y(L)$ and $\Phi(L)$ contain all measured input–output data $\{u(t), y(t){:}t = 1, 2, \ldots, L\}$. From Eq. (3.79), we have

$$\boldsymbol{Y}(L) = \boldsymbol{\Phi}(L)\boldsymbol{\theta} + \boldsymbol{V}(L).$$

Define a quadratic criterion function,

$$J(\boldsymbol{\theta}) := \sum_{j=1}^{L} \left[y(j) - \varphi^{\mathrm{T}}(j)\boldsymbol{\theta}\right]^2 = \left\| \boldsymbol{Y}(L) = \boldsymbol{\Phi}(L)\boldsymbol{\theta} \right\|^2.$$

Minimizing the criterion function $J(\theta)$ and letting its partial derivative with respect to $\theta$ be zero, we can obtain the least squares algorithm for estimating the parameter vector $\theta$:

$$\hat{\boldsymbol{\theta}} = [\boldsymbol{\Phi}^{\mathrm{T}}(L)\boldsymbol{\Phi}(L)]^{-1}\boldsymbol{\Phi}^{\mathrm{T}}(L)\boldsymbol{Y}(L). \tag{3.80}$$

Because the stacked information matrix $\Phi(L)$ contains the unknown terms $x(t-i)(i = 1, 2, \ldots, m)$. The above approach cannot be used to calculate $\hat{\theta}$, the solution is to let $\hat{\varphi}_k(t)$ denote the information vector by replacing $x(t-i)$ in (3.77) with their estimates $\hat{x}_{k-1}(t-i)$ at iteration $k-1$, and $\hat{\Phi}_k(L)$ denotes the stacked information matrix obtained by replacing $\varphi(t-i)$ in $\Phi(L)$ with $\hat{\varphi}_k(t-i)$.

Replacing $\Phi(L)$ in (3.80) with $\hat{\Phi}_k(L)$, we can obtain the least squares-based iterative identification (LSI) algorithm for estimating $\theta$:

$$\hat{\boldsymbol{\theta}}_k = \left[\hat{\boldsymbol{\Phi}}_k^{\mathrm{T}}(L)\hat{\boldsymbol{\Phi}}_k(L)\right]^{-1}\hat{\boldsymbol{\Phi}}_k^{\mathrm{T}}(L)\boldsymbol{Y}(L), k = 1, 2, 3, \cdots, \tag{3.81}$$

$$\hat{\boldsymbol{\Phi}}_k(L) = \left[\hat{\varphi}_k(L), \hat{\varphi}_k(L-1), \cdots, \hat{\varphi}_k(1)\right]^{\mathrm{T}}, \tag{3.82}$$

$$\boldsymbol{Y}(L) = [y(L), y(L-1), \cdots, y(1)]^{\mathrm{T}}, \tag{3.83}$$

$$\hat{\varphi}_k(t) = [-\hat{x}_{k-1}(t-1), -\hat{x}_{k-1}(t-2), \cdots, -\hat{x}_{k-1}(t-m), u(t-1), u(t-2), \cdots, u(t-nr)]^{\mathrm{T}}, \tag{3.84}$$

$$\hat{x}_k(t) = \hat{\varphi}_k^{\mathrm{T}}(t)\hat{\boldsymbol{\theta}}_k, \ t = 1, 2, \cdots, L. \tag{3.85}$$

### 3.3.3 Example

Consider the following state-space system with two-step delay:

$$\boldsymbol{x}(t+1) = \begin{bmatrix} 0 & 1 \\ -1 & -0.07 \end{bmatrix} \boldsymbol{x}(t) + \begin{bmatrix} 0.0075 & -0.01 \\ 0 & 1.2 \end{bmatrix} \boldsymbol{x}(t-1) + \begin{bmatrix} 1.00 \\ -1.00 \end{bmatrix} u(t),$$
$$y(t) = [1, 0]\boldsymbol{x}(t) + v(t).$$

Thus, we have

$$\begin{aligned} y(t) &= \frac{\beta(z)}{\alpha(z)} u(t) + v(t) \\ &= \frac{\beta_1 z^{-1} + \beta_2 z^{-2} + \beta_3 z^{-3}}{1 + \alpha_1 z^{-1} + \alpha_2 z^{-2} + \alpha_3 z^{-3} + \alpha_4 z^{-4}} u(t) + v(t) \\ &= \frac{z^{-1} - 0.93 z^{-2} - 1.19 z^{-3}}{1 + 0.07 z^{-1} - 0.2075 z^{-2} - 0.01052 z^{-3} + 0.009 z^{-4}} u(t) + v(t), \end{aligned}$$

$$\begin{aligned} \boldsymbol{\theta} &= [\alpha_1, \alpha_2, \alpha_3, \alpha_4, \beta_1, \beta_2, \beta_3]^{\mathrm{T}} \\ &= [0.07, -0.2075, -0.01052, 0.009, 1, -0.93, -1.19]^{\mathrm{T}}. \end{aligned}$$

In simulation, the input $\{u(t)\}$ is taken as an uncorrelated persistent excitation signal sequence with zero mean and unit variance and $\{v(t)\}$ as a white noise sequence with zero mean and variance $\sigma^2 = 0.10^2$, $\sigma^2 = 0.50^2$, and $\sigma^2 = 1.00^2$, respectively. Apply the LSI algorithm to estimate the parameters of this example system. The parameter estimates and their errors with different data lengths $t = L = 1000$, 2000, and 3000 are shown in Tables 3.4–3.6, and the parameter estimation errors versus $k$ are shown in Figs. 3.6–3.8.

From Tables 3.4–3.6 and Figs. 3.6–3.8, we can draw the following conclusions: (1) the parameter estimation errors given by the LSI algorithm become small as the iteration $k$ increases; (2) a longer data length $L$ leads to a smaller estimation error under the same noise level; (3) for large data length, the parameter estimation errors of the LSI algorithm converge a positive constant.

**Table 3.4** The LSI estimates and errors versus iteration $k$ ($L = 1000$).

| $\sigma^2$ | $\kappa$ | $\alpha_1$ | $\alpha_2$ | $\alpha_3$ | $\alpha_4$ | $\beta_1$ | $\beta_2$ | $\beta_3$ | $\delta$ (%) |
|---|---|---|---|---|---|---|---|---|---|
| $0.10^2$ | 1 | −0.00302 | −0.00067 | −0.00985 | −0.00225 | 1.01865 | −0.99244 | −0.91775 | 19.50105 |
| | 2 | 0.18477 | −0.28106 | 0.02194 | −0.03725 | 0.99973 | −0.81250 | −1.38383 | 14.82316 |
| | 5 | 0.07358 | −0.20841 | −0.01070 | 0.00869 | 0.99967 | −0.92725 | −1.19678 | 0.45010 |
| | 10 | 0.07392 | −0.20865 | −0.01061 | 0.00861 | 0.99967 | −0.92690 | −1.19736 | 0.49227 |
| $0.50^2$ | 1 | −0.02111 | 0.01008 | −0.01639 | 0.00170 | 1.01761 | −0.99608 | −0.92560 | 19.78452 |
| | 2 | 0.19101 | −0.28260 | 0.02726 | −0.04069 | 0.99835 | −0.8096 1 | −1.39776 | 15.67879 |
| | 5 | 0.08746 | −0.21182 | −0.01139 | 0.00740 | 0.99833 | −0.91670 | −1.22324 | 2.20026 |
| | 10 | 0.08835 | −0.21242 | −0.01116 | 0.00720 | 0.99833 | −0.91581 | −1.22475 | 2.30995 |
| $1.00^2$ | 1 | −0.04373 | 0.02352 | −0.02456 | 0.00663 | 1.01630 | −1.00064 | −0.93542 | 20.25402 |
| | 2 | 0.19937 | −0.28534 | 0.03431 | −0.04522 | 0.99661 | −0.80545 | −1.41630 | 16.84650 |
| | 5 | 0.10078 | −0.21344 | −0.01308 | 0.00646 | 0.99666 | −0.90756 | −1.24972 | 3.90524 |
| | 10 | 0.10412 | −0.21567 | −0.01218 | 0.00568 | 0.99666 | −0.90425 | −1.25535 | 4.31200 |
| True values | | 0.07000 | −0.20750 | −0.01052 | 0.00900 | 1.00000 | −0.93000 | −1.19000 | |

**Table 3.5** The LSI estimates and errors versus iteration $k$ ($L = 2000$).

| $\sigma^2$ | $k$ | $\alpha_1$ | $\alpha_2$ | $\alpha_3$ | $\alpha_4$ | $\beta_1$ | $\beta_2$ | $\beta_3$ | $\delta$ (%) |
|---|---|---|---|---|---|---|---|---|---|
| $0.10^2$ | 1 | 0.00300 | −0.00193 | −0.00830 | −0.00412 | 1.00553 | −0.99509 | −0.90914 | 19.76656 |
| | 2 | 0.19192 | −0.28337 | 0.02125 | −0.03803 | 0.99786 | −0.80821 | −1.38650 | 15.23719 |
| | 5 | 0.06937 | −0.20549 | −0.01187 | 0.00887 | 0.99788 | −0.93199 | −1.18446 | 0.36920 |
| | 10 | 0.06970 | −0.20571 | −0.01179 | 0.00880 | 0.99788 | −0.93166 | −1.18501 | 0.33373 |
| $0.50^2$ | 1 | −0.00021 | −0.00033 | −0.01855 | 0.00129 | 0.99742 | −1.00025 | −0.89836 | 20.35744 |
| | 2 | 0.17490 | −0.26697 | 0.01152 | −0.03678 | 0.98941 | −0.83170 | −1.34107 | 12.22066 |
| | 5 | 0.06863 | −0.19856 | −0.01674 | 0.00792 | 0.98942 | −0.93818 | −1.16526 | 1.65567 |
| | 10 | 0.06905 | −0.19883 | −0.01664 | 0.00783 | 0.98942 | −0.93776 | −1.16596 | 1.61162 |
| $1.00^2$ | 1 | −0.00421 | 0.00166 | −0.03138 | 0.00806 | 0.98728 | −1.00670 | −0.88488 | 21.14400 |
| | 2 | 0.15975 | −0.24995 | 0.00154 | −0.03604 | 0.97886 | −0.85460 | −1.29415 | 9.33488 |
| | 5 | 0.06873 | −0.19043 | −0.02246 | 0.00640 | 0.97885 | −0.94490 | −1.14298 | 3.15888 |
| | 10 | 0.06942 | −0.19088 | −0.02229 | 0.00626 | 0.97885 | −0.94422 | −1.14413 | 3.08874 |
| True values | | 0.07000 | −0.20750 | −0.01052 | 0.00900 | 1.00000 | −0.93000 | −1.19000 | |

**Table 3.6** The LSI estimates and errors versus iteration $k$ ($L = 3000$).

| $\sigma^2$ | $\kappa$ | $\alpha_1$ | $\alpha_2$ | $\alpha_3$ | $\alpha_4$ | $\beta_1$ | $\beta_2$ | $\beta_3$ | $\delta$ (%) |
|---|---|---|---|---|---|---|---|---|---|
| $0.10^2$ | 1 | −0.00412 | −0.00272 | −0.00420 | 0.00036 | 1.00396 | −0.99980 | −0.91295 | 19.69904 |
| | 2 | 0.18577 | −0.28068 | 0.02174 | −0.03505 | 0.99874 | −0.81435 | −1.37764 | 14.53396 |
| | 5 | 0.06922 | −0.20590 | −0.01143 | 0.00960 | 0.99875 | −0.93174 | −1.18591 | 0.27764 |
| | 10 | 0.06937 | −0.20600 | −0.01139 | 0.00957 | 0.99875 | −0.93159 | −1.18616 | 0.26023 |
| $0.50^2$ | 1 | −0.01012 | −0.00359 | −0.01023 | −0.00107 | 0.99900 | −1.00348 | −0.90669 | 20.04502 |
| | 2 | 0.18292 | −0.27561 | 0.01875 | −0.03357 | 0.99368 | −0.82171 | −1.36198 | 13.58056 |
| | 5 | 0.06700 | −0.20013 | −0.01481 | 0.01180 | 0.99375 | −0.93778 | −1.17112 | 1.28024 |
| | 10 | 0.06716 | −0.20023 | −0.01477 | 0.01176 | 0.99375 | −0.93763 | −1.17138 | 1.26250 |
| $1.00^2$ | 1 | −0.01763 | −0.00467 | −0.01776 | −0.00287 | 0.99280 | −1.00807 | −0.89887 | 20.50070 |
| | 2 | 0.17980 | −0.26933 | 0.01524 | −0.03169 | 0.98736 | −0.83043 | −1.34299 | 12.45934 |
| | 5 | 0.06485 | −0.19333 | −0.01882 | 0.01439 | 0.98750 | −0.94470 | −1.15373 | 2.45897 |
| | 10 | 0.06504 | −0.19346 | −0.01878 | 0.01434 | 0.98750 | −0.94451 | −1.15405 | 2.43707 |
| True values | | 0.07000 | −0.20750 | −0.01052 | 0.00900 | 1.00000 | −0.93000 | −1.19000 | |

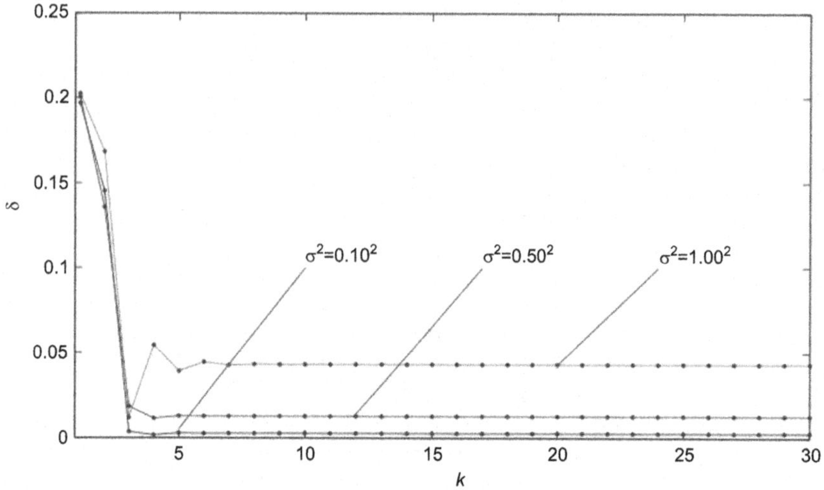


**Figure 3.6** The LSI estimation errors $\delta$ versus $k$ with different $\sigma^2(L = 1000)$.

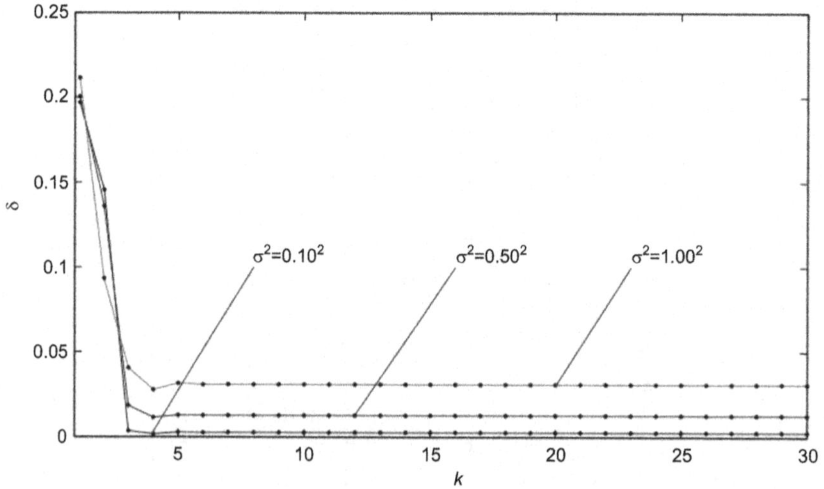


**Figure 3.7** The LSI estimation errors $\delta$ versus $k$ with different $\sigma^2(L = 2000)$.

### 3.3.4 Conclusions

This chapter presents a least squares-based iterative parameter estimation algorithm for linear systems based on the state-space model with time delay. The proposed algorithm performs a hierarchical computational process at each iteration and provides more accurate estimation of system parameters than the recursive approaches. The simulation results show that the

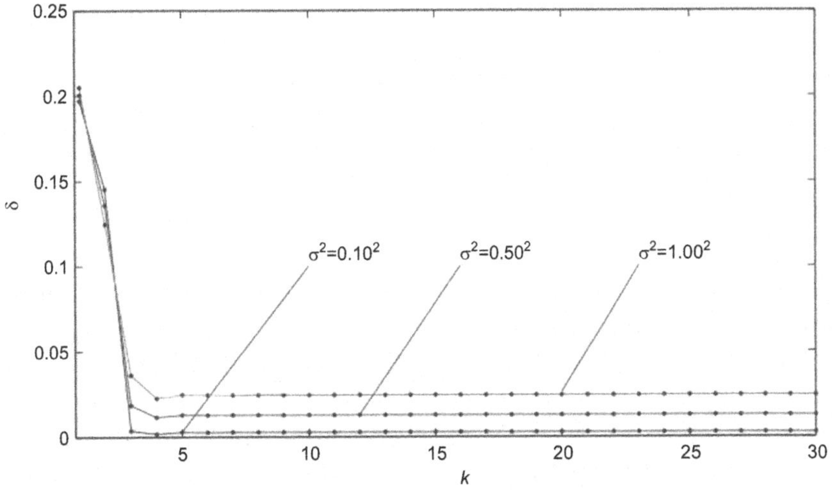


**Figure 3.8** The LSI estimation errors $\delta$ versus $k$ with different $\sigma^2(L = 3000)$.

proposed algorithm has fast convergence rates and can generate highly accurate parameter estimates after only several iterations. The proposed techniques can be extended to identify Hammerstein nonlinear models, time-varying systems, and nonuniformly sampled multirate systems.

## 3.4 Two-stage least squares-based iterative parameter identification algorithm for time-delay systems

The state-space models represented by differential/difference equations have wide range of applications in adaptive control, system analysis, and parameter estimation and therefore have received a lot of research attention in parameter and state estimation in the past several decades. Various state estimation methods have been presented for linear systems and nonlinear systems.

The ultimate purpose of the system identification is to find a model that is very close to a real system basic to the given input–output data. The general optimization algorithm can be used to get the parameters and a signal model from the observation data. For system analysis and system identification, the basic is to obtain the parameters of a system. The iterative methods can be used for solving not only some matrix equations, but also parameter estimation problems. Different from the recursive algorithms, the iterative methods can make full use of all the observed information to improve the accuracy of parameter estimation.

Time delays are widely exist in control systems and industrial processes due to the signal interruption and material transport. They are encountered in process control, fault diagnosis and detection, and signal processing. It is extremely difficult for the control system to respond to the input changes in time delays. Therefore, the identification of the time-delay systems has always been a hot research topic. In view of the linear regression equation, a large amount of literature has been established in the analysis of time delays in the field of control and identification. The influence of time delays on the function of control systems has attracted many researchers in the industrial process.

It combines the iterative identification principle and the least squares identification idea to present a two-stage least squares-based iterative algorithm for state-space models with time delays. To solve the problem of the information matrix containing unmeasurable items, the unmeasurable items are substituted for their iterative residuals, which are calculated by the preceding parameter estimates. Compared with the iterative algorithm, the two-stage least squares iterative algorithm can produce faster convergence rate and higher accurate parameter estimation.

### 3.4.1 The system description and input–output representation

Consider the following state-space system with time delays,

$$\begin{aligned} \boldsymbol{x}(t+1) = \boldsymbol{A}_1\boldsymbol{x}(t) + \boldsymbol{A}_2\boldsymbol{x}(t-1) + \boldsymbol{A}_3\boldsymbol{x}(t-2) + \cdots \\ + \boldsymbol{A}_r\boldsymbol{x}(t-r+1) + \boldsymbol{f}u(t), \end{aligned} \tag{3.86}$$

$$y(t) = \boldsymbol{c}\boldsymbol{x}(t) + v(t), \tag{3.87}$$

where $x(t) = [x_1(t), x_2(t), \ldots, x_n(t)]^{\mathrm{T}} \in R^n$ represents the state vector, $u(t) \in R$ represents the system input, $y(t) \in R$ stands for the system output, $v(t) \in R$ denotes random noise with zero mean, $A_i \in R^{n\times n} (i = 1, 2, \ldots, r)$, $f \in R^n$, and $c \in R^{1\times n}$ are the system parameter matrices/vectors.

Let us recommend some notation. "$A =: X$" or "$X := A$" stands for "$A$ is defined as $X$"; the symbol $I(I_n)$ stands for unit matrix of appropriate size $(n \times n)$; $z$ is a unit forward shift operator: $zx(t) = x(t+1)$ and $z^{-1}x(t) = x(t-1)$; the superscript T is the matrix/vector transpose; $\hat{\theta}(t)$ represents the estimate of $\theta$ at time $t$; adj$[X]$ stands for the adjoint matrix of the square matrix $X$ : adj$[X] = \det[X]X^{-1}$; $\det[X] = |X|$ represents the determinant of the square matrix $X$.

The purpose of this chapter is to develop a two-stage least squares-based iterative parameter estimation algorithm and a least squares-based iterative estimation algorithm. The existing input–output measurement data $\{u(t), y(t)\}$ are used to estimate the unknown parameters $(A_i, f)$. Without loss of generality, assume that the order $m$ is given and $u(t)=0, y(t)=0$, and $v(t)=0$ for $t\leq 0$. The following transforms the time-delay state-space model in (3.86) and (3.87) into an input–output representation and gives its identification model, simultaneously.

**Lemma 3.2.**: For the state-space model in (3.86) and (3.87), the transfer function from the input $u(t)$ to the output $y(t)$ having the form

$$G(z) := \frac{\boldsymbol{c}\mathrm{adj}\left[z^r\boldsymbol{I} - z^{r-1}\boldsymbol{A}_1 - z^{r-2}\boldsymbol{A}_2 - \cdots - \boldsymbol{A}_r\right]\boldsymbol{f}z^{r-1}}{\det[z^r\boldsymbol{I} - z^{r-1}\boldsymbol{A}_1 - z^{r-2}\boldsymbol{A}_2 - \cdots - \boldsymbol{A}_r]}.$$

**Proof** The first method is to prove this lemma by using the property of the shift operator $z$.

Writing for the sake of convenience, Eq. (3.86) can be represented by

$$\begin{aligned} z\boldsymbol{x}(t) = {} & \boldsymbol{A}_1\boldsymbol{x}(t) + z^{-1}\boldsymbol{A}_2\boldsymbol{x}(t) + z^{-2}\boldsymbol{A}_3\boldsymbol{x}(t) + \cdots \\ & + z^{-r+1}\boldsymbol{A}_r\boldsymbol{x}(t) + \boldsymbol{f}u(t). \end{aligned} \tag{3.88}$$

Multiplying both sides by $z^{r-1}$, we get the alternative forms

$$\begin{aligned} z^r\boldsymbol{x}(t) = {} & z^{r-1}\boldsymbol{A}_1\boldsymbol{x}(t) + z^{r-2}\boldsymbol{A}_2\boldsymbol{x}(t) + z^{r-3}\boldsymbol{A}_3\boldsymbol{x}(t) + \cdots \\ & + \boldsymbol{A}_r\boldsymbol{x}(t) + z^{r-1}\boldsymbol{f}u(t), \end{aligned} \tag{3.89}$$

or

$$\boldsymbol{x}(t) = \left(z^r\boldsymbol{I} - z^{r-1}\boldsymbol{A}_1 - z^{r-2}\boldsymbol{A}_2 - \cdots - \boldsymbol{A}_r\right)^{-1}\boldsymbol{f}z^{r-1}u(t).$$

Substituting $x(t)$ into Eq. (3.87) can be expressed in the output equation:

$$y(t) = \boldsymbol{c}\left(z^r\boldsymbol{I} - z^{r-1}\boldsymbol{A}_1 - z^{r-2}\boldsymbol{A}_2 - \cdots - \boldsymbol{A}_r\right)^{-1}\boldsymbol{f}z^{r-1}u(t) + v(t) =: G(z)u(t) + v(t).$$

Thus, we have the transfer function of the system from the input $u(t)$ to the output $y(t)$:

$$\begin{aligned} G(z) &= \boldsymbol{c}\left(z^r\boldsymbol{I} - z^{r-1}\boldsymbol{A}_1 - z^{r-2}\boldsymbol{A}_2 - \cdots - \boldsymbol{A}_r\right)^{-1}\boldsymbol{f}z^{r-1} \\ &= \frac{\boldsymbol{c}\mathrm{adj}\left[z^r\boldsymbol{I} - z^{r-1}\boldsymbol{A}_1 - z^{r-2}\boldsymbol{A}_2 - \cdots - \boldsymbol{A}_r\right]\boldsymbol{f}z^{r-1}}{\det[z^r\boldsymbol{I} - z^{r-1}\boldsymbol{A}_1 - z^{r-2}\boldsymbol{A}_2 - \cdots - \boldsymbol{A}_r]} \\ &=: \frac{\beta(z)}{\alpha(z)}, \end{aligned}$$

where $\alpha(z)$ and $\beta(z)$ are polynomials in a unit backward shift operator $z^{-1}[z^{-1}y(t) = y(t-1)]$, and

$$\begin{aligned}\alpha(z) &:= z^{-m}\det\left[z^{r}\boldsymbol{I} - \boldsymbol{A}_1 z^{r-1} - z^{r-2} - \cdots - \boldsymbol{A}_r\right] \\ &= z^{-m}\left(z^{m} + \alpha_1 z^{m-1} + \alpha_2 z^{m-2} + \cdots + \alpha_m\right) \\ &= 1 + \alpha_1 z^{-1} + \alpha_2 z^{-2} + \cdots + \alpha_m z^{-m},\end{aligned} \tag{3.90}$$

$$\begin{aligned}\beta(z) &:= z^{-m}\mathrm{cadj}\left[z^{r}\boldsymbol{I} - \boldsymbol{A}_1 z^{r-1} - z^{r-2} - \cdots - \boldsymbol{A}_r\right]\boldsymbol{f} z^{r-1} \\ &= \beta_1 z^{-1} + \beta_2 z^{-2} + \cdots + \beta_m z^{-m}.\end{aligned} \tag{3.91}$$

We assume that the order $m$ is given and can be achieved by trial and error. Normally, if the model is used as the prediction, the order should be selected to be large; otherwise if the model is for control, the order should not be too large for simplifying controller design.

Define the parameter vector $\theta$, $\theta_1$, and $\theta_2$ and the information vectors $\varphi(t)$, $\varphi_1(t)$, and $\varphi_2(t)$ as

$$\begin{aligned}&\boldsymbol{\theta} := \left[\boldsymbol{\theta}_1^{\mathrm{T}}, \boldsymbol{\theta}_2^{\mathrm{T}}\right]^{\mathrm{T}} \in \mathbb{R}^{2m}, \\ &\boldsymbol{\theta}_1 := [\alpha_1, \alpha_2, \cdots, \alpha_m]^{\mathrm{T}} \in \mathbb{R}^{m}, \\ &\boldsymbol{\theta}_2 := \left[\beta_1, \beta_2, \cdots, \beta_m\right]^{\mathrm{T}} \in \mathbb{R}^{m}, \\ &\boldsymbol{\varphi}(t) := \left[\boldsymbol{\varphi}_1^{\mathrm{T}}(t), \boldsymbol{\varphi}_2^{\mathrm{T}}(t)\right]^{\mathrm{T}} \in \mathbb{R}^{2m}, \\ &\boldsymbol{\varphi}_1(t) := [-x(t-1), -x(t-2), \cdots, -x(t-m)]^{\mathrm{T}} \in \mathbb{R}^{m}, \\ &\boldsymbol{\varphi}_2(t) := [u(t-1), -u(t-2), \cdots, -u(t-m)]^{\mathrm{T}} \in \mathbb{R}^{m}.\end{aligned}$$

Define an intermediate variable:

$$x(t) := \frac{\beta(z)}{\alpha(z)} u(t). \tag{3.92}$$

Substituting (3.90) and (3.91) into (3.92) can be described by

$$x(t) = \frac{\beta_1 z^{-1} + \beta_2 z^{-2} + \cdots + \beta_m z^{-m}}{1 + \alpha_1 z^{-1} + \alpha_2 z^{-2} + \cdots + \alpha_m z^{-m}} u(t),$$

or

$$\begin{aligned}x(t) &= -\alpha_1 x(t-1) - \alpha_2 x(t-2) - \cdots - \alpha_m x(t-m) \\ &\quad + \beta_1 u(t-1) + \beta_2 u(t-2) + \cdots + \beta_m u(t-m) \\ &= \boldsymbol{\varphi}_1^{\mathrm{T}}(t)\boldsymbol{\theta}_1 + \boldsymbol{\varphi}_2^{\mathrm{T}}(t)\boldsymbol{\theta}_2 \\ &= \boldsymbol{\varphi}^{\mathrm{T}}(t)\boldsymbol{\theta}.\end{aligned}$$

Thus, the identification model of the state-space system with time delays in (3.86) and (3.87) is of the form

$$y(t) = \boldsymbol{\varphi}^{\mathrm{T}}(t)\boldsymbol{\theta} + v(t). \tag{3.93}$$

### 3.4.2 The two-stage least squares-based iterative algorithm

This section derives a two-stage least squares-based iterative parameter estimation algorithm using the decomposition technique. The basic method is to decompose the model in (3.93) into two subsystems, which contains one parameter vector each other, and to identify each subsystem one by one.

Define two intermediate variables,

$$y_1(t) := y(t) - \boldsymbol{\varphi}_2^{\mathrm{T}}(t)\boldsymbol{\theta}_2, \tag{3.94}$$

$$y_2(t) := y(t) - \boldsymbol{\varphi}_1^{\mathrm{T}}(t)\boldsymbol{\theta}_1. \tag{3.95}$$

The identification model in (3.93) can be decomposed into two fictitious subsystems,

$$y_1(t) = \boldsymbol{\varphi}_1^{\mathrm{T}}(t)\boldsymbol{\theta}_1 + v(t), \tag{3.96}$$

$$y_2(t) = \boldsymbol{\varphi}_2^{\mathrm{T}}(t)\boldsymbol{\theta}_2 + v(t). \tag{3.97}$$

Consider the data from $t=1$ to $t=L(L\rangle 2m)$ and define the stacked output vectors $Y(L)$, $Y_1(L)$, and $Y_2(L)$, the stacked information matrices $\Phi_1(L)$, $\Phi_2(L)$ and the stacked white noise vector $V(L)$ as

$$\boldsymbol{Y}(L) := \begin{bmatrix} y(1) \\ y(2) \\ \vdots \\ y(L) \end{bmatrix} \in \mathbb{R}^L, \qquad \boldsymbol{V}(L) := \begin{bmatrix} v(1) \\ v(2) \\ \vdots \\ v(L) \end{bmatrix} \in \mathbb{R}^L,$$

$$\boldsymbol{Y}_1(L) := \begin{bmatrix} y_1(1) \\ y_1(2) \\ \vdots \\ y_1(L) \end{bmatrix} \in \mathbb{R}^L, \qquad \boldsymbol{Y}_2(L) := \begin{bmatrix} y_2(1) \\ y_2(2) \\ \vdots \\ y_2(L) \end{bmatrix} \in \mathbb{R}^L,$$

$$\Phi_1(L) := \begin{bmatrix} \boldsymbol{\varphi}_1^{\mathrm{T}}(1) \\ \boldsymbol{\varphi}_1^{\mathrm{T}}(2) \\ \vdots \\ \boldsymbol{\varphi}_1^{\mathrm{T}}(L) \end{bmatrix} \in \mathbb{R}^{L\times m},$$

$$\Phi_2(L) := \begin{bmatrix} \boldsymbol{\varphi}_2^{\mathrm{T}}(1) \\ \boldsymbol{\varphi}_2^{\mathrm{T}}(2) \\ \vdots \\ \boldsymbol{\varphi}_2^{\mathrm{T}}(L) \end{bmatrix} \in \mathbb{R}^{L\times m}.$$

Define two criterion functions,

$$J_1(\boldsymbol{\theta}_1) := \| \boldsymbol{Y}_1(L) - \boldsymbol{\Phi}_1(L)\boldsymbol{\theta}_1 \|^2,$$
$$J_2(\boldsymbol{\theta}_2) := \| \boldsymbol{Y}_2(L) - \boldsymbol{\Phi}_2(L)\boldsymbol{\theta}_2 \|^2.$$

For the two optimization problems, letting the partial derivatives of $J_1(\theta_1)$ and $J_2(\theta_2)$ with regard to $\theta_1$ and $\theta_2$ be zero becomes

$$\frac{\partial J_1(\boldsymbol{\theta}_1)}{\partial \boldsymbol{\theta}_1} = -2\boldsymbol{\Phi}_1^{\mathrm{T}}(L)[\boldsymbol{Y}_1(L) - \boldsymbol{\Phi}_1(L)\boldsymbol{\theta}_1] = \mathbf{0},$$
$$\frac{\partial J_2(\boldsymbol{\theta}_2)}{\partial \boldsymbol{\theta}_2} = -2\boldsymbol{\Phi}_2^{\mathrm{T}}(L)[\boldsymbol{Y}_2(L) - \boldsymbol{\Phi}_2(L)\boldsymbol{\theta}_2] = \mathbf{0}.$$

We can acquire the following least squares estimates of the parameter vectors $\theta_1$ and $\theta_2$:

$$\begin{aligned}
\hat{\boldsymbol{\theta}}_1 &= \left[\boldsymbol{\Phi}_1^{\mathrm{T}}(L)\boldsymbol{\Phi}_1(L)\right]^{-1}\boldsymbol{\Phi}_1^{\mathrm{T}}(L)\boldsymbol{Y}_1(L) \\
&= \left[\boldsymbol{\Phi}_1^{\mathrm{T}}(L)\boldsymbol{\Phi}_1(L)\right]^{-1}\boldsymbol{\Phi}_1^{\mathrm{T}}(L)[\boldsymbol{Y}(L) - \boldsymbol{\Phi}_2(L)\boldsymbol{\theta}_2], \\
\hat{\boldsymbol{\theta}}_2 &= \left[\boldsymbol{\Phi}_2^{\mathrm{T}}(L)\boldsymbol{\Phi}_2(L)\right]^{-1}\boldsymbol{\Phi}_2^{\mathrm{T}}(L)\boldsymbol{Y}_2(L) \\
&= \left[\boldsymbol{\Phi}_2^{\mathrm{T}}(L)\boldsymbol{\Phi}_2(L)\right]^{-1}\boldsymbol{\Phi}_2^{\mathrm{T}}(L)[\boldsymbol{Y}(L) - \boldsymbol{\Phi}_1(L)\boldsymbol{\theta}_1].
\end{aligned}$$

Based on the hierarchical identification principle: let $k = 1, 2, 3, \ldots$ be an iterative variable, $\hat{\theta}(k) := [\hat{\theta}_1^{\mathrm{T}}(k), \hat{\theta}_2^{\mathrm{T}}(k)]^{\mathrm{T}}$ be the iterative estimate of $\theta(k) := [\theta_1^{\mathrm{T}}(k), \theta_2^{\mathrm{T}}(k)]^{\mathrm{T}}$ at iteration $k$, and $\hat{x}_k(t-i)$ be the estimate of $x(t-i)$ at iteration $k$, and define

$$\begin{aligned}
&\hat{\varphi}_k(t) := \begin{bmatrix} \hat{\varphi}_{1k}(t) \\ \varphi_2(t) \end{bmatrix} \in \mathbb{R}^{2m}, \\
&\hat{\varphi}_{1k}(t) := [-\hat{x}_{k-1}(t-1), -\hat{x}_{k-1}(t-2), \cdots, -\hat{x}_{k-1}(t-m)]^{\mathrm{T}} \in \mathbb{R}^m, \\
&\hat{\boldsymbol{\Phi}}_1(L) := \left[\hat{\varphi}_{1k}(1), \hat{\varphi}_{1k}(2), \cdots, \hat{\varphi}_{1k}(L)\right] \in \mathbb{R}^{L \times m}.
\end{aligned}$$

Thus, we can summarize the two-stage least squares-based iterative (TS-LSI) identification algorithm for estimating $\theta_i$ as follows:

$$\hat{\boldsymbol{\theta}}_{1k} = \left[\hat{\boldsymbol{\Phi}}_1^{\mathrm{T}}(L)\hat{\boldsymbol{\Phi}}_1(L)\right]^{-1}\hat{\boldsymbol{\Phi}}_1^{\mathrm{T}}(L)\left[\boldsymbol{Y}(L) - \boldsymbol{\Phi}_2(L)\hat{\boldsymbol{\theta}}_{2,k-1}\right], k = 1, 2, 3, \cdots, \tag{3.98}$$

$$\hat{\boldsymbol{\theta}}_{2k} = \left[\hat{\boldsymbol{\Phi}}_2^{\mathrm{T}}(L)\boldsymbol{\Phi}_2(L)\right]^{-1}\boldsymbol{\Phi}_2^{\mathrm{T}}(L)\left[\boldsymbol{Y}(L) - \hat{\boldsymbol{\Phi}}_1(L)\hat{\boldsymbol{\theta}}_{1k}\right], \tag{3.99}$$

$$\mathbf{Y}(L) = [y(1), y(2), \cdots, y(L)]^{\mathrm{T}}, \tag{3.100}$$

$$\hat{\mathbf{\Phi}}_1(L) = \left[\hat{\varphi}_{1k}(1), \hat{\varphi}_{1k}(2), \cdots, \hat{\varphi}_{1k}(L)\right]^{\mathrm{T}}, \tag{3.101}$$

$$\mathbf{\Phi}_2(L) = \left[\varphi_2(1), \varphi_2(2), \cdots, \varphi_2(L)\right]^{\mathrm{T}}, \tag{3.102}$$

$$\hat{\varphi}_{1k}(t) = [-\hat{x}_{k-1}(t-1), \cdots, -\hat{x}_{k-1}(t-m)]^{\mathrm{T}}, \tag{3.103}$$

$$\varphi_2(t) = [u(t-1), u(t-2), \cdots, u(t-m)]T, \tag{3.104}$$

$$\hat{\varphi}_k(t) = \left[\hat{\varphi}_{1k}^{\mathrm{T}}(t), \hat{\varphi}_2^{\mathrm{T}}(t)\right]^{\mathrm{T}}, \tag{3.105}$$

$$\hat{\boldsymbol{\theta}}_k = \left[\hat{\boldsymbol{\theta}}_{1k}^{\mathrm{T}}, \hat{\boldsymbol{\theta}}_{2k}^{\mathrm{T}}\right]^{\mathrm{T}}, \tag{3.106}$$

$$\hat{x}_k(t) = \hat{\varphi}_{1k}^{\mathrm{T}}(t)\hat{\boldsymbol{\theta}}_{1k} + \hat{\varphi}_2^{\mathrm{T}}(t)\hat{\boldsymbol{\theta}}_{2k}. \tag{3.107}$$

The steps of carrying out the TS-LSI algorithm in (3.98–3.107) to calculate the parameter estimates $\theta_1(k)$ and $\theta_2(k)$ are in the following list.

- Collect the input–output data $\{u(t), y(t){:}i = 1, 2, \ldots, L\}$ ($L$ is the data length), and form $Y(L)$ by (3.100), $\varphi_2(t)$ by (3.104), and $\Phi_2(L)$ by (3.102).
- To initialize, let $k = 1, \theta_1(0) = 1_m/p_0, \theta_2(0) = 1_m/p_0, p_0 = 10^6$, $\hat{x}_0(t) =$ random number for all $t$.
- Form $\hat{\varphi}_{1k}(t)$ by (3.103), $\hat{\Phi}_1(L)$ by (3.101), and $\varphi_k(t)$ by (3.105).
- Update $\theta_1(k)$ and $\theta_2(k)$ by (3.98) and (3.99), respectively.
- Form $\theta_k$ by (3.106).
- Compute $\hat{x}_k(t)$ by (3.107).
- If $||\hat{\theta}_1(k) - \hat{\theta}_1(k-1)|| + ||\hat{\theta}_2(k) - \hat{\theta}_2(k-1)|| \leq \varepsilon$, obtain the parameter estimates $\hat{\theta}_1(k)$ and $\hat{\theta}_2(k)$; otherwise, increase $k$ by 1 and go to Step 3.

## 3.4.3 The least squares-based iterative algorithm

In order to compare with the proposed TS-LSI algorithm, this section gives the least squares-based iterative (LSI) algorithm.

The parameter vector $\theta$ and the information vector $\varphi(t)$ are defined as

$$\boldsymbol{\theta} := \left[\alpha_1, \alpha_2 \cdots, \alpha_m, \beta_1, \beta_2, \cdots, \beta_m\right]^{\mathrm{T}} \in \mathbb{R}^{2m}, \tag{3.108}$$

$$\varphi(t) := [-x(t-1), -x(t-2), \cdots, -x(t-m), u(t-1), u(t-2), \cdots, u(t-m)]^{\mathrm{T}} \in \mathbb{R}^{2m}. \tag{3.109}$$

Eqs. (3.86) and (3.87) can be stated as follows

$$y(t) = \varphi^{\mathrm{T}}(t)\boldsymbol{\theta} + v(t).$$

Obviously, the parameter vector $\theta$ contains all the parameters to be estimated. The available measurement data are $u(t)$, and $y(t), x(t)$ is the inner variable and unmeasured. Thus, let $\hat{x}_k(t-i)(i = 1, 2, \ldots, m)$ be the estimate of $x(t-i)$ at iteration $k$, and $\hat{\varphi}_k(t)$ denotes the information vector obtained by replacing $x(t-i)$ in (3.109) with $\hat{x}_{k-1}(t-i)$. For the identification model in (3.93), we can obtain the following least squares-based iterative identification algorithm for estimating $\theta$:

$$\hat{\boldsymbol{\theta}}_k = \left[\hat{\boldsymbol{\Phi}}_k^{\mathrm{T}}(L)\hat{\boldsymbol{\Phi}}_k(L)\right]^{-1}\hat{\boldsymbol{\Phi}}_k^{\mathrm{T}}(L)\mathbf{Y}(L), \quad k = 1, 2, 3, \tag{3.110}$$

$$\hat{\boldsymbol{\Phi}}_k(L) = \left[\hat{\varphi}_k(1), \hat{\varphi}_k(2), \cdots, \hat{\varphi}_k(L)\right]^{\mathrm{T}}, \tag{3.111}$$

$$\mathbf{Y}(L) = [y(1), y(2), \cdots, y(L)]^{\mathrm{T}}, \tag{3.112}$$

$$\hat{\varphi}_k(t) = [-\hat{x}_{k-1}(t-1), \cdots, -\hat{x}_{k-1}(t-m), u(t-1), u(t-2), \cdots, u(t-m)]^{\mathrm{T}}, \tag{3.113}$$

$$\hat{x}_k(t) = \hat{\varphi}_k^{\mathrm{T}}(t)\hat{\boldsymbol{\theta}}_k, \ t = 1, 2, \cdots, L. \tag{3.114}$$

To initialize, let $k = 1$, $\hat{x}_0(t) =$ random number for all $t$. From (3.110–3.114), we can see that the LSI algorithm performs a hierarchical interactive process: when computing the parameter estimate $\hat{\theta}(k)$, the unknown inner variables $x(t-i)$ in the information vector are replaced with their corresponding estimates $\hat{x}_{k-1}(t-i)$ at iteration $k-1$, and the estimates $\hat{x}_{k-1}(t-i)$ are obtained by using the parameter estimates $\hat{\theta}(k)$.

### 3.4.4 Example

Consider the following state-space system with time delay:

$$\begin{aligned}
\boldsymbol{x}(t+1) &= \begin{bmatrix} 0 & 1 \\ -1 & 0.07 \end{bmatrix} \boldsymbol{x}(t) \\
&\quad + \begin{bmatrix} -0.02 & -0.03 \\ 0 & 1.15 \end{bmatrix} \boldsymbol{x}(t-1) \\
&\quad + \begin{bmatrix} 1.00 \\ -1.00 \end{bmatrix} u(t), \\
y(t) &= [1, 0]\boldsymbol{x}(t) + v(t).
\end{aligned}$$

The corresponding input–output representation is expressed as

$$\begin{aligned}
y(t) &= \frac{\beta(z)}{\alpha(z)} u(t) + v(t) \\
&= \frac{\beta_1 z^{-1} + \beta_2 z^{-2} + \beta_3 z^{-3}}{1 + \alpha_1 z^{-1} + \alpha_2 z^{-2} + \alpha_3 z^{-3} + \alpha_4 z^{-4}} u(t) + v(t) \\
&= \frac{z^{-1} - 1.07 z^{-2} - 1.12 z^{-3}}{1 - 0.07 z^{-1} - 0.13 z^{-2} - 0.0314 z^{-3} - 0.023 z^{-4}} u(t) + v(t),
\end{aligned}$$

$$\begin{aligned}
\boldsymbol{\theta} &= \left[\alpha_1, \alpha_2, \alpha_3, \alpha_4, \beta_1, \beta_2, \beta_3\right]^{\mathrm{T}} \\
&= [-0.07, -0.13, -0.0314, -0.023, 1, -1.07, -1.12]^{\mathrm{T}}.
\end{aligned}$$

In simulation, the input $\{u(t)\}$ is taken as an uncorrelated persistent excitation signal sequence with zero mean and unit variance, and $\{v(t)\}$ as a white noise sequence with zero mean and variance $\sigma^2 = 0.20^2$ and $\sigma^2 = 0.50^2$. Apply the TS-LSI algorithm and the LSI algorithm to estimate the parameters of this example system. The parameter estimates and their errors with data length $t = L = 2000$, 3000, and 4000 are shown in Tables 3.7–3.9, and the parameter estimation errors versus $k$ are shown in Figs. 3.9–3.10.

From Tables 3.7–3.11 and Figs. 3.9 and 3.10, the following conclusions can be drawn: (1) comparison with the LSI algorithm, the parameter estimation errors given by the TS-LSI algorithm become smaller along with the iteration $k$ increases; (2) for large data length, the parameter estimation errors of the TS-LSI algorithm converge to a positive constant.

## 3.5 Conclusions

This chapter presents a two-stage least squares-based iterative parameter estimation algorithm for linear systems based on the state-space model

**Table 3.7** The two-stage LSI parameter estimates and errors ($L = 2000$).

| $\sigma^2$ | $\kappa$ | $\alpha_1$ | $\alpha_2$ | $\alpha_3$ | $\alpha_4$ | $\beta_1$ | $\beta_2$ | $\beta_3$ | $\delta$ (%) |
|---|---|---|---|---|---|---|---|---|---|
| $0.20^2$ | 1 | −0.03970 | 0.07162 | 0.06860 | −0.00933 | 1.00208 | −1.00052 | −1.05031 | 13.39852 |
| | 2 | −0.07548 | −0.11888 | −0.04737 | −0.05070 | 1.00704 | −1.07343 | −1.10054 | 2.17289 |
| | 3 | −0.07992 | −0.12878 | −0.03866 | −0.02501 | 1.00662 | −1.07790 | −1.10341 | 1.25363 |
| | 4 | −0.07993 | −0.12911 | −0.03677 | −0.02400 | 1.00660 | −1.07777 | −1.10373 | 1.20636 |
| | 5 | −0.07975 | −0.12923 | −0.03663 | −0.02413 | 1.00660 | −1.07759 | −1.10404 | 1.18456 |
| | 6 | −0.07957 | −0.12936 | −0.03657 | −0.02418 | 1.00660 | −1.07741 | −1.10435 | 1.16379 |
| | 7 | −0.07940 | −0.12949 | −0.03652 | −0.02422 | 1.00660 | −1.07723 | −1.10466 | 1.14340 |
| | 8 | −0.07922 | −0.12961 | −0.03647 | −0.02426 | 1.00660 | −1.07705 | −1.10496 | 1.12340 |
| | 9 | −0.07905 | −0.12974 | −0.03641 | −0.02430 | 1.00660 | −1.07688 | −1.10526 | 1.10378 |
| | 10 | −0.07888 | −0.12986 | −0.03636 | −0.02433 | 1.00660 | −1.07671 | −1.10555 | 1.08454 |
| $0.50^2$ | 1 | −0.0:!009 | 0.07647 | 0.07353 | −0.01260 | 1.01171 | −0.99631 | −1.04025 | 13.94617 |
| | 2 | −0.07936 | −0.12459 | −0.05334 | −0.05718 | 1.01691 | −1.07405 | −1.09438 | 2.82174 |
| | 3 | −0.08393 | −0.13495 | −0.04361 | −0.02868 | 1.01647 | −1.07873 | −1.09750 | 1.91354 |
| | 4 | −0.08399 | −0.13549 | −0.04139 | −0.02781 | 1.01645 | −1.07863 | −1.09800 | 1.85492 |
| | 5 | −0.08375 | −0.13568 | −0.041 18 | −0.02796 | 1.01645 | −1.07838 | −1.09845 | 1.83035 |
| | 6 | −0.08349 | −0.13586 | −0.04111 | −0.02802 | 1.01645 | −1.07811 | −1.09889 | 1.80714 |
| | 7 | −0.08324 | −0.13604 | −0.04103 | −0.02808 | 1.01645 | −1.07786 | −1.09933 | 1.78448 |
| | 8 | −0.08299 | −0.13622 | −0.04096 | −0.02813 | 1.01645 | −1.07760 | −1.09976 | 1.76238 |
| | 9 | −0.08274 | −0.13640 | −0.0:!088 | −0.02819 | 1.01645 | −1.07735 | −1.10018 | 1.74085 |
| | 10 | −0.08250 | −0.13657 | −0.04081 | −0.02824 | 1.01644 | −1.07710 | −1.10060 | 1.71988 |
| True values | | −0.07000 | −0.13000 | −0.03140 | −0.02300 | 1.00000 | −1.07000 | −1.12000 | |

**Table 3.8** The two-stage LSI parameter estimates and errors ($L = 3000$).

| $\sigma^2$ | $\kappa$ | $\alpha_1$ | $\alpha_2$ | $\alpha_3$ | $\alpha_4$ | $\beta_1$ | $\beta_2$ | $\beta_3$ | $\delta$ (%) |
|---|---|---|---|---|---|---|---|---|---|
| $0.20^2$ | 1 | −0.02045 | 0.05989 | 0.03424 | −0.00895 | 1.00102 | −1.00144 | −1.05285 | 12.35342 |
| | 2 | −0.07376 | −0.11689 | −0.05007 | −0.04424 | 1.00422 | −1.07221 | −1.10147 | 1.98735 |
| | 3 | −0.07856 | −0.12478 | −0.03834 | −0.02282 | 1.00489 | −1.07602 | −1.10388 | 1.17034 |
| | 4 | −0.07863 | −0.12495 | −0.03662 | −0.02231 | 1.00480 | −1.07622 | −1.10427 | 1.12936 |
| | 5 | −0.07850 | −0.12506 | −0.03649 | −0.02245 | 1.00479 | −1.07608 | −1.10451 | 1.11153 |
| | 6 | −0.07837 | −0.12516 | −0.03645 | −0.02248 | 1.00479 | −1.07595 | −1.10473 | 1.09542 |
| | 7 | −0.07824 | −0.12525 | −0.03641 | −0.02251 | 1.00479 | −1.07582 | −1.10495 | 1.07963 |
| | 8 | −0.07811 | −0.12534 | −0.03637 | −0.02254 | 1.00479 | −1.07569 | −1.10517 | 1.06411 |
| | 9 | −0.07798 | −0.12543 | −0.03633 | −0.02257 | 1.00479 | −1.07556 | −1.10539 | 1.04885 |
| | 10 | −0.07786 | −0.12552 | −0.03630 | −0.02260 | 1.00479 | −1.07544 | −1.10560 | 1.03386 |
| $0.50^2$ | 1 | −0.02003 | 0.05970 | 0.04142 | −0.00869 | 1.00798 | −0.99724 | −1.04861 | 12.61233 |
| | 2 | −0.07413 | −0.11715 | −0.05346 | −0.04596 | 1.01128 | −1.06892 | −1.09820 | 2.29307 |
| | 3 | −0.07946 | −0.12580 | −0.04266 | −0.02464 | 1.01201 | −1.07322 | −1.10093 | 1.48499 |
| | 4 | −0.07956 | −0.12598 | −0.04073 | −0.02403 | 1.01192 | −1.07345 | −1.10134 | 1.42781 |
| | 5 | −0.07943 | −0.12610 | −0.04059 | −0.02418 | 1.01192 | −1.07332 | −1.10157 | 1.41234 |
| | 6 | −0.07930 | −0.12619 | −0.04055 | −0.02422 | 1.01192 | −1.07319 | −1.10179 | 1.39894 |
| | 7 | −0.07917 | −0.12628 | −0.04051 | −0.02425 | 1.01192 | −1.07305 | −1.1020 I | 1.38583 |
| | 8 | −0.07905 | −0.12637 | −0.04047 | −0.02427 | 1.01191 | −1.07293 | −1.10223 | 1.37299 |
| | 9 | −0.07892 | −0.12646 | −0.04043 | −0.02430 | 1.01191 | −1.07280 | −1.10244 | 1.36040 |
| | 10 | −0.07880 | −0.12655 | −0.04039 | −0.02433 | 1.01191 | −1.07267 | −1.10265 | 1.34806 |
| True values | | −0.07000 | −0.13000 | −0.03140 | −0.02300 | 1.00000 | −1.07000 | −1.12000 | |

**Table 3.9** The parameter estimates and errors with $\sigma^2 = 0.50^2 (L = 4000)$.

| Algorithms | $\kappa$ | $\alpha_1$ | $\alpha_2$ | $\alpha_3$ | $\alpha_4$ | $\beta_1$ | $\beta_2$ | $\beta_3$ | $\delta$ (%) |
|---|---|---|---|---|---|---|---|---|---|
| TS–LSI | 1 | −0.01675 | 0.05559 | 0.01741 | −0.00067 | 1.01385 | −0.98957 | −1.04851 | 12.31792 |
| | 2 | −0.07255 | −0.11637 | −0.05274 | −0.04506 | 1.01056 | −1.06455 | −1.09896 | 2.24070 |
| | 3 | −0.07849 | −0.12599 | −0.04216 | −0.02685 | 1.00997 | −1.06970 | −1.10178 | 1.37826 |
| | 4 | −0.07865 | −0.12607 | −0.04017 | −0.02608 | 1.00994 | −1.06989 | −1.10178 | 1.3321 1 |
| | 5 | −0.07868 | −0.12606 | −0.04009 | −0.02620 | 1.00993 | −1.06992 | −1.10174 | 1.33372 |
| | 6 | −0.07871 | −0.12604 | −0.04010 | −0.02620 | 1.00993 | −1.06995 | −1.10169 | 1.33635 |
| | 7 | −0.07873 | −0.12602 | −0.04011 | −0.02620 | 1.00993 | −1.06998 | −1.10165 | 1.33891 |
| | 8 | −0.07876 | −0.12600 | −0.04011 | −0.02619 | 1.00993 | −1.07000 | −1.10160 | 1.34143 |
| | 9 | −0.07878 | −0.12599 | −0.04012 | −0.02618 | 1.00993 | −1.07003 | −1.10156 | 1.34393 |
| | 10 | −0.07881 | −0.12597 | −0.04013 | −0.02618 | 1.00993 | −1.07005 | −1.10152 | 1.34639 |
| LSI | 1 | −0.00129 | 0.00027 | 0.00536 | 0.00674 | 1.01532 | −0.99026 | −1.04990 | 10.17503 |
| | 2 | −0.04041 | −0.16755 | −0.04148 | −0.05864 | 1.00980 | −1.03177 | −1.18268 | 5.16876 |
| | 3 | −0.08997 | −0.11600 | −0.04257 | −0.02050 | 1.00996 | −1.08117 | −1.08018 | 2.72201 |
| | 4 | −0.07992 | −0.12541 | −0.04059 | −0.02628 | 1.00994 | −1.07121 | −1.09988 | 1.45022 |
| | 5 | −0.08061 | −0.12465 | −0.04066 | −0.02574 | 1.00994 | −1.07189 | −1.09839 | 1.53258 |
| | 6 | −0.08049 | −0.12478 | −0.04064 | −0.02581 | 1.00994 | −1.07176 | −1.09864 | 1.51792 |
| | 7 | −0.08050 | −0.12476 | −0.04064 | −0.02581 | 1.00994 | −1.07177 | −1.09862 | 1.51935 |
| | 8 | −0.08050 | −0.12477 | −0.04064 | −0.02581 | 1.00994 | −1.07177 | −1.09862 | 1.51914 |
| | 9 | −0.08050 | −0.12477 | −0.04064 | −0.02581 | 1.00994 | −1.07177 | −1.09862 | 1.51916 |
| | 10 | −0.08050 | −0.12477 | −0.04064 | −0.02581 | 1.00994 | −1.07177 | −1.09862 | 1.51916 |
| True values | | −0.07000 | −0.13000 | −0.03140 | −0.02300 | 1.00000 | −1.07000 | −1.12000 | |

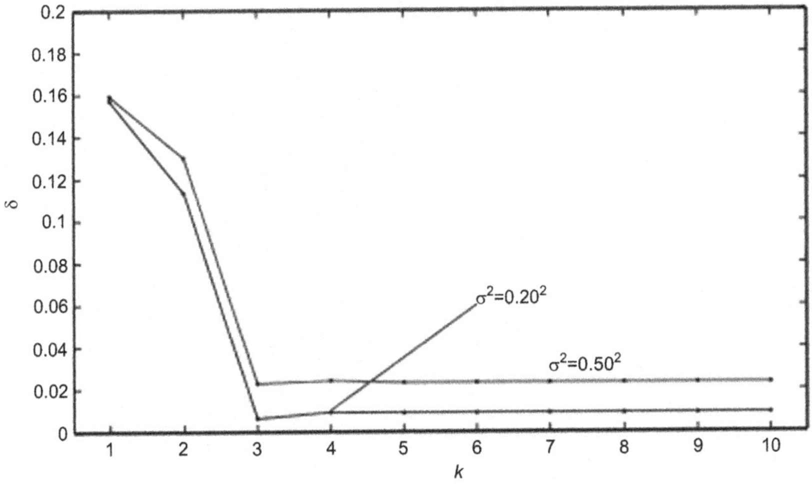


**Figure 3.9** The LSI estimation errors $\delta$ versus $k$ with different $\sigma^2(L = 2000)$.

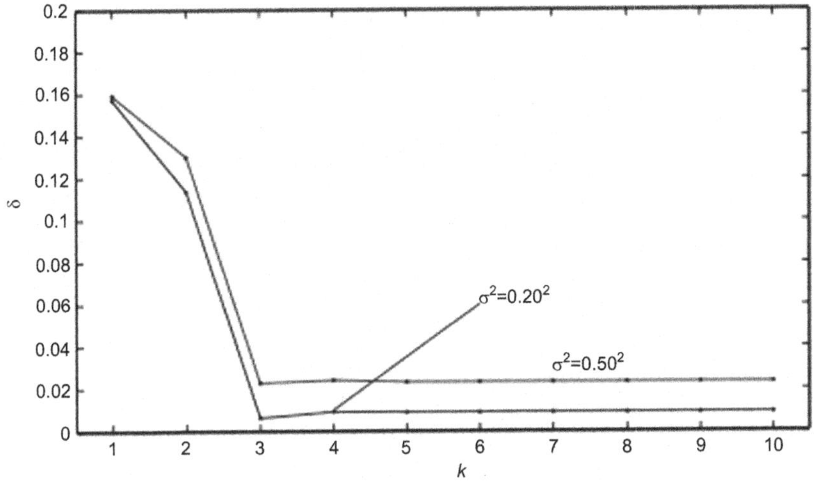


**Figure 3.10** The two-stage LSI estimation errors $\delta$ versus $k$ with different $\sigma^2(L = 3000)$.

with time delay. The given algorithm carries out a hierarchical computational process at each iteration and offers accurate parameter estimates. The simulation results show that the proposed algorithm has fast convergence rate, and the high precision parameter estimation can be obtained after several iterations. The given techniques can be used for identifying nonlinear models, time-varying systems, and nonuniformly sampled multirate systems.

**Table 3.10** The LSI parameter estimates and errors ($L = 3000, \sigma^2 = 0.20^2$).

| $\kappa$ | $\alpha_1$ | $\alpha_2$ | $\alpha_3$ | $\alpha_4$ | $\beta_1$ | $\beta_2$ | $\beta_3$ | $\delta$ (%) |
|---|---|---|---|---|---|---|---|---|
| 1 | 0.00409 | 0.00199 | −0.00494 | 0.00120 | 0.99325 | −0.99979 | −0.90504 | 15.58489 |
| 2 | 0.14668 | −0.18323 | −0.00524 | −0.03296 | 0.99261 | −0.85527 | −1.23563 | 9.40149 |
| 3 | 0.07353 | −0.14079 | −0.02818 | −0.01308 | 0.99241 | −0.92818 | −1.11984 | 0.49731 |
| 4 | 0.07465 | −0.14152 | −0.02799 | −0.01339 | 0.99241 | −0.92707 | −1.12161 | 0.55864 |
| 5 | 0.07458 | −0.14148 | −0.02801 | −0.01338 | 0.99241 | −0.92714 | −1.12150 | 0.55386 |
| 6 | 0.07458 | −0.14148 | −0.02801 | −0.01338 | 0.99241 | −0.92714 | −1.12150 | 0.55377 |
| 7 | 0.07458 | −0.14148 | −0.0280 I | −0.01338 | 0.99241 | −0.92714 | −1.12150 | 0.55383 |
| 8 | 0.07458 | −0.14148 | −0.02801 | −0.01338 | 0.99241 | −0.92714 | −1.12150 | 0.55382 |
| 9 | 0.07458 | −0.14148 | −0.02801 | −0.01338 | 0.99241 | −0.92714 | −1.12150 | 0.55382 |
| 10 | 0.07458 | −0.14148 | −0.02801 | −0.01338 | 0.99241 | −0.92714 | −1.12150 | 0.55382 |
| True values | 0.07000 | −0.14000 | −0.02930 | −0.01 150 | 1.00000 | −0.93000 | −1.12000 | |

**Table 3.11** The LSI parameter estimates and errors ($L = 3000, \sigma^2 = 0.50^2$).

| $\kappa$ | $\alpha_1$ | $\alpha_2$ | $\alpha_3$ | $\alpha_4$ | $\beta_1$ | $\beta_2$ | $\beta_3$ | $\delta$ (%) |
|---|---|---|---|---|---|---|---|---|
| 1 | 0.00988 | 0.00657 | −0.00829 | 0.00506 | 0.98198 | −1.00170 | −0.89975 | 15.92850 |
| 2 | 0.14044 | −0.17662 | −0.00770 | −0.03402 | 0.98124 | −0.86479 | −1.21649 | 8.20910 |
| 3 | 0.08492 | −0.14574 | −0.02495 | −0.01698 | 0.98100 | −0.91958 | −1.12926 | 1.65333 |
| 4 | 0.08213 | −0.14406 | −0.02588 | −0.01630 | 0.98101 | −0.92233 | −1.12475 | 1.42687 |
| 5 | 0.08204 | −0.14401 | −0.02592 | −0.01630 | 0.9810 I | −0.92242 | −1.12461 | 1.42063 |
| 6 | 0.08205 | −0.14402 | −0.02591 | −0.01630 | 0.98!01 | −0.92241 | −1.12463 | 1.42137 |
| 7 | 0.08205 | −0.14402 | −0.02591 | −0.01630 | 0.98101 | −0.92241 | −1.12463 | 1.42141 |
| 8 | 0.08205 | −0.14402 | −0.02591 | −0.01630 | 0.98101 | −0.92241 | −1.12463 | 1.42141 |
| 9 | 0.08205 | −0.14402 | −0.02591 | −0.01630 | 0.98101 | −0.92241 | −1.12463 | 1.42141 |
| 10 | 0.08205 | −0.14402 | −0.02591 | −0.01630 | 0.98101 | −0.92241 | −1.12463 | 1.42141 |
| True values | 0.07000 | −0.14000 | −0.02930 | −0.01150 | 1.00000 | −0.93000 | −1.12000 | |

## References

Hidayat, E., Medvedev, A., 2012. Laguerre domain identification of continuous linear time-delay systems from impulse response data. Automatica 48, 2902–2907.

Kim, J.H., 2011. Note on stability of linear systems with time-varying delay. Automatica 47, 2118–2121.

Qin, J., Zheng, W., Gao, H., 2011. Consensus of multiple second-order vehicles with a time-varying reference signal under directed topology. Automatica 47, 1983–1991.

Shi, Y., Yu, B., 2011. Robust mixed H-2/H-infinity control of networked control systems with random time delays in both forward and backward communication links. Automatica 47, 754–760.

CHAPTER 4

# Multivariable time-delay system identification

## 4.1 Parameter estimation for a multivariable state space system with d-step delay

The state space model which originates from the state variable method described by differential equations has become a significant tool for identifying and analyzing dynamic systems. There exist a variety of identification methods for state space models, for example, the maximum likelihood methods, the subspace identification methods.

The hierarchical identification principle is based on the decomposition technique and is widely used in state and parameter estimation, for example, the least squares-based parameter identification algorithms, gradient-based parameter estimation algorithms, and the iterative estimation algorithms (Sui et al., 2018). The identification of the time-delay systems is significant for system control and system analysis and has received considerable attention from many researchers in industrial processes.

It combines the least squares theory and the hierarchical identification principle and presents the state and parameter estimate algorithms for time-delay systems. The proposed method deals with the identification problem for the state space model with its corresponding input–output representation. Compared with the recursive least squares algorithm, the hierarchical identification algorithm has less computation burden and is easier to implement.

### 4.1.1 The canonical state space model for state delay systems

Let us introduce some notation. "$A =: X$" or "$X := A$" stands for "$A$ is defined as $X$"; the symbol $I(I_n)$ stands for an identity matrix of appropriate size ($n \times n$); $z$ represents a unit forward shift operator: $zx(t) = x(t+1)$ and $z^{-1}x(t) = x(t-1)$; the superscript $T$ denotes the matrix/vector transpose; $\hat{\theta}(t)$ denotes the estimate of $\theta$ at time $t$; $e_i$ stands for the ith column of an identity matrix. Consider the following state space system with d-step delay depicted in Fig. 4.1,

*State Space Systems With Time-Delays Analysis, Identification, and Applications*
DOI: https://doi.org/10.1016/B978-0-323-91768-1.00016-2

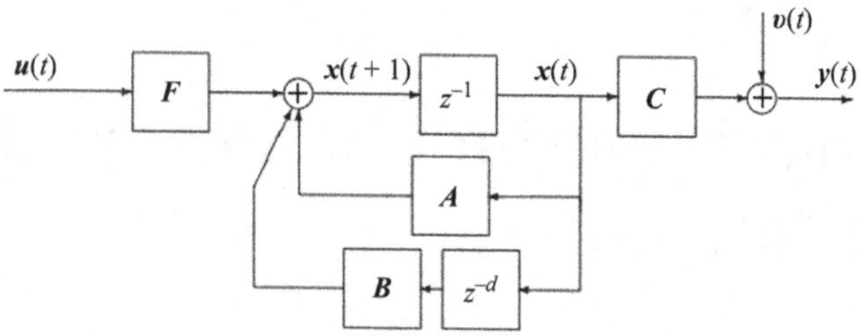


**Figure 4.1** The diagram of the state space model with d-step delay.

$$\boldsymbol{x}(t+1) = \boldsymbol{A}\boldsymbol{x}(t) + \boldsymbol{B}\boldsymbol{x}(t-d) + \boldsymbol{F}\boldsymbol{u}(t), \tag{4.1}$$

$$\boldsymbol{y}(t) = \boldsymbol{C}\boldsymbol{x}(t) + \boldsymbol{v}(t), \tag{4.2}$$

$$\boldsymbol{A} := \begin{bmatrix} \boldsymbol{A}_{11} & \boldsymbol{A}_{12} & \cdots & \boldsymbol{A}_{1m} \\ \boldsymbol{A}_{21} & \boldsymbol{A}_{22} & \cdots & \boldsymbol{A}_{2m} \\ \vdots & \vdots & & \vdots \\ \boldsymbol{A}_{m1} & \boldsymbol{A}_{m2} & \cdots & \boldsymbol{A}_{mm} \end{bmatrix} \in \mathbb{R}^{n\times n}, \quad \boldsymbol{A}_{ii} := \begin{bmatrix} 0 & 1 & \vdots & 0 \\ \vdots & 0 & \ddots & \vdots \\ 0 & \cdots & 0 & 1 \\ a_{ii,1} & a_{ii,2} & \cdots & a_{ii,n_i} \end{bmatrix} \in \mathbb{R}^{n_i\times n_i},$$

$$\boldsymbol{A}_{ij} := \begin{bmatrix} 0 & \cdots & 0 \\ \vdots & & \vdots \\ 0 & \cdots & 0 \\ a_{ij,1} & \cdots & a_{ij,n_j} \end{bmatrix} \in \mathbb{R}^{n_i\times n_i}, \quad i \neq j,$$

$$\boldsymbol{B} := \begin{bmatrix} \boldsymbol{B}_1 \\ \boldsymbol{B}_2 \\ \vdots \\ \boldsymbol{B}_m \end{bmatrix} \in \mathbb{R}^{n\times n}, \quad \boldsymbol{B}_i := \begin{bmatrix} \boldsymbol{b}_{i1} \\ \boldsymbol{b}_{i2} \\ \vdots \\ \boldsymbol{b}_{in_i} \end{bmatrix} \in \mathbb{R}^{n_i\times n}, \quad \boldsymbol{b}_{ij} \in \mathbb{R}^{1\times n},$$

$$\boldsymbol{F} := \begin{bmatrix} \boldsymbol{F}_1 \\ \boldsymbol{F}_2 \\ \vdots \\ \boldsymbol{F}_m \end{bmatrix} \in \mathbb{R}^{n\times r}, \quad \boldsymbol{F}_i := \begin{bmatrix} \boldsymbol{f}_{i1} \\ \boldsymbol{f}_{i2} \\ \vdots \\ \boldsymbol{f}_{in_i} \end{bmatrix} \in \mathbb{R}^{n_i\times n^r}, \quad \boldsymbol{f}_{ij} \in \mathbb{R}^{1\times r},$$

$$\boldsymbol{C} := \operatorname{diag}\left[\boldsymbol{e}_1^{\mathrm{T}},\ \boldsymbol{e}_2^{\mathrm{T}}, \cdots, \boldsymbol{e}_m^{\mathrm{T}}\right] \in \mathbb{R}^{m\times m}, \qquad \boldsymbol{e}_i := [1, 0, \cdots, 0]^{\mathrm{T}} \in \mathbb{R}^{n_i},$$

where $x(t) = [x_1(t), x_2(t), \ldots, x_m(t)]^{\mathrm{T}} \in R^n$ is the state vector, $u(t) \in R^r$ and $y(t) = [y_1(t), y_2(t), \ldots, y_m(t)]^{\mathrm{T}} \in R^m$ denote the input and output vectors of the system, $v(t) = [v_1(t), v_2(t), \ldots, v_m(t)]^T \in R^m$ is a random noise with zero mean, $A \in R^{n\times n}$, $B \in R^{n\times n}$, $F \in R^{n\times r}$, and $C \in R^{m\times n}$ are the system parameter matrices, and $n_i \geq 1$ is the observability structure index with $n_1 + n_2 + \cdots + n_m = n$.

Assume that $(C, A)$ is observable and $u(t) = 0$ and $y(t) = 0$ for $t \leq 0$. The system matrices $A$, $B$, and $F$ are the unknown parameters to be estimated from input–output data $\{u(t), y(t)\}$.

### 4.1.2 The identification model

This section derives the identification model of the canonical state space system in (4.1) and (4.2). Eqs. (4.1) and (4.2) can be equivalently written as

$$\begin{bmatrix} \boldsymbol{x}_1(t+1) \\ \boldsymbol{x}_2(t+1) \\ \vdots \\ \boldsymbol{x}_m(t+1) \end{bmatrix} = \begin{bmatrix} \boldsymbol{A}_{11} & \boldsymbol{A}_{12} & \cdots & \boldsymbol{A}_{1m} \\ \boldsymbol{A}_{21} & \boldsymbol{A}_{22} & \cdots & \boldsymbol{A}_{2m} \\ \vdots & \vdots & & \vdots \\ \boldsymbol{A}_{m1} & \boldsymbol{A}_{m1} & \cdots & \boldsymbol{A}_{mm} \end{bmatrix} \begin{bmatrix} \boldsymbol{x}_1(t) \\ \boldsymbol{x}_2(t) \\ \vdots \\ \boldsymbol{x}_m(t) \end{bmatrix} + \begin{bmatrix} \boldsymbol{B}_1 \\ \boldsymbol{B}_1 \\ \vdots \\ \boldsymbol{B}_m \end{bmatrix} x(t-d) + \begin{bmatrix} \boldsymbol{F}_1 \\ \boldsymbol{F}_1 \\ \vdots \\ \boldsymbol{F}_m \end{bmatrix} \boldsymbol{u}(t), \tag{4.3}$$

$$\begin{bmatrix} y_1(t) \\ y_2(t) \\ \vdots \\ y_m(t) \end{bmatrix} = \mathrm{diag}\left[\boldsymbol{e}_1^{\mathrm{T}},\ \boldsymbol{e}_2^{\mathrm{T}}, \cdots, \boldsymbol{e}_m^{\mathrm{T}}\right] \begin{bmatrix} \boldsymbol{x}_1(t) \\ \boldsymbol{x}_2(t) \\ \vdots \\ \boldsymbol{x}_m(t) \end{bmatrix} + \begin{bmatrix} v_1(t) \\ v_2(t) \\ \vdots \\ v_m(t) \end{bmatrix}, \tag{4.4}$$

where $x_i(t) = [x_{i1}(t), x_{i2}(t), \ldots, x_{im}(t)]^{\mathrm{T}} \in R^n$ is the state vector.

Eqs. (4.3) and (4.4) can be decomposed into $m$ subsystems:

$$\begin{aligned} \begin{bmatrix} x_{i1}(t+1) \\ x_{i2}(t+1) \\ \vdots \\ x_{in_i}(t+1) \end{bmatrix} &= \begin{bmatrix} 0 & 1 & \cdots & 0 \\ \vdots & 0 & \ddots & \vdots \\ 0 & \cdots & 0 & 1 \\ a_{ii,1} & a_{ii,2} & \cdots & a_{ii,n_i} \end{bmatrix} \begin{bmatrix} x_{i1}(t) \\ x_{i2}(t) \\ \vdots \\ x_{in_i}(t) \end{bmatrix} \\ &+ \sum_{j\neq i}^{m} \begin{bmatrix} 0 & \cdots & 0 \\ \vdots & & \vdots \\ 0 & \cdots & 0 \\ a_{ij,1} & \cdots & a_{ij,n_j} \end{bmatrix} \begin{bmatrix} x_{j1}(t) \\ x_{j2}(t) \\ \vdots \\ x_{jn_j}(t) \end{bmatrix} \\ &+ \begin{bmatrix} \boldsymbol{b}_{i1} \\ \boldsymbol{b}_{i1} \\ \vdots \\ \boldsymbol{b}_{in_i} \end{bmatrix} \boldsymbol{x}(t-d) + \begin{bmatrix} \boldsymbol{f}_{i1} \\ \boldsymbol{f}_{i2} \\ \vdots \\ \boldsymbol{f}_{in_i} \end{bmatrix} \boldsymbol{u}(t), \end{aligned} \tag{4.5}$$

$$\begin{aligned} y_i(t) &= \boldsymbol{e}_i^{\mathrm{T}} \boldsymbol{x}_i(t) + v_i(t) \\ &= x_{i1}(t) + v_i(t). \end{aligned} \tag{4.6}$$

Expanding (4.5) gives

$$\begin{cases} x_{i1}(t+1) = x_{i2}(t) + \boldsymbol{b}_{i1}\boldsymbol{x}(t-d) + \boldsymbol{f}_{i1}\boldsymbol{u}(t), \\ x_{i2}(t+1) = x_{i3}(t) + \boldsymbol{b}_{i2}\boldsymbol{x}(t-d) + \boldsymbol{f}_{i2}\boldsymbol{u}(t), \\ \quad\vdots \\ x_{in_i}(t+1) = \boldsymbol{a}_i^{\mathrm{T}}\boldsymbol{x}(t) + \boldsymbol{b}_{in_i}\boldsymbol{x}(t-d) + \boldsymbol{f}_{in_i}\boldsymbol{u}(t). \end{cases} \tag{4.7}$$

Using the properties of the shift operator $z$ and multiplying the $j$th equation of (4.7) by $z^{-j}(j=1,2,\ldots,n_i)$ give

$$\begin{cases} z^{-1}x_{i1}(t+1) & = z^{-1}x_{i2}(t)+z^{-1}\boldsymbol{b}_{i1}\boldsymbol{x}(t-d)+z^{-1}\boldsymbol{f}_{i1}\boldsymbol{u}(t), \\ z^{-2}x_{i2}(t+1) & \quad z^{-2}x_{i3}(t)+z^{-2}\boldsymbol{b}_{i2}\boldsymbol{x}(t-d)+z^{-2}\boldsymbol{f}_{i2}\boldsymbol{u}(t), \\ \vdots & \\ z^{1-n_i}x_{i,n_{i-1}}(t+1) & = z^{1-n_i}x_{in_i}(t)+z^{1-n_i}\boldsymbol{b}_{i,n_i-1}\boldsymbol{x}(t-d)+z^{1-n_i}\boldsymbol{f}_{i,n_{i-1}}\boldsymbol{u}(t), \\ z^{-n_i}x_{in_i}(t+1) & = z^{-n_i}\boldsymbol{a}_i^{\mathrm{T}}\boldsymbol{x}(t)+z^{-n_i}\boldsymbol{b}_{in_i}\boldsymbol{x}(t-d)+z^{-n_i}\boldsymbol{f}_{in_i}\boldsymbol{u}(t), \end{cases}$$

or

$$\begin{cases} x_{i1}(t) & = x_{i2}(t-1)+\boldsymbol{b}_{i1}\boldsymbol{x}(t-d-1)+\boldsymbol{f}_{i1}\boldsymbol{u}(t-1), \\ x_{i2}(t-1) & = x_{i3}(t-2)+\boldsymbol{b}_{i2}\boldsymbol{x}(t-d-2)+\boldsymbol{f}_{i2}\boldsymbol{u}(t-2), \\ \vdots & \\ x_{i,n_i-1}(t-n_i+2) & = x_{in_i}(t-n_i+1)+\boldsymbol{b}_{i,n_i-1}\boldsymbol{x}(t-d-n_i+1)+\boldsymbol{f}_{i,n_i-1}\boldsymbol{u}(t-n_i+1), \\ x_{in_i}(t-n_i+1) & = \boldsymbol{a}_i^{\mathrm{T}}\boldsymbol{x}(t-n_i)+\boldsymbol{b}_{in_i}\boldsymbol{x}(t-d-n_i)+\boldsymbol{f}_{in_i}\boldsymbol{u}(t-n_i), \end{cases} \tag{4.8}$$

where $a_i=[a_{i1,1},a_{i1,2},\ldots,a_{i1,n_1},\ldots,a_{im,1},a_{im,2},\ldots,a_{im,n_m}]^{\mathrm{T}}\in R^n$.

When $d\leq n_i-1$, define the information vector $\phi_i(t)$ and the parameter vector $\theta_i$:

$$\varphi_i(t):=\left[\boldsymbol{x}^{\mathrm{T}}(t-d-1),\boldsymbol{x}^{\mathrm{T}}(t-d-2),\cdots,\boldsymbol{x}^{\mathrm{T}}(t-n_i),\cdots,\boldsymbol{x}^{\mathrm{T}}(t-n_i-d),\right.$$
$$\left.\boldsymbol{u}^{\mathrm{T}}(t-1),\boldsymbol{u}^{T}(t-2),\cdots,\boldsymbol{u}^{\mathrm{T}}(t-n_i)\right]^{\mathrm{T}}\in\mathbb{R}^{n_i\times(n+r)},$$

$$\boldsymbol{\theta}_i:=\left[\boldsymbol{b}_{i1},\boldsymbol{b}_{i2},\cdots,\boldsymbol{b}_{i,n_i-d-1},\boldsymbol{a}_i+\boldsymbol{b}_{i,n_i-d},\boldsymbol{b}_{i,n_i-d+1},\cdots,\boldsymbol{b}_{in_i},\boldsymbol{F}_i^{\mathrm{T}}\right]^{\mathrm{T}}\in\mathbb{R}^{n_i\times(n+r)}.$$

When $d\geq n_i$, define the information vector $\phi_i(t)$ and the parameter vector $\theta_i$:

$$\varphi_i(t):=\left[\boldsymbol{x}^{\mathrm{T}}(t-n_i),\boldsymbol{x}^{\mathrm{T}}(t-d-1),\boldsymbol{x}^{\mathrm{T}}(t-d-2),\cdots,\boldsymbol{x}^{\mathrm{T}}(t-d-n_i+1),\right.$$
$$\left.\boldsymbol{x}^{\mathrm{T}}(t-d-n_i),\boldsymbol{u}^{\mathrm{T}}(t-1),\boldsymbol{u}^{\mathrm{T}}(t-2),\cdots,\boldsymbol{u}^{\mathrm{T}}(t-n_i)\right]^{\mathrm{T}}\in\mathbb{R}^{n+n_i\times(n+r)},$$

$$\boldsymbol{\theta}_i:=\left[\boldsymbol{a}_i,\boldsymbol{b}_{i1},\boldsymbol{b}_{i2},\cdots,\boldsymbol{b}_{i,n_i-1},\boldsymbol{b}_{in_i},\boldsymbol{F}_i^{\mathrm{T}}\right]^{\mathrm{T}}\in\mathbb{R}^{n+n_i\times(n+r)}.$$

Adding all equations of (4.8) gives

$$\begin{aligned} x_{i1}(t) &= \boldsymbol{a}_i\boldsymbol{x}(t-n_i)+\boldsymbol{b}_{i1}\boldsymbol{x}(t-2)+\boldsymbol{b}_{i2}\boldsymbol{x}(t-3)+\cdots+\boldsymbol{b}_{i,n_i-1}\boldsymbol{x}(t-n_i) \\ &\quad +\boldsymbol{b}_{in_i}\boldsymbol{x}(t-n_i-1)+\boldsymbol{f}_{i1}\boldsymbol{u}(t-1)+\boldsymbol{f}_{i2}\boldsymbol{u}(t-2)+\cdots \\ &\quad +\boldsymbol{f}_{i,n_i-1}\boldsymbol{u}(t-n_i+1)+\boldsymbol{f}_{in_i}\boldsymbol{u}(t-n_i) \\ &= \boldsymbol{\varphi}_i^{\mathrm{T}}(t)\boldsymbol{\theta}_i. \end{aligned}$$

From (4.6), we have

$$y_i(t) = x_{i1}(t) + v_i(t) = \boldsymbol{\varphi}_i^{\mathrm{T}}(t)\boldsymbol{\theta}_i + v_i(t). \tag{4.9}$$

This is the identification model of the state space system with d-step delay.

### 4.1.3 The parameter and state estimation algorithm

This section derives a hierarchical identification algorithm to estimate the parameter vector $\theta_i$ in (4.9) and uses the Kalman filtering theory to estimate the state vector $x(t+1)$ of the system.

From the identification expression in (4.9), difficulties arise in that the information vector $\varphi_i(t)$ contains the unknown state variable $x(t-i)$ $(i = 1 + d, 2 + d, \ldots, n_i + d)$ and the parameter vector $\theta_i$ is unknown. We adopt the hierarchical identification principle, when estimating the parameter vector $\theta_i$, replacing the unknown state vector $x(t-i)$ with its corresponding estimate $\hat{x}(t-i)$; similarly, when estimating the state vector $x(t+1)$, replacing the unknown parameter vector $\theta_i$ with its estimate $\hat{\theta}_i(t)$ at time $t$.

Minimizing the criterion function,

$$J(\boldsymbol{\theta}_i) := \sum_{j=1}^{t} \left[y_i(j) - \boldsymbol{\varphi}_i^{\mathrm{T}}(j)\boldsymbol{\theta}_i\right]^2.$$

We can obtain the following least squares parameter estimation algorithm based on the hierarchical identification principle:

$$\hat{\boldsymbol{\theta}}_i(t) = \hat{\boldsymbol{\theta}}_i(t-1) + \boldsymbol{L}_i(t)\left[y_i(t) - \hat{\boldsymbol{\varphi}}_i^{\mathrm{T}}(t)\hat{\boldsymbol{\theta}}_i(t-1)\right], \tag{4.10}$$

$$\boldsymbol{L}_i(t) = \frac{\boldsymbol{P}_i(t-1)\hat{\boldsymbol{\varphi}}_i(t)}{1 + \hat{\boldsymbol{\varphi}}_i^{\mathrm{T}}(t)\boldsymbol{P}_i(t-1)\hat{\boldsymbol{\varphi}}_i(t)}, \tag{4.11}$$

$$\boldsymbol{P}_i(t) = \boldsymbol{P}_i(t-1) - \frac{\boldsymbol{P}_i(t-1)\hat{\boldsymbol{\varphi}}_i(t)\hat{\boldsymbol{\varphi}}_i^{\mathrm{T}}(t)\boldsymbol{P}_i(t-1)}{1 + \hat{\boldsymbol{\varphi}}_i^{\mathrm{T}}(t)\boldsymbol{P}_i(t-1)\hat{\boldsymbol{\varphi}}_i(t)}, \tag{4.12}$$

$$\hat{\boldsymbol{\varphi}}_i(t) = \left[\hat{\boldsymbol{x}}^{\mathrm{T}}(t-d-1), \hat{\boldsymbol{x}}^{\mathrm{T}}(t-d-2), \cdots, \hat{\boldsymbol{x}}^{\mathrm{T}}(t-n_i), \cdots, \hat{\boldsymbol{x}}^{\mathrm{T}}(t-n_i-d),\right.$$

$$\left.\boldsymbol{u}^{\mathrm{T}}(t-1), \boldsymbol{u}^{\mathrm{T}}(t-2), \cdots, \boldsymbol{u}^{\mathrm{T}}(t-n_i)\right]^{\mathrm{T}}, \quad d \leqslant n_i - 1, \tag{4.13}$$

$$\hat{\varphi}_i(t) = \left[\hat{\boldsymbol{x}}^{\mathrm{T}}(t-n_i), \hat{\boldsymbol{x}}^{\mathrm{T}}(t-d-1), \hat{x}^{\mathrm{T}}(t-d-2), \cdots, \hat{\boldsymbol{x}}^{\mathrm{T}}(t-d-n_i+1),\right.$$

$$\left.\hat{x}^{\mathrm{T}}(t-d-n_i), \boldsymbol{u}^{\mathrm{T}}(t-1), \boldsymbol{u}^{\mathrm{T}}(t-2), \cdots, \boldsymbol{u}^{\mathrm{T}}(t-n_i)\right]^{\mathrm{T}}, d \geqslant n_i. \quad (4.14)$$

We can use the Kalman filter theory to estimate the state, and the algorithm is as follows:

$$\hat{\boldsymbol{x}}(t+1) = \hat{\boldsymbol{A}}(t)\hat{\boldsymbol{x}}(t) + \hat{\boldsymbol{B}}(t)\hat{\boldsymbol{x}}(t-d) + \hat{\boldsymbol{F}}(t)\boldsymbol{u}(t) + \boldsymbol{L}(t)\left[y(t) - \hat{\varphi}^{\mathrm{T}}(t)\hat{\boldsymbol{\theta}}(t)\right], \quad (4.15)$$

$$\boldsymbol{L}(t) = \hat{\boldsymbol{A}}(t)\boldsymbol{P}(t)\boldsymbol{C}^{T}\left[\boldsymbol{I} + \boldsymbol{C}\boldsymbol{P}(t)\boldsymbol{C}^{\mathrm{T}}\right]^{-1} \quad (4.16)$$

$$\boldsymbol{P}(t+1) = \hat{\boldsymbol{A}}(t)\boldsymbol{P}(t)\hat{\boldsymbol{A}}^{\mathrm{T}}(t) - \hat{\boldsymbol{A}}(t)\boldsymbol{P}(t)\boldsymbol{C}^{\mathrm{T}}\left[\boldsymbol{I} + \boldsymbol{C}\boldsymbol{P}(t)\boldsymbol{C}^{\mathrm{T}}\right]^{-1}\boldsymbol{C}\boldsymbol{P}(t)\hat{\boldsymbol{A}}^{\mathrm{T}}(t), \quad (4.17)$$

$$\hat{\boldsymbol{\theta}}_i(t) = \left[\hat{\boldsymbol{b}}_{i1}(t), \hat{\boldsymbol{b}}_{i2}(t), \cdots, \hat{\boldsymbol{b}}_{i,n_i-d-1}(t), \hat{\boldsymbol{a}}_i(t) + \hat{\boldsymbol{b}}_{i,n_i-d}(t),\right.$$

$$\left.\hat{\boldsymbol{b}}_{i,n_i-d+1}(t), \cdots, \hat{\boldsymbol{b}}_{in_i}(t), \boldsymbol{F}_i^{\mathrm{T}}(t)\right]^{\mathrm{T}}, d \leqslant n_i - 1,$$

$$\hat{\boldsymbol{\theta}}_i(t) = \left[\hat{\boldsymbol{a}}_i(t),\ \hat{\boldsymbol{b}}_{i1}(t),\ \hat{\boldsymbol{b}}_{i2}(t), \cdots, \hat{\boldsymbol{b}}_{i,n_i-1}(t),\ \hat{\boldsymbol{b}}_{in_i}(t),\ \boldsymbol{F}_i^{T}(t)\right]^{T},\ d \geqslant n_i, \quad (4.18)$$

$$\hat{\boldsymbol{a}}_i(t) = \left[\hat{a}_{i1,1}(t), \hat{a}_{i1,2}(t), \cdots, \hat{a}_{i1,n_1}(t), \cdots, \hat{a}_{im,1}(t), \hat{a}_{im,2}(t), \cdots, \hat{a}_{im,n_m}(t)\right]^{\mathrm{T}},$$

$$\hat{\boldsymbol{A}}(t) = \begin{bmatrix} \hat{\boldsymbol{A}}_{11}(t) & \hat{\boldsymbol{A}}_{12}(t) & \cdots & \hat{\boldsymbol{A}}_{1m}(t) \\ \hat{\boldsymbol{A}}_{21}(t) & \hat{\boldsymbol{A}}_{22}(t) & \cdots & \hat{\boldsymbol{A}}_{2m}(t) \\ \vdots & \vdots & & \vdots \\ \hat{\boldsymbol{A}}_{m1}(t) & \hat{\boldsymbol{A}}_{m2}(t) & \cdots & \hat{\boldsymbol{A}}_{mm}(t) \end{bmatrix},$$

$$\hat{\boldsymbol{A}}_{ii}(t) = \begin{bmatrix} 0 & 1 & \cdots & 0 \\ \vdots & 0 & \ddots & \vdots \\ 0 & \cdots & 0 & 1 \\ \hat{a}_{ii,1}(t) & \hat{a}_{ii,2}(t) & \cdots & \hat{a}_{ii,n_i}(t) \end{bmatrix},\ \hat{\boldsymbol{A}}_{ij}(t) = \begin{bmatrix} 0 & \cdots & 0 \\ \vdots & & \vdots \\ 0 & \cdots & 0 \\ \hat{a}_{ij,1}(t) & \cdots & \hat{a}_{ij,n_j}(t) \end{bmatrix}, \quad (4.19)$$

$$\hat{\boldsymbol{B}}(t) = \begin{bmatrix} \hat{\boldsymbol{B}}_1(t) \\ \hat{\boldsymbol{B}}_2(t) \\ \vdots \\ \hat{\boldsymbol{B}}_m(t) \end{bmatrix}, \quad \hat{\boldsymbol{B}}_i(t) = \begin{bmatrix} \hat{\boldsymbol{b}}_{i1}(t) \\ \hat{\boldsymbol{b}}_{i2}(t) \\ \vdots \\ \hat{\boldsymbol{b}}_{in_i}(t) \end{bmatrix}, \quad \hat{\boldsymbol{F}}(t) = \begin{bmatrix} \hat{\boldsymbol{F}}_1(t) \\ \hat{\boldsymbol{F}}_2(t) \\ \vdots \\ \hat{\boldsymbol{F}}_m(t) \end{bmatrix},$$

$$\hat{\boldsymbol{F}}_i(t) = \begin{bmatrix} \hat{\boldsymbol{f}}_{i1}(t) \\ \hat{\boldsymbol{f}}_{i2}(t) \\ \vdots \\ \hat{\boldsymbol{f}}_{in_i}(t) \end{bmatrix}. \tag{4.20}$$

The steps of computing the parameter estimate $\hat{\theta}_i(t)$ in (4.10)–(4.14) and the state estimate in (4.15)–(4.20) are listed in the following.

- Let $t = 1$; set the initial values $\hat{\theta}_i(0) = 1_{n^2+n}/p_0$, $P_i(0) = p_0 I$, $P(1) = I$, $p_0 = 10^6$, $\hat{x}(1) = 0$.
- Collect the input–output data $u(t)$ and $y(t)$, and form $\hat{\phi}_i(t)$ by (4.13) or (4.14).
- Compute the gain vector $L_i(t)$ by (4.11) and the covariance matrix $P_i(t)$ by (4.12).
- Update the parameter estimation vector $\hat{\theta}_i(t)$ by (4.10).
- Read $\hat{a}_i(t)$, $\hat{b}_{ij}(t)$, and $\hat{F}_i(t)$ from $\hat{\theta}_i(t)$ according to the definition of $\hat{\theta}_i(t)$.
- Form $\hat{A}(t)$, $\hat{B}(t)$, and $\hat{F}(t)$ by (4.19) and (4.20).
- Compute the gain vector $L(t)$ by (4.16) and the covariance matrix $P(t)$ by (4.17).
- Compute the state estimation vector $\hat{x}(t+1)$ by (4.15).
- Increase $t$ by 1 and go to step 2, continue the recursive calculation.

The flowchart of computing the parameter estimate $\hat{\theta}_i(t)$ in (4.10)–(4.14) and the state estimate $\hat{x}(t+1)$ in (4.15)–(4.20) are shown in Fig. 4.2.

### 4.1.4 Example

Consider the following state space system with two-step delay:

$$\boldsymbol{x}(t+1) = \begin{bmatrix} 0 & 1 & 0 & 0 \\ 0.32 & 0.49 & 0.55 & 0.01 \\ 0 & 0 & 0 & 1 \\ -0.40 & -0.86 & -0.60 & -0.65 \end{bmatrix}$$

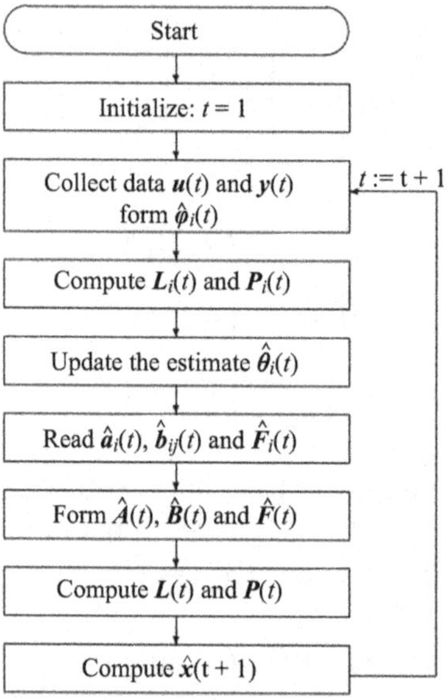


**Figure 4.2** The flowchart of computing the parameter estimate $\hat{\theta}_i(t)$ and the state estimate $\hat{x}(t+1)$.

$$\boldsymbol{x}(t) + \begin{bmatrix} 0.20 & -0.25 & 0.01 & 0.01 \\ 0.10 & 0.10 & 0.23 & 0.14 \\ 0.30 & 0.01 & 0.20 & 0.10 \\ 0.10 & 0.50 & 1.10 & 0.20 \end{bmatrix} \boldsymbol{x}(t-2) + \begin{bmatrix} 0.70 & 0.10 \\ 0.50 & 0.14 \\ 0.15 & 0.60 \\ 0.01 & 0.50 \end{bmatrix} \boldsymbol{u}(t),$$

$$\boldsymbol{y}(t) = \begin{bmatrix} 1 & 0 & 0 & 0 \\ 0 & 0 & 1 & 0 \end{bmatrix} \boldsymbol{x}(t) + \boldsymbol{v}(t).$$

Since the input $u(t)$ and output $y(t)$ are two-dimensional vectors, the system can be decomposed into two subsystems. The information vectors and parameter vectors are

$$\begin{aligned} \boldsymbol{\varphi}_i(t) = [&x_i(t-2),\ x_2(t-2),\ x_3(t-2),\ x_4(t-2),\ x_1(t-3),\ x_2(t-3), \\ &x_3(t-3),\ x_4(t-3), x_1(t-4),\ x_2(t-4),\ x_3(t-4),\ x_4(t-4), \\ &u_1(t-1),\ u_2(t-1),\ u_1(t-2),\ u_2(t-2)]^{\mathrm{T}}, \quad i = 1, 2, \end{aligned}$$

$$\begin{aligned} \boldsymbol{\theta}_1 = &\left[a_{21},\ a_{22},\ a_{23},\ a_{24},\ b_{11},\ b_{12},\ b_{13},\ b_{14},\ b_{21},\ b_{22},\ b_{23},\ b_{24},\ f_{11}, \right. \\ &\left. f_{12},\ f_{21},\ f_{22}\right]^{\mathrm{T}} \\ = &[0.32,\ 0.49,\ 0.55,\ 0.01,\ 0.20, -0.25,\ 0.01,\ 0.01,\ 0.10,\ 0.10, \\ &0.22,\ 0.14,\ 0.70,\ 0.10,\ 0.50,\ 0.14]^{\mathrm{T}}, \end{aligned}$$

**Table 4.1** The parameter estimates and errors with $\sigma^2 = 0.20^2$ for subsystem 1.

| $t$ | 100 | 200 | 500 | 1000 | 2000 | 3000 |
|---|---|---|---|---|---|---|
| $a_{21} = 0.32$ | 0.34172 | 0.30684 | 0.30307 | 0.31499 | 0.31846 | 0.32556 |
| $a_{22} = 0.49$ | 0.47293 | 0.51905 | 0.52173 | 0.50641 | 0.50087 | 0.48805 |
| $a_{23} = 0.55$ | 0.52494 | 0.49729 | 0.53975 | 0.53397 | 0.53724 | 0.54645 |
| $a_{24} = 0.01$ | −0.00094 | 0.05598 | 0.02233 | 0.02766 | 0.02078 | 0.01006 |
| $b_{11} = 0.20$ | 0.25448 | 0.19555 | 0.18444 | 0.19939 | 0.20129 | 0.20970 |
| $b_{12} = -0.25$ | −0.35660 | −0.20399 | −0.22150 | −0.22897 | −0.24132 | −0.26105 |
| $b_{13} = 0.01$ | 0.07983 | 0.01279 | −0.01952 | −0.00029 | 0.00281 | 0.00597 |
| $b_{14} = 0.01$ | −0.06088 | 0.09800 | 0.04511 | 0.04560 | 0.03409 | 0.01509 |
| $b_{21} = 0.10$ | 0.07412 | 0.06607 | 0.09251 | 0.09415 | 0.09992 | 0.10817 |
| $b_{22} = 0.10$ | 0.03564 | 0.14311 | 0.12825 | 0.11306 | 0.10382 | 0.09307 |
| $b_{23} = 0.23$ | 0.20842 | 0.22802 | 0.23300 | 0.23111 | 0.23201 | 0.23417 |
| $b_{24} = 0.14$ | 0.02243 | 0.16633 | 0.17926 | 0.16193 | 0.15301 | 0.14344 |
| $f_{11} = 0.70$ | 0.69817 | 0.69047 | 0.69976 | 0.70163 | 0.70169 | 0.70242 |
| $f_{12} = 0.10$ | 0.08291 | 0.09771 | 0.09933 | 0.09907 | 0.09824 | 0.09756 |
| $f_{21} = 0.50$ | 0.48362 | 0.49099 | 0.49890 | 0.50035 | 0.50066 | 0.50277 |
| $f_{22} = 0.14$ | 0.12468 | 0.14667 | 0.14775 | 0.14454 | 0.14161 | 0.13927 |
| $\delta$(%) | 16.83778 | 11.06949 | 6.67615 | 4.58945 | 2.84750 | 1.70786 |

**Table 4.2** The parameter estimates and errors with $\sigma^2 = 0.20^2$ for subsystem 2.

| $t$ | 100 | 200 | 500 | 1000 | 2000 | 3000 |
|---|---|---|---|---|---|---|
| $a_{41} = -0.40$ | −0.40585 | −0.43720 | −0.43619 | −0.42690 | −0.41518 | −0.40337 |
| $a_{42} = -0.86$ | −0.88077 | −0.80238 | −0.80656 | −0.82305 | −0.83690 | −0.85153 |
| $a_{43} = -0.60$ | −0.52392 | −0.64200 | −0.60299 | −0.60503 | −0.60532 | −0.59738 |
| $a_{44} = -0.65$ | −0.75095 | −0.60135 | −0.63830 | −0.63613 | −0.64159 | −0.65298 |
| $b_{31} = 0.30$ | 0.20968 | 0.26528 | 0.27934 | 0.28729 | 0.30032 | 0.30454 |
| $b_{32} = 0.01$ | 0.00691 | 0.09789 | 0.04594 | 0.03849 | 0.01991 | 0.00004 |
| $b_{33} = 0.20$ | 0.17897 | 0.21795 | 0.17049 | 0.18872 | 0.18804 | 0.19039 |
| $b_{34} = 0.10$ | −0.08016 | 0.16219 | 0.12453 | 0.12120 | 0.11509 | 0.09609 |
| $b_{41} = 0.10$ | 0.09018 | 0.06746 | 0.10945 | 0.10064 | 0.10392 | 0.10779 |
| $b_{42} = 0.50$ | 0.43403 | 0.50384 | 0.50190 | 0.50580 | 0.49674 | 0.49300 |
| $b_{43} = 0.10$ | 0.05355 | 0.08340 | 0.10646 | 0.10515 | 0.10582 | 0.10438 |
| $b_{44} = 0.20$ | 0.10850 | 0.17601 | 0.22398 | 0.21032 | 0.20983 | 0.20362 |
| $f_{31} = 0.15$ | 0.15213 | 0.14366 | 0.14746 | 0.14800 | 0.14876 | 0.15037 |
| $f_{32} = 0.60$ | 0.61174 | 0.61131 | 0.60750 | 0.60375 | 0.60050 | 0.59858 |
| $f_{41} = 0.01$ | 0.00338 | 0.00943 | 0.00907 | 0.00856 | 0.00862 | 0.01030 |
| $f_{42} = 0.50$ | 0.48059 | 0.50340 | 0.50116 | 0.50372 | 0.49987 | 0.49781 |
| $\delta$(%) | 16.25509 | 9.32823 | 5.47162 | 3.83092 | 2.32346 | 1.31101 |

**Table 4.3** The parameter estimates and errors with $\sigma^2 = 0.50^2$ for subsystem 1.

| $t$ | 100 | 200 | 500 | 1000 | 2000 | 3000 |
|---|---|---|---|---|---|---|
| $a_{21} = 0{-}32$ | 0.45155 | 0.32711 | 0.29386 | 0.31572 | 0.32035 | 0.33660 |
| $a_{22} = 0.49$ | 0.31347 | 0.49821 | 0.54299 | 0.51776 | 0.51036 | 0.48076 |
| $a_{23} = 0{-}55$ | 0.49036 | 0.41788 | 0.52600 | 0.51010 | 0.51837 | 0.54133 |
| $a_{24} = 0.01$ | 0.01683 | 0.14440 | 0.04690 | 0.05770 | 0.03862 | 0.01123 |
| $b_{11} = 0.20$ | 0.36913 | 0.20853 | 0.16924 | 0.20342 | 0.20556 | 0.22577 |
| $b_{12} = -0.25$ | −0.49308 | −0.12747 | −0.17737 | −0.19695 | −0.22790 | −0.27737 |
| $b_{13} = 0.01$ | 0.32485 | 0.09592 | −0.03349 | −0.00043 | −0.00014 | 0.00497 |
| $b_{14} = 0.01$ | −0.19134 | 0.21043 | 0.08842 | 0.09562 | 0.06814 | 0.02134 |
| $b_{21} = 0.10$ | −0.02323 | −0.01591 | 0.06994 | 0.07995 | 0.09684 | 0.11850 |
| $b_{22} = 0.10$ | −0.03139 | 0.21173 | 0.17289 | 0.13380 | 0.11019 | 0.08303 |
| $b_{23} = 0.23$ | 0.14009 | 0.20254 | 0.22980 | 0.22937 | 0.23314 | 0.23919 |
| $b_{24} = 0.14$ | −0.23669 | 0.15008 | 0.21743 | 0.18494 | 0.16734 | 0.14524 |
| $f_{11} = 0.70$ | 0.69941 | 0.67727 | 0.69990 | 0.70423 | 0.70428 | 0.70612 |
| $f_{12} = 0.10$ | 0.08028 | 0.10406 | 0.10197 | 0.09941 | 0.09648 | 0.09449 |
| $f_{21} = 0.50$ | 0.46075 | 0.47770 | 0.49763 | 0.50102 | 0.50177 | 0.50704 |
| $f_{22} = 0.14$ | 0.12086 | 0.16464 | 0.16215 | 0.15254 | 0.14463 | 0.13861 |
| $\delta(\%)$ | 53.69499 | 28.05292 | 14.18717 | 10.76650 | 6.67045 | 4.21919 |

**Table 4.4** The parameter estimates and errors with $\sigma^2 = 0.50^2$ for subsystem 2.

| $t$ | 100 | 200 | 500 | 1000 | 2000 | 3000 |
|---|---|---|---|---|---|---|
| $a_{41} = -0.40$ | −0.32353 | −0.44392 | −0.46928 | −0.45680 | −0.43253 | −0.40477 |
| $a_{42} = -0.86$ | −1.05760 | −0.79280 | −0.75995 | −0.78439 | −0.81077 | −0.84453 |
| $a_{43} = -0.60$ | −0.40338 | −0.70115 | −0.60520 | −0.61123 | −0.61277 | −0.59308 |
| $a_{44} = -0.65$ | −0.86862 | −0.50935 | −0.61384 | −0.61208 | −0.62718 | −0.65625 |
| $b_{31} = 0.30$ | 0.13434 | 0.24383 | 0.26009 | 0.27409 | 0.30374 | 0.31329 |
| $b_{32} = 0.01$ | −0.01385 | 0.22497 | 0.09798 | 0.08030 | 0.03453 | −0.01502 |
| $b_{33} = 0.20$ | 0.31646 | 0.33512 | 0.16492 | 0.19088 | 0.17983 | 0.18251 |
| $b_{34} = 0.10$ | −0.39511 | 0.23168 | 0.14895 | 0.14675 | 0.13476 | 0.08819 |
| $b_{41} = 0.10$ | 0.02484 | −0.00781 | 0.11095 | 0.09539 | 0.10645 | 0.11718 |
| $b_{42} = 0.50$ | 0.34453 | 0.51108 | 0.50697 | 0.51552 | 0.49250 | 0.48290 |
| $b_{43} = 0.10$ | −0.04767 | 0.04014 | 0.10766 | 0.10870 | 0.11232 | 0.10941 |
| $b_{44} = 0.20$ | −0.14197 | 0.07893 | 0.23312 | 0.21262 | 0.21793 | 0.20456 |
| $f_{31} = 0.15$ | 0.15765 | 0.13455 | 0.14402 | 0.14513 | 0.14694 | 0.15096 |
| $f_{32} = 0.60$ | 0.62127 | 0.62294 | 0.61659 | 0.60829 | 0.60075 | 0.59610 |
| $f_{41} = 0.01$ | −0.00323 | 0.00937 | 0.00810 | 0.00661 | 0.006699 | 0.01086 |
| $f_{42} = 0.50$ | 0.46032 | 0.51276 | 0.50444 | 0.50994 | 0.50000 | 0.49474 |
| $\delta$(%) | 46.15140 | 23.47298 | 10.54793 | 8.30891 | 5.02747 | 2.90838 |

$$\boldsymbol{\theta}_2 = [a_{41},\ a_{42},\ a_{43},\ a_{44},\ b_{31},\ b_{32},\ b_{33},\ b_{34},\ b_{41},\ b_{42},\ b_{43},\ b_{44},\ f_{31},\ f_{32},\ f_{41},\ f_{42}]^{\mathrm{T}}$$
$$= [-0.40, -0.86, -0.60, -0.65,\ 0.30,\ 0.01,\ 0.20,\ 0.10,\ 0.10,\ 0.50,\ 0.10,\ 0.20,\ 0.15,\ 0.60,\ 0.01,\ 0.50]^{\mathrm{T}}.$$

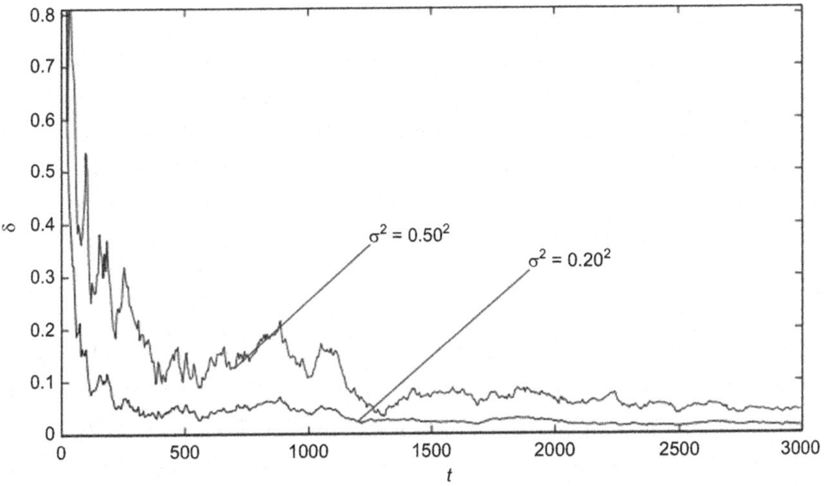


**Figure 4.3** The parameter estimates errors $\delta$ versus $t$ with $\sigma^2 = 0.20^2$ and $\sigma^2 = 0.50^2$ for subsystem 1.

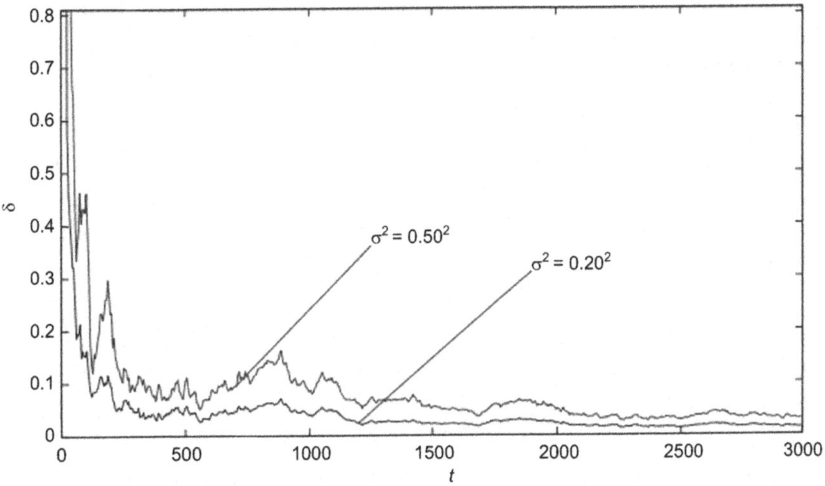


**Figure 4.4** The parameter estimates errors $\delta$ versus $t$ with $\sigma^2 = 0.20^2$ and $\sigma^2 = 0.50^2$ for subsystem 2.

In simulation, the input $\{u(t)\}$ is taken as an uncorrelated persistent excitation signal vector sequence with zero mean and unit variance and $\{v(t)\}$ as a white noise vector sequence with zero mean and variance $\sigma^2 = 0.20^2$ and $\sigma^2 = 0.50^2$, respectively. Apply the least squares parameter estimation algorithm to estimate the parameters of this example system. The parameter estimates and their estimation errors are shown in Tables 4.1–4.4 with $\sigma^2 = 0.20^2$ and $\sigma^2 = 0.50^2$, respectively, and the parameter estimation errors $\delta_i$ versus $t$ are shown in Figs. 4.3 and 4.4, where $\delta_i := ||\hat{\theta}_i(t) - \theta_i||/||\theta_i||$.

From the simulation results in Tables 4.1–4.4 and Figs. 4.3 and 4.4, we can draw the following conclusions: (1) a lower noise level leads to a faster convergence rate of the parameter estimates to the true parameters; (2) the parameter estimation errors $\delta_i$ become smaller (in general) with $t$ increases—see Figs. 4.3 and 4.4; (3) the parameter estimation accuracy under the same data lengths becomes higher—see Tables 4.1–4.4.

### 4.1.5 Conclusions

This chapter presents a least squares parameter estimation algorithm based on the hierarchical identification principle for canonical state space systems with d-step delay. Compared with the recursive least squares methods, the proposed algorithms can not only identify the system unknown parameters, but also identify the unknown system states. The simulation results show that the proposed algorithm is effective.

## 4.2 State filtering and parameter estimation for multivariable system

Parameter identification and state identification are the core problems of system identification, signal filtering, and controller design (Ding et al., 2020). The mathematical model is the basic property of the system, and the system identification makes use of statistical algorithms to develop the mathematical model of the dynamic systems from the known information (Chen et al., 2018). The iterative algorithms and the recursive algorithms are the typical parameter identification algorithms, which have a wide range of applications in seeking the roots of the equation and achieving parameter estimation methods. In this work, the differential equation is extended to the two-input two-output model with time delay, which is difficult from the point of view of identification, and the time-delay system is generally ubiquitous in industry.

With the development of control theory and industrial technology, many controlled objects were abstracted into state space models. State space models have had important applications in system modeling, system control and system analysis and thus have received much research attention in parameter and state estimation in several decades. Hence, the system identification based on the state space model has important theoretical and practical value. In engineering, the states of some systems are unknown. Therefore, the system state estimation plays an important role in control design and system identification (Gan et al., 2015).

The complexity and uncertainty of the analysis or sampling often accompany by the outputs with uncertain delays. The existence of time delay makes it difficult for the control system to respond to timely changes in input. Besides, the time delay can lead to instability and poor performance of controlled processes. The parameter and delay estimation of delay system is a challenging problem, which has attracted a lot of attention, especially in the case of measurement noise. Some useful techniques have been introduced in this aspect. Stojanovic discussed the robust finite-time stability of discrete time systems with interval time-varying delay and nonlinear perturbations.

In this chapter, a residual-based least squares identification algorithm is investigated to simultaneously estimate states and parameters of a two-input two-output system. Based on the thought of decomposition in the identification model, the two-input two-output system is decomposed into two less dimension and variables two-input single-output subsystems, again to identify each subsystem. To solve the difficulty of the information matrix including unmeasurable noise terms, the unknown noise terms are replaced with their estimated residuals, which are computed through the preceding parameter estimates. The simulation results indicate that the proposed algorithm works well. The contributions lie in three aspects:

- The two-input two-output model with time delay is decomposed into two two-input single-output models with few dimensions and few variables based on the idea of identification model decomposition.
- The presented algorithm can make full use of all data to generate highly accurate parameter estimates.
- The deducing process of the identification model is simplified to reduce the computational load of multivariable system identification.

### 4.2.1 The problem formulation

Consider the following two-input two-output state space system with time delay,

$$\boldsymbol{x}(t+1) = \boldsymbol{A}\boldsymbol{x}(t) + \boldsymbol{B}\boldsymbol{x}(t-d) + \boldsymbol{F}\boldsymbol{u}(t), \tag{4.21}$$

$$\boldsymbol{y}(t) = \boldsymbol{C}\boldsymbol{x}(t) + \boldsymbol{v}(t). \tag{4.22}$$

where $x(t) \in R^n$ is the unmeasurable state vector, $u(t) = [u_1(t), u_2(t)]^{\mathrm{T}} \in R^2$ is the input vector, $y(t) = [y_1(t), y_2(t)]^{\mathrm{T}} \in R^2$ is the output vector, and $v(t) = [v_1(t), v_2(t)]^{\mathrm{T}} \in R^2$ is the uncorrelated stochastic noise with zero mean. Assume that $n$ is known, and $u(t) = 0$ for $t \leq 0$. Since it is a multivariable system with coupling, the matrices $A \in R^{n \times n}$, $B \in R^{n \times n}$, $F \in R^{n \times 2}$, and $c \in R^{2 \times n}$ are the system parameter matrices to be identified.

$$\boldsymbol{A} := \begin{bmatrix} \boldsymbol{A}_1 & \boldsymbol{A}_{12} \\ \boldsymbol{A}_{21} & \boldsymbol{A}_2 \end{bmatrix} \in \mathbb{R}^{n \times n},$$

$$\boldsymbol{A}_i := \begin{bmatrix} 0 & 1 & 0 & \cdots & 0 \\ 0 & 0 & 1 & \ddots & \vdots \\ \vdots & \vdots & & \ddots & 0 \\ 0 & 0 & \cdots & 0 & 1 \\ a_i(1) & a_i(2) & a_i(3) & \cdots & a_i(n_i) \end{bmatrix} \in \mathbb{R}^{n_i \times n_i}, \qquad i,j = 1,2, \qquad i \neq j,$$

$$\boldsymbol{A}_{ij} := \begin{bmatrix} 0 & \cdots & & & \cdots & 0 \\ \vdots & & & & & \vdots \\ 0 & \cdots & & & \cdots & 0 \\ a_{ij}(1) & a_{ij}(2) & \cdots & a_{ij}(n_{ij}) & 0 & \cdots & 0 \end{bmatrix} \in \mathbb{R}^{n_i \times n_j},$$

$$n_{ij} \leqslant \begin{cases} n_i + 1, & i > j, \\ n_i, & i \leqslant j, \end{cases}$$

$$\boldsymbol{B} := \left[\boldsymbol{B}_1^{\mathrm{T}}, \boldsymbol{B}_2^{\mathrm{T}}\right]^{\mathrm{T}} \in \mathbb{R}^{n \times n},$$

$$\boldsymbol{B}_i := \left[\boldsymbol{b}_{i1}^{\mathrm{T}}, \boldsymbol{b}_{i2}^{\mathrm{T}}, \cdots, \boldsymbol{b}_{in_i}^{\mathrm{T}}\right]^{\mathrm{T}} \in \mathbb{R}^{n_i \times n}, \ \boldsymbol{b}_{ik} \in \mathbb{R}^{1 \times n},$$

$$\boldsymbol{F} := \left[\boldsymbol{F}_1^{\mathrm{T}}, \boldsymbol{F}_2^{\mathrm{T}}\right]^{\mathrm{T}} \in \mathbb{R}^{n \times 2},$$

$$\boldsymbol{F}_i := \left[\boldsymbol{f}_{i1}^{\mathrm{T}}, \boldsymbol{f}_{i2}^{\mathrm{T}}, \cdots, \boldsymbol{f}_{in_i}^{\mathrm{T}}\right]^{\mathrm{T}} \in \mathbb{R}^{n_i \times 2}, \boldsymbol{f}_{ik} \in \mathbb{R}^{1 \times 2},$$

$$\mathbf{C} := \begin{bmatrix} \boldsymbol{e}_1^{\mathrm{T}} & 0 \\ 0 & \boldsymbol{e}_2^{\mathrm{T}} \end{bmatrix} \in \mathbb{R}^{2 \times n}, \ \boldsymbol{e}_i = [1, 0, \cdots, 0]^{\mathrm{T}} \in \mathbb{R}^{n_i}.$$

Here, $n_i \geq 1$ are the observability indices, satisfying $n_1 + n_2 = n$. Let $x(t) = [x_1^{\mathrm{T}}(t), x_2^{\mathrm{T}}(t)]^{\mathrm{T}} \in R^n$, $x_i(t) \in R^{n_i}$.

The models (4.21) and (4.22) are rewritten as

$$\begin{bmatrix} \boldsymbol{x}_1(t+1) \\ \boldsymbol{x}_2(t+1) \end{bmatrix} = \begin{bmatrix} \boldsymbol{A}_1 & \boldsymbol{A}_{12} \\ \boldsymbol{A}_{21} & \boldsymbol{A}_2 \end{bmatrix} \begin{bmatrix} \boldsymbol{x}_1(t) \\ \boldsymbol{x}_2(t) \end{bmatrix} + \begin{bmatrix} \boldsymbol{B}_1 \\ \boldsymbol{B}_2 \end{bmatrix} x(t-d) + \begin{bmatrix} \boldsymbol{F}_1 \\ \boldsymbol{F}_2 \end{bmatrix} \boldsymbol{u}(t), \tag{4.23}$$

$$\begin{bmatrix} y_1(t) \\ y_2(t) \end{bmatrix} = \begin{bmatrix} \boldsymbol{e}_1^{\mathrm{T}} & \\ & \boldsymbol{e}_2^{\mathrm{T}} \end{bmatrix} \begin{bmatrix} \boldsymbol{x}_1(t) \\ \boldsymbol{x}_2(t) \end{bmatrix} + \begin{bmatrix} v_1(t) \\ v_2(t) \end{bmatrix}. \tag{4.24}$$

In this chapter, it is a multivariable system, here the two-input and two-output model is rewritten as two dual-input single-output subsystems with few dimensions and few variables, and each subsystem is identified. Then, the model in (4.23) and (4.24) gives

$$\boldsymbol{x}_1(t+1) = \boldsymbol{A}_1\boldsymbol{x}_1(t) + \boldsymbol{A}_{12}\boldsymbol{x}_2(t) + \boldsymbol{B}_1\boldsymbol{x}(t-d) + \boldsymbol{F}_1\boldsymbol{u}(t), \tag{4.25}$$

$$y_1(t) = \boldsymbol{e}_1^{\mathrm{T}}\boldsymbol{x}_1(t) + v_1(t), \tag{4.26}$$

$$\boldsymbol{x}_2(t+1) = \boldsymbol{A}_2\boldsymbol{x}_2(t) + \boldsymbol{A}_{21}\boldsymbol{x}_1(t) + \boldsymbol{B}_2\boldsymbol{x}(t-d) + \boldsymbol{F}_2\boldsymbol{u}(t), \tag{4.27}$$

$$y_2(t) = \boldsymbol{e}_2^{\mathrm{T}}\boldsymbol{x}_2(t) + v_2(t). \tag{4.28}$$

Because the models (4.23) and (4.24) contain the unknown parameter vectors/matrices of the system and unmeasurable state vectors, which is the difficulty of identification, the idea of this paper is to replace the state vector with a measurable input and output. First analyze the first subsystem, and from (4.25) and (4.26), we have

$$\begin{aligned} y_1(t+i) = {} & \boldsymbol{e}_1^{\mathrm{T}}\boldsymbol{A}_1^{i}\boldsymbol{x}_1(t) + \boldsymbol{e}_1^{\mathrm{T}}\boldsymbol{A}_1^{i-1}\boldsymbol{A}_{12}\boldsymbol{x}_2(t) + \boldsymbol{e}_1^{\mathrm{T}}\boldsymbol{A}_1^{i-2}\boldsymbol{A}_{12}\boldsymbol{x}_2(t+1) + \cdots \\ & + \boldsymbol{e}_1^{\mathrm{T}}\boldsymbol{A}_{12}\boldsymbol{x}_2(t+i-1) + \boldsymbol{e}_1^{\mathrm{T}}\boldsymbol{A}_1^{i-1}\boldsymbol{B}_1\boldsymbol{x}(t-d) + \boldsymbol{e}_1^{T}\boldsymbol{A}_1^{i-2}\boldsymbol{B}_1\boldsymbol{x}(t-d+1) \\ & + \cdots + \boldsymbol{e}_1^{\mathrm{T}}\boldsymbol{B}_1\boldsymbol{x}(t-d+i-1) + \boldsymbol{e}_1^{\mathrm{T}}\boldsymbol{A}_1^{i-1}\boldsymbol{F}_1\boldsymbol{u}(t) + \cdots \\ & + \boldsymbol{e}_1^{\mathrm{T}}\boldsymbol{F}_1\boldsymbol{u}(t+i-1) + v_1(t+i), \quad i = 0, 1, \ldots, n_1 - 1, \end{aligned} \tag{4.29}$$

$$\begin{aligned} y_i(t+n_1) = {} & \boldsymbol{e}_1^{\mathrm{T}}\boldsymbol{A}_1^{n1}\boldsymbol{x}_1(t) + \boldsymbol{e}_1^{\mathrm{T}}\boldsymbol{A}_1^{n_1-1}\boldsymbol{A}_{12}\boldsymbol{x}_2(t) + \boldsymbol{e}_1^{\mathrm{T}}\boldsymbol{A}_1^{n_1-2}\boldsymbol{A}_{12}\boldsymbol{x}_2(t+1) + \cdots \\ & + \boldsymbol{e}_1^{\mathrm{T}}\boldsymbol{A}_{12}\boldsymbol{x}_2(t+n_1-1) + \boldsymbol{e}_1^{\mathrm{T}}\boldsymbol{A}_1^{n_1-1}\boldsymbol{B}_1\boldsymbol{x}(t-d) \\ & + \boldsymbol{e}_1^{\mathrm{T}}\boldsymbol{A}_1^{n_1-2}\boldsymbol{B}_1\boldsymbol{x}(t-d+1) + \cdots + \boldsymbol{e}_1^{\mathrm{T}}\boldsymbol{B}_1\boldsymbol{x}(t-d+n_1-1) \\ & + \boldsymbol{e}_1^{\mathrm{T}}\boldsymbol{A}_1^{n_1-1}\boldsymbol{F}_1\boldsymbol{u}(t) + \boldsymbol{e}_1^{\mathrm{T}}\boldsymbol{A}_1^{n_1-2}\boldsymbol{F}_1\boldsymbol{u}(t+1) + \cdots \\ & + \boldsymbol{e}_1^{\mathrm{T}}\boldsymbol{F}_1\boldsymbol{u}(t+n_1-1) + v_1(t+n_1). \end{aligned} \tag{4.30}$$

Since the models (4.21) and (4.22) are in the observable canonical form, the decomposed subsystem is still the observable canonical model. According to the special structure of the matrix $A_1$ and $e_1^{\mathrm{T}}$, it is known that the observability matrix T is an identity matrix:

$$\boldsymbol{T}=\begin{bmatrix} \boldsymbol{e}_1^{\mathrm{T}} \\ \boldsymbol{e}_1^{\mathrm{T}}\boldsymbol{A}_1 \\ \vdots \\ \boldsymbol{e}_1^{\mathrm{T}}\boldsymbol{A}_1^{n_1-1} \end{bmatrix}=\boldsymbol{I}_{n_1}. \tag{4.31}$$

Postmultiplying both sides of the above equation by the matrix $A_{12}$ yields:

$$\begin{bmatrix} \boldsymbol{e}_1^{\mathrm{T}}\boldsymbol{A}_{12} \\ \boldsymbol{e}_1^{\mathrm{T}}\boldsymbol{A}_1\boldsymbol{A}_{12} \\ \vdots \\ \boldsymbol{e}_1^{\mathrm{T}}\boldsymbol{A}_1^{n_1-1}\boldsymbol{A}_{12} \end{bmatrix}=\boldsymbol{A}_{12}. \tag{4.32}$$

Observing the structure of matrix $A_{12}$, we get

$$\boldsymbol{e}_1^{\mathrm{T}}\boldsymbol{A}_1^{k-1}\boldsymbol{A}_{12}=0, \quad k=1,2,\ldots,n_1-1.$$

Observing Eq. (4.32), $e_1^{\mathrm{T}}A_1^{k-1}A_{12}$ is the last row of matrix $A_{12}$, which is

$$\boldsymbol{e}_1^{\mathrm{T}}\boldsymbol{A}_1^{n_1-1}\boldsymbol{A}_{12}=[a_{12}(1),\ a_{12}(2),\ldots,a_{12}(n_{12}),0,\ldots,0].$$

Define

$$\begin{aligned}
\boldsymbol{Y}_i(t+n_1) &:= [y_i(t),\ y_i(t+1),\ldots,y_i(t+n_i-1)]^{\mathrm{T}} \quad \in\mathbb{R}^{n_i},\\
\boldsymbol{U}_i(t+n_1) &:= \left[\boldsymbol{u}^{\mathrm{T}}(t),\boldsymbol{u}^{\mathrm{T}}(t+1),\ldots,\boldsymbol{u}^{\mathrm{T}}(t+n_i-1)\right]^{\mathrm{T}} \quad \in\mathbb{R}^{2n_i},\\
\boldsymbol{X}(t-d+n_i) &:= \left[\boldsymbol{x}^{\mathrm{T}}(t-d),\boldsymbol{x}^{\mathrm{T}}(t-d+1),\ldots,\boldsymbol{x}^{\mathrm{T}}(t-d+n_i-1)\right]^{\mathrm{T}} \quad \in\mathbb{R}^{nn_i},\\
\boldsymbol{V}_i(t+n_i) &:= [v_i(t),\ v_i(t+1),\ldots,v_i(t+n_i-1)]^{\mathrm{T}} \quad \in\mathbb{R}^{n_i},
\end{aligned}$$

$$\boldsymbol{M}_i:=\begin{bmatrix} 0 & 0 & \cdots & 0 & 0 \\ \boldsymbol{e}_i^{\mathrm{T}}\boldsymbol{B}_i & 0 & \cdots & 0 & 0 \\ \boldsymbol{e}_i^{\mathrm{T}}\boldsymbol{A}_i\boldsymbol{B}_i & \boldsymbol{e}_i^{\mathrm{T}}\boldsymbol{B}_i & & \vdots & \vdots \\ \vdots & \vdots & & 0 & 0 \\ \boldsymbol{e}_i^{\mathrm{T}}\boldsymbol{A}_i^{n_i-2}\boldsymbol{B}_i & \boldsymbol{e}_i^{\mathrm{T}}\boldsymbol{A}_i^{n_i-3}\boldsymbol{B}_i & \cdots & \boldsymbol{e}_i^{\mathrm{T}}\boldsymbol{B}_i & 0 \end{bmatrix} \in\mathbb{R}^{n_i\times(nn_i)},$$

$$\boldsymbol{Q}_i:=\begin{bmatrix} 0 & 0 & \cdots & 0 & 0 \\ \boldsymbol{e}_i^{\mathrm{T}}\boldsymbol{F}_i & 0 & \cdots & 0 & 0 \\ \boldsymbol{e}_i^{\mathrm{T}}\boldsymbol{A}_i\boldsymbol{F}_i & \boldsymbol{e}_i^{\mathrm{T}}\boldsymbol{F}_i & & \vdots & \vdots \\ \vdots & \vdots & & 0 & 0 \\ \boldsymbol{e}_i^{\mathrm{T}}\boldsymbol{A}_i^{n_i-2}\boldsymbol{F}_i & \boldsymbol{e}_i^{\mathrm{T}}\boldsymbol{A}_i^{n_i-3}\boldsymbol{F}_i & \cdots & \boldsymbol{e}_i^{\mathrm{T}}\boldsymbol{F}_i & 0 \end{bmatrix} \in\mathbb{R}^{n_i\times(2n_i)}.$$

From Eqs. (4.29) and (4.30), we have

$$\boldsymbol{Y}_1(t+n_1)=\boldsymbol{T}\boldsymbol{x}_1(t)+\boldsymbol{M}_1\boldsymbol{X}(t-d+n_1)+\boldsymbol{Q}_1\boldsymbol{U}_1(t+n_1)+\boldsymbol{V}_1(t+n_1).$$

Combining the observable matrix in (4.31) and the above equation gives

$$\boldsymbol{x}_1(t) = \boldsymbol{Y}_1(t+n_1) - \boldsymbol{M}_1\boldsymbol{X}(t-d+n_1) - \mathbf{Q}_1\boldsymbol{U}_1(t+n_1) - \boldsymbol{V}_1(t+n_1). \tag{4.33}$$

For the second subsystem, (4.27) and (4.28) are rewritten as

$$\boldsymbol{x}_2(t) = \boldsymbol{Y}_2(t+n_2) - \boldsymbol{M}_2\boldsymbol{X}(t-d+n_2) - \mathbf{Q}_2\boldsymbol{U}_2(t+n_2) - \boldsymbol{V}_2(t+n_2). \tag{4.34}$$

For $n_{12} \leq n_1$ and $n_{12} \leq \min\{n_1, n_2\}$, the number of nonzero elements of $e_1^{\mathrm{T}} A_1^{n_1-1} A_{12} Q_2 \in R^{1\times(2n_2)}$ is $2n_{12}$, and now construct a vector associated with it:

$$l_1 := \begin{cases} \left[e_1^{\mathrm{T}}\boldsymbol{A}_1^{n_1-1}\boldsymbol{A}_{12}\mathbf{Q}_2, 0, \cdots, 0\right] \in \mathbb{R}^{1\times(2n_1)}, n_1 \geqslant n_2 \\ \left[e_1^{\mathrm{T}}\boldsymbol{A}_1^{n_1-1}\boldsymbol{A}_{12}\mathbf{Q}_2\right]\begin{bmatrix}\boldsymbol{I}_{2n1} \\ 0\end{bmatrix} \in \mathbb{R}^{1\times(2n_1)}, n_1 < n_2, \end{cases}$$

similarly,

$$\boldsymbol{h}_1 := \begin{cases} \left[e_1^{\mathrm{T}}\boldsymbol{A}_1^{n_1-1}\boldsymbol{A}_{12}\boldsymbol{M}_2, 0, \cdots, 0\right] \in \mathbb{R}^{1\times(nn_1)}, & n_1 \geqslant n_2 \\ \left[e_1^{\mathrm{T}}\boldsymbol{A}_1^{n_1-1}\boldsymbol{A}_{12}\boldsymbol{M}_2\right]\begin{bmatrix}\boldsymbol{I}_{2n_1} \\ 0\end{bmatrix} \in \mathbb{R}^{1\times(nn_1)}, & n_1 < n_2. \end{cases}$$

In order to get the parameter estimates, define the information set at time $t$ by $\varphi_1(t)$ and the parameter vector $\theta_1$ as

$$\begin{aligned}
\varphi_1(t+n_1) &:= \left[\varphi_{11}^{\mathrm{T}}(t+n_1), \boldsymbol{U}_1^{\mathrm{T}}(t+n_1), \boldsymbol{X}^{\mathrm{T}}(t-d+n_1)\right]^{\mathrm{T}} \in \mathbb{R}^{n+2n_1+nn_1}, \\
\varphi_{11}(t+n_1) &:= \begin{bmatrix}\boldsymbol{Y}_1(t+n_1) - \boldsymbol{V}_1(t+n_1) \\ \boldsymbol{Y}_2(t+n_2) - \boldsymbol{V}_2(t+n_2)\end{bmatrix} \in \mathbb{R}^n, \\
\boldsymbol{\theta}_1 &:= \left[\boldsymbol{\theta}_{11}^{\mathrm{T}}, \boldsymbol{\theta}_{12}^{\mathrm{T}}, \boldsymbol{\theta}_{13}^{\mathrm{T}}\right]^{\mathrm{T}} \in \mathbb{R}^{n+2n_1+nn_1},
\end{aligned}$$

$$\boldsymbol{\theta}_{11} := \left[e_1^{\mathrm{T}}\boldsymbol{A}_1^{n_1}, e_1^{\mathrm{T}}\boldsymbol{A}_1^{n_1-1}\boldsymbol{A}_{12}\right]^{\mathrm{T}} \in \mathbb{R}^n, \tag{4.35}$$

$$\boldsymbol{\theta}_{12} := \left[-e_1^{\mathrm{T}}\boldsymbol{A}_1^{n_1}\mathbf{Q}_1 - l_1 + \left[e_1^{\mathrm{T}}\boldsymbol{A}_1^{n_1-1}\boldsymbol{F}_1, e_1^{\mathrm{T}}\boldsymbol{A}_1^{n_1-2}\boldsymbol{F}_1, \cdots, e_1^{\mathrm{T}}\boldsymbol{F}_1\right]\right]^{\mathrm{T}} \in \mathbb{R}^{2n_1}, \tag{4.36}$$

$$\boldsymbol{\theta}_{13} := \left[-e_1^{\mathrm{T}}\boldsymbol{A}_1^{n_1}\boldsymbol{M}_1 - \boldsymbol{h}_1 + \left[e_1^{\mathrm{T}}\boldsymbol{A}_1^{n_1-1}\boldsymbol{B}_1, e_1^{\mathrm{T}}\boldsymbol{A}_1^{n_1-2}\boldsymbol{B}_1, \cdots, e_1^{\mathrm{T}}\boldsymbol{B}_1\right]\right]^{\mathrm{T}} \in \mathbb{R}^{nn_1}. \tag{4.37}$$

Substituting the state variable in (4.33) and (4.34) into (4.30) gives

$$\begin{aligned}
y_1(t+n_1) &= \boldsymbol{e}_1^{\mathrm{T}}\boldsymbol{A}_1^{n_1}\boldsymbol{x}_1(t) + \boldsymbol{e}_1^{\mathrm{T}}\boldsymbol{A}_1^{n_1-1}\boldsymbol{A}_{12}\boldsymbol{x}_2(t) \\
&\quad + \boldsymbol{e}_1^{\mathrm{T}}\boldsymbol{A}_1^{n_1-1}\boldsymbol{B}_1\boldsymbol{x}(t-d) + \boldsymbol{e}_1^{\mathrm{T}}\boldsymbol{A}_1^{n_1-2}\boldsymbol{B}_1\boldsymbol{x}(t-d+1) + \cdots + \boldsymbol{e}_1^{\mathrm{T}}\boldsymbol{B}_1\boldsymbol{x}(t-d+n_1-1) \\
&\quad + \boldsymbol{e}_1^{\mathrm{T}}\boldsymbol{A}_1^{n_1-1}\boldsymbol{F}_1\boldsymbol{u}(t) + \boldsymbol{e}_1^{\mathrm{T}}\boldsymbol{A}_1^{n_1-2}\boldsymbol{F}_1\boldsymbol{u}(t+1) + \cdots + \boldsymbol{e}_1^{\mathrm{T}}\boldsymbol{F}_1\boldsymbol{u}(t+n_1-1) + v_1(t+n_1) \\
&= \left[\boldsymbol{\varphi}_{11}^{\mathrm{T}}(t+n_1)\boldsymbol{\theta}_{11}\boldsymbol{U}_1^{\mathrm{T}}(t+n_1)\boldsymbol{\theta}_{12} + \boldsymbol{X}^{\mathrm{T}}(t-d+n_1)\boldsymbol{\theta}_{13} + v_1(t+n_1)\right] \\
&= \left[\boldsymbol{\varphi}_{11}^{\mathrm{T}}(t+n_1),\ \boldsymbol{U}_1^{\mathrm{T}}(t+n_1), \boldsymbol{X}^{\mathrm{T}}(t-d+n_1)\right]\begin{bmatrix}\boldsymbol{\theta}_{11}\\ \boldsymbol{\theta}_{12}\\ \boldsymbol{\theta}_{13}\end{bmatrix} + v_1(t+n_1) \\
&= \boldsymbol{\varphi}_1^{\mathrm{T}}(t+n_1)\boldsymbol{\theta}_1 + v_1(t+n_1).
\end{aligned} \tag{4.38}$$

Replacing $t$ in (4.38) with $t-n_1$ can be simplified as the following regression model,

$$y_1(t) = \boldsymbol{\varphi}_1^{\mathrm{T}}(t)\boldsymbol{\theta}_1 + v_1(t). \tag{4.39}$$

**Remark 1**: The above equation is the identification model of the two-input two-output state space system with time delay. For research convenience, assume $t$ is the current moment, $\{u(t), y(t){:}t = 0, 1, 2, \ldots\}$ is the measurable input–output information, $y(t)$ and $\varphi(t)$ are the current information, and $\{y(t-i), \varphi(t-i){:}i = 1, 2, \ldots, p-1\}$ are the past information.

### 4.2.2 The parameter estimation algorithm

According to the least squares theory for Eq. (4.39), minimizing the criterion function,

$$J_1(\boldsymbol{\theta}_1) := \sum_{k=1}^{t}\left[y_1(k) - \boldsymbol{\varphi}_1^{\mathrm{T}}(k)\boldsymbol{\theta}_1\right]^2.$$

We get the following recursive least square (RLS) algorithm to estimate the parameter vector $\theta_1$:

$$\hat{\boldsymbol{\theta}}_1(t) = \hat{\boldsymbol{\theta}}_1(t-1) + \boldsymbol{L}_1(t)\left[y_1(t) - \hat{\boldsymbol{\varphi}}_1^{\mathrm{T}}(t)\hat{\boldsymbol{\theta}}_1(t-1)\right], \tag{4.40}$$

$$\boldsymbol{L}_1(t) = \boldsymbol{P}_1(t)\boldsymbol{\varphi}_1(t) = \frac{\boldsymbol{P}_1(t-1)\boldsymbol{\varphi}_1(t)}{1 + \boldsymbol{\varphi}_1^{\mathrm{T}}(t)\boldsymbol{P}_1(t-1)\boldsymbol{\varphi}_1(t)}, \tag{4.41}$$

$$\boldsymbol{P}_1(t) = \boldsymbol{P}_1(t-1) - \frac{\boldsymbol{P}_1(t-1)\boldsymbol{\varphi}_1(t)\boldsymbol{\varphi}_1^{\mathrm{T}}(t)\boldsymbol{P}_1(t-1)}{1 + \boldsymbol{\varphi}_1^{\mathrm{T}}(t)\boldsymbol{P}_1(t-1)\boldsymbol{\varphi}_1(t)}, \tag{4.42}$$

where $L_1(t) \in R^{n+2n_1+nn_1}$ is the gain vector, and $P_1(t) \in R^{(n+2n_1+nn_1)\times(n+2n_1+nn_1)}$ is the covariance matrix.

**Remark 2**: For the information vector $\varphi_1(t)$ in (4.40) and (4.42) containing the unknown noise item $v_1(t-i)$ and the state vector $x(t-d-i)$, the above algorithm cannot realize, which is the difficulty in identification. This section adopts the basic idea of replacing the unknown noise item $v_1(t-i)$ and the state vector $x(t-d-i)$ in $\varphi_1(t)$ with the estimated residual $\hat{v}_1(t-i)$ and the estimated state vector $\hat{x}(t-d-i)$.

Use the estimates $\hat{v}_i(t)$ and $\hat{x}(t)$ of $v_i(t)$ and $x(t)$ to construct the estimates $\hat{V}_i(t)$ and $\hat{X}(t)$ of $V_i(t)$ and $X(t)$:

$$
\begin{aligned}
\hat{\varphi}_1(t) &:= \left[\hat{\varphi}_{11}^{\mathrm{T}}(t), \boldsymbol{U}_1^{\mathrm{T}}(t), \hat{\boldsymbol{X}}^{\mathrm{T}}(t-d)\right]^{\mathrm{T}} \in \mathbb{R}^{n+2n_1+nn_1}, \\
\hat{\varphi}_{11}(t) &:= \begin{bmatrix} \boldsymbol{Y}_1(t) - \hat{\boldsymbol{V}}_1(t) \\ \boldsymbol{Y}_2(t) - \hat{\boldsymbol{V}}_2(t) \end{bmatrix} \in \mathbb{R}^n, \\
\hat{\boldsymbol{X}}(t-d) &:= \left[\hat{\boldsymbol{x}}^{\mathrm{T}}(t-n_1-d),\ \hat{\boldsymbol{x}}^{\mathrm{T}}(t-n_1-d+1), \ldots, \hat{\boldsymbol{x}}(t-d-1)\right]^{\mathrm{T}} \in \mathbb{R}^{nn_1}, \\
\hat{\boldsymbol{V}}_1(t) &:= \left[\hat{v}_1^{\mathrm{T}}(t-n_1), \hat{v}_1^{\mathrm{T}}(t-n_1+1), \ldots, \hat{v}_1^{\mathrm{T}}(t-1)\right]^{\mathrm{T}} \in \mathbb{R}^{n_1}, \\
\hat{\boldsymbol{V}}_2(t) &:= \left[\hat{v}_2^{\mathrm{T}}(t-n_2), \hat{v}_2^{\mathrm{T}}(t-n_2+1), \ldots, \hat{v}_2^{\mathrm{T}}(t-1)\right]^{\mathrm{T}} \in \mathbb{R}^{n_2}.
\end{aligned}
$$

Define $\hat{\theta}_1(t) := [\hat{\theta}_{11}^{\mathrm{T}}(t), \hat{\theta}_{12}^{\mathrm{T}}(t), \hat{\theta}_{13}^{\mathrm{T}}(t)]^{\mathrm{T}}$ is the estimate of $\theta_1 = [\theta_{11}^{\mathrm{T}}, \theta_{12}^{\mathrm{T}}, \theta_{13}^{\mathrm{T}}]^{\mathrm{T}}$ at time $t$. According to Eq. (4.39), the estimate of $v_1(t)$ is calculated as $\hat{v}_1(t) = y_1(t) - \hat{\varphi}_1^{\mathrm{T}}(t)\hat{\theta}_1(t)$. Thus, replacing the unknown variable $\varphi_1(t)$ on the right-hand sides of algorithm (4.40)–(4.42) with its corresponding estimate $\hat{\varphi}_1(t)$ and replacing the unknown $\theta_1(t)$ with its estimate $\hat{\theta}_1(t-1)$ at the previous time $t-1$, we obtain the following parameter estimation-based recursive least squares algorithm to calculate $\theta_1$:

$$\hat{\boldsymbol{\theta}}_1(t) = \hat{\boldsymbol{\theta}}_1(t-1) + \boldsymbol{L}_1(t)\left[y_1(t) - \hat{\varphi}_1^{\mathrm{T}}(t)\hat{\boldsymbol{\theta}}_1(t-1)\right], \tag{4.43}$$

$$\boldsymbol{L}_1(t) = \boldsymbol{P}_1(t)\hat{\varphi}_1(t) = \frac{\boldsymbol{P}_1(t-1)\hat{\varphi}_1(t)}{1 + \hat{\varphi}_1^{\mathrm{T}}(t)\boldsymbol{P}_1(t-1)\hat{\varphi}_1(t),} \tag{4.44}$$

$$\boldsymbol{P}_1(t) = \left[\boldsymbol{I} - \boldsymbol{L}_1(t)\hat{\varphi}_1^{\mathrm{T}}(t)\right]\boldsymbol{P}_1(t-1),\ \boldsymbol{P}_1(0) = p_0\boldsymbol{I}, \tag{4.45}$$

$$\hat{v}_1(t) = y_1(t) - \hat{\varphi}_1^{\mathrm{T}}(t)\hat{\boldsymbol{\theta}}_1(t), \tag{4.46}$$

$$\hat{\varphi}_1(t) = [y_1(t-n_1) - \hat{v}_1(t-n_1), y_1(t-n_1+1) - \hat{v}_1(t-n_1+1), \cdots,\\ y_1(t-1) - \hat{v}_1(t-1) y_2(t-n_2) - \hat{v}_2(t-n_2), y_2(t-n_2+1)\\ - \hat{v}_2(t-n_2+1), \cdots, y_2(t-1) - \hat{v}_2(t-1), \hat{\boldsymbol{x}}(t-n_1-d),\\ \hat{\boldsymbol{x}}(t-n_1-d+1), \cdots, \hat{\boldsymbol{x}}(t-d-1), \boldsymbol{u}^{\mathrm{T}}(t-n_1), \cdots,\\ \boldsymbol{u}^{\mathrm{T}}(t-1)]. \tag{4.47}$$

Similar to the derivation process of the parameter vector $\theta_1$, the second subsystem is obtained as follows.

Define $l_2$ and $h_2$ as

$$l_2 := \begin{cases} \left[\boldsymbol{e}_2^{\mathrm{T}} \boldsymbol{A}_2^{n_2-1} \boldsymbol{A}_{21} \mathbf{Q}_1, 0, \cdots, 0\right] \in \mathbb{R}^{1 \times (2n_2)}, n_2 \geqslant n_1 \\ \left[\boldsymbol{e}_2^{T} \boldsymbol{A}_2^{n_2-1} \boldsymbol{A}_{21} \mathbf{Q}_1\right] \begin{bmatrix} \boldsymbol{I}_{2n1} \\ 0 \end{bmatrix} \in \mathbb{R}^{1 \times (2n_2)}, n_2 < n_1, \end{cases}$$

$$\boldsymbol{h}_2 := \begin{cases} \left[\boldsymbol{e}_2^{\mathrm{T}} \boldsymbol{A}_2^{n_1-1} \boldsymbol{A}_{21} \boldsymbol{M}_1, 0, \cdots, 0\right] \in \mathbb{R}^{1 \times (nn_2)}, n_2 \geqslant n_1 \\ \left[\boldsymbol{e}_2^{\mathrm{T}} \boldsymbol{A}_2^{n_1-1} \boldsymbol{A}_{21} \boldsymbol{M}_1\right] \begin{bmatrix} I_{2n_1} \\ 0 \end{bmatrix} \in \mathbb{R}^{1 \times (nn_2)}, n_2 \geqslant n_1. \end{cases}$$

According to the least squares principle, minimize the cost function. When computing one parameter vector, others are replaced with their estimates; then, we get the parameter estimation-based recursive least squares algorithm as follows:

$$\hat{\boldsymbol{\theta}}_2(t) = \hat{\boldsymbol{\theta}}_2(t-1) + \boldsymbol{L}_2(t)\left[y_2(t) - \hat{\varphi}_2^{\mathrm{T}}(t)\hat{\boldsymbol{\theta}}_2(t-1)\right], \tag{4.48}$$

$$\boldsymbol{L}_2(t) = \boldsymbol{P}_2(t)\hat{\varphi}_2(t) = \frac{\boldsymbol{P}_2(t-1)\hat{\varphi}_2(t)}{1 + \hat{\varphi}_2^{\mathrm{T}}(t)\boldsymbol{P}_2(t-1)\hat{\varphi}_2(t)}, \tag{4.49}$$

$$\boldsymbol{P}_2(t) = \left[\boldsymbol{I} - \boldsymbol{L}_2(t)\hat{\varphi}_2^{\mathrm{T}}(t)\right]\boldsymbol{P}_2(t-1),\ \boldsymbol{P}_2(0) = \boldsymbol{P}_2(0) = p_0\boldsymbol{I}, \tag{4.50}$$

$$\hat{v}_2(t) = y_2(t) - \hat{\varphi}_2^{\mathrm{T}}(t)\hat{\boldsymbol{\theta}}_2(t), \tag{4.51}$$

$$\hat{\varphi}_2(t) = [y_1(t-n_1) - \hat{v}_1(t-n_1), y_1(t-n_1+1) - \hat{v}_1(t-n_1+1), \cdots,\\ y_1(t-1) - \hat{v}_1(t-1), y_2(t-n_2) - \hat{v}_2(t-n_2), y_2(t-n_2+1)\\ - \hat{v}_2(t-n_2+1), \cdots, y_2(t-1) - \hat{v}_2(t-1), \hat{\boldsymbol{x}}(t-n_2-d),\\ \hat{\boldsymbol{x}}(t-n_2-d+1), \cdots, \hat{\boldsymbol{x}}(t-d-1),\\ \boldsymbol{u}^{\mathrm{T}}(t-n_2), \cdots, \boldsymbol{u}^{\mathrm{T}}(t-1)]^{\mathrm{T}}, \tag{4.52}$$

where $\hat{\theta}_2(t) := [\hat{\theta}_{21}^{\mathrm{T}}(t), \hat{\theta}_{22}^{\mathrm{T}}(t), \hat{\theta}_{23}^{\mathrm{T}}(t)]^{\mathrm{T}}$ is the estimate of $\theta_2 = [\theta_{21}^T, \theta_{22}^T, \theta_{23}^T]^T$ at time $t$, the gain matrix $L_2(t) \in R^{n+2n_2+nn_2}$, and the covariance matrix $P_2(t) \in R^{(n+2n_2+nn_2)\times(n+2n_2+nn_2)}$.

Eqs. (4.43)−(4.47) and (4.48)−(4.52) form the recursive least squares algorithm whose initial values $\hat{\theta}_1(0)$ and $\hat{\theta}_2(0)$ are taken as zero vectors of appropriate sizes, $\hat{v}_i(j), \hat{x}(j), \hat{\varphi}_1(j), \hat{\varphi}_2(j), u(j)$ and $y(j)$ as zero vectors or zero matrices of appropriate sizes for $j \leq 0$, and $P_1(0) = p_0 I, P_2(0) = p_0 I$, $p_0 = 10^6$, $I$ is an identity matrix of appropriate dimensions.

**Remark 3**: Since we consider a multivariable system, the coupling of the system needs to be analyzed during the decomposition, which is to realize the decoupling of the system: the multivariable system that makes the input and output are correlated to each other realizes that each output is only controlled by the corresponding input.

**Theorem 1**: For the system in (4.21) and (4.22), the identification model in (4.39), and the least squares algorithm in (4.43)−(4.52), suppose that $\{v(t)\}$ is a white noise sequence with zero mean and variance $\sigma^2$, i.e., $\mathrm{E}[v(t)] = 0, \mathrm{E}[v^2(t)] = \sigma^2, \mathrm{E}[v(t)v(s)] = 0, s \neq t$, the input $u(t)$ is deterministic, and the following least squares parameter estimate $\hat{\theta}_i$ is an unbiased estimate of $\theta_i$,

$$\begin{aligned}\mathrm{E}\left[\hat{\boldsymbol{\theta}}_i\right] &= \boldsymbol{\theta}_i + \left[\left[\sum_{k=1}^{t}\varphi_i(k)\varphi_i^{\mathrm{T}}(k)\right]^{-1}\sum_{k=1}^{t}\varphi_i(k)v(k)\right]\\ &= \boldsymbol{\theta}_i + \left[\sum_{k=1}^{t}\varphi_i(k)\varphi_i^{\mathrm{T}}(k)\right]^{-1}\sum_{k=1}^{t}\varphi_i(k)\mathrm{E}[v(k)]\\ &= \boldsymbol{\theta}_i.\end{aligned}$$

### 4.2.3 The state estimation algorithm

The relationship between the parameter vector $\theta_1$ and the matrices/vector $A_1, A_{12}, B_1, e_1^{\mathrm{T}}$ has been established, and postmultiplying the observable matrix in (4.31) on both sides by $B_1$ gives

$$\begin{bmatrix} e_1^{\mathrm{T}}\boldsymbol{B}_1 \\ e_1^{\mathrm{T}}\boldsymbol{A}_1\boldsymbol{B}_1 \\ \vdots \\ e_1^{\mathrm{T}}\boldsymbol{A}_1^{n_1-1}\boldsymbol{B}_1 \end{bmatrix} = \boldsymbol{B}_1. \tag{4.53}$$

From Eq. (4.53) and the definition of $B_1$, we have

$$e_1^{\mathrm{T}} A_1^{k-1} B_1 = b_{1k}, \quad k = 1, 2, \cdots, n_1.$$

Using the above equation, the matrix $M_1$ and the similar matrix $M_2$ are simplified as

$$M_i = \begin{bmatrix} 0 & 0 & \cdots & 0 & 0 \\ b_{i1} & 0 & \cdots & 0 & 0 \\ b_{i2} & b_{i1} & \ddots & \vdots & \vdots \\ \vdots & \vdots & \ddots & 0 & 0 \\ b_{i,n_i-1} & b_{i,n_i-2} & \cdots & b_{i1} & 0 \end{bmatrix}, \quad i = 1, 2.$$

Observing the structure of $A_{12}$, from the last line of the formula (4.32) gives

$$e_1^{\mathrm{T}} A_1^{n_1-1} A_{12} = [a_{12}(1), a_{12}(2), \ldots, a_{12}(n_{12}), 0, \ldots, 0], \tag{4.54}$$

$$e_1^{\mathrm{T}} A_1^{n_1} = [a_1(1), a_1(2), \ldots, a_1(n_1), 0, \ldots, 0]. \tag{4.55}$$

Postmultiplying Eq. (4.54) on both sides with the matrix $M_2$ and postmultiplying Eq. (4.55) on both sides with the matrix $M_1$, we have

$$\begin{aligned} e_1^{\mathrm{T}} A_1^{n_1-1} A_{12} M_2 = \big[ & a_{12}(2) b_{21} + a_{12}(3) b_{22} + \cdots + a_{12}(n_{12}) b_{2,(n_{12}-1)}, \\ & a_{12}(3) b_{21} + a_{12}(4) b_{22} + \cdots + a_{12}(n_{12}) b_{2,(n_{12}-2)}, \ldots, 0, \end{aligned} \tag{4.56}$$

$$\begin{aligned} e_1^{\mathrm{T}} A_1^{n_1} M_1 = \big[ & a_1(2) b_{11} + a_1(3) b_{12} + \cdots + a_1(n_1) b_{1,(n_1-1)}, \\ & a_1(3) b_{11} + a_1(4) b_{12} + \cdots + a_1(n_1) b_{1,(n_1-2)}, \ldots, \\ & a_1(n_1) b_{11}, 0]. \end{aligned} \tag{4.57}$$

From Eq. (4.56), we get that the number of nonzero elements of $e_1^T A_1^{n_1-1} A_{12}$ is $2n_{12}$; according to the above equations, the parameters of $\theta_{11}$ in (4.35), $\theta_{12}$ in (4.36), and $\theta_{13}$ in (4.37) are expressed as

$$\begin{aligned} \theta_{11} &= \left[ e_1^{\mathrm{T}} A_1^{n_1}, e_1^{\mathrm{T}} A_1^{n_1-1} A_{12} \right]^{\mathrm{T}} \\ &= [a_1(1), a_1(2), \ldots, a_1(n_1), a_{12}(1), a_{12}(2), \ldots, a_{12}(n_{12}), 0, \ldots, 0]^{\mathrm{T}}, \end{aligned} \tag{4.58}$$

$$\begin{aligned} \theta_{12} &= \left[ -e_1^{\mathrm{T}} A_1^{n_1} Q_1 - l_1 + \left[ e_1^{\mathrm{T}} A_1^{n_1-1} F_1, e_1^{\mathrm{T}} A_1^{n_1-2} F_1, \ldots, e_1^{\mathrm{T}} F_1 \right] \right] \\ &= \Big[ -a_1(2) f_{11} - a_1(3) f_{12} - \cdots - a_1(n_1) f_{1,(n_1-1)} + f_{1,n_1} - a_{12}(2) f_{21} \\ &\quad - a_{12}(3) f_{22} - \cdots - a_{12}(n_{12}) f_{2,(n_{12}-1)}, - a_1(3) f_{11} - a_1(4) f_{12} - \cdots \\ &\quad - a_1(n_1) f_{1,(n_1-2)} + f_{1,(n_1-1)} - a_{12}(3) f_{21} - a_{12}(4) f_{22} - \cdots \\ &\quad - a_{12}(n_{12}) f_{2,(n_{12}-2)}, \ldots, - a_1(n_1) f_{11} + f_{12}, f_{11} \Big]^{\mathrm{T}}, \end{aligned} \tag{4.59}$$

$$\begin{aligned}\boldsymbol{\theta}_{13} &= \left[-\boldsymbol{e}_1^{\mathrm{T}}\boldsymbol{A}_1^{n_1}\boldsymbol{M}_1 - \boldsymbol{h}_1 + \left[\boldsymbol{e}_1^{\mathrm{T}}\boldsymbol{A}_1^{n_1-1}\boldsymbol{B}_1, \boldsymbol{e}_1^{\mathrm{T}}\boldsymbol{A}_1^{n_1-2}\boldsymbol{B}_1, \cdots, \boldsymbol{e}_1^{\mathrm{T}}\boldsymbol{B}_1\right]\right]^{\mathrm{T}} \\ &= \left[-a_1(2)\boldsymbol{b}_{11} - a_1(3)\boldsymbol{b}_{12} - \cdots - a_1(n_1)\boldsymbol{b}_{1,(n_1-1)} + \boldsymbol{b}_{1,n_1} - a_{12}(2)\boldsymbol{b}_{21}\right. \\ &\quad - a_{12}(3)\boldsymbol{b}_{22} - \cdots - a_{12}(n_{12})\boldsymbol{b}_{2,(n_{12}-1)}, - a_1(3)\boldsymbol{b}_{11} - a_1(4)\boldsymbol{b}_{12} - \cdots \\ &\quad - a_1(n_1)\boldsymbol{b}_{1,(n_1-2)} + \boldsymbol{b}_{1,(n_1-1)} - a_{12}(3)\boldsymbol{b}_{21} - a_{12}(4)\boldsymbol{b}_{22} - \cdots \\ &\quad \left. - a_{12}(n_{12})\boldsymbol{b}_{2,(n_{12}-2)}, \ldots, -a_1(n_1)\boldsymbol{b}_{11} + \boldsymbol{b}_{12}, \boldsymbol{b}_{11}\right]^{\mathrm{T}}. \end{aligned} \tag{4.60}$$

Let $\theta_{12}^{\mathrm{T}} = :[g_1, g_2, \ldots, g_{n_1}] \in R^{1\times(2n_1)}$.

Eq. (4.59) is converted into the following form:

$$\left[\begin{array}{ccccc|cccc} -a_1(2) & -a_1(3) & \cdots & -a_1(n_1) & 1 & -a_{12}(2) & \cdots & -a_{12}(n_{12}) & 0 \\ -a_1(3) & -a_1(4) & \cdots & 1 & & -a_{12}(3) & \cdots & 0 & \\ \vdots & \vdots & \ddots & & & \vdots & & & \\ & & & & & -a_{12}(n_{12}) & 0 & & \\ -a_1(n_1) & 1 & & & & 0 & & & \\ 1 & & & & & 0 & & & \end{array}\right]$$

$$\boldsymbol{F} = \begin{bmatrix} \boldsymbol{g}_1 \\ \boldsymbol{g}_2 \\ \vdots \\ \boldsymbol{g}_{n_1-1} \\ \boldsymbol{g}_{n_1} \end{bmatrix}. \tag{4.61}$$

Define $\theta_{13}^{T} =: [r_1, r_2, \ldots, r_{n_1}] \in R^{1\times(nn_1)}$.

By arranging Eq. (4.60), the following matrix equation is obtained.

$$\left[\begin{array}{ccccc|cccc} -a_1(2) & -a_1(3) & \cdots & -a_1(n_1) & 1 & -a_{12}(2) & \cdots & -a_{12}(n_{12}) & 0 \\ -a_1(3) & -a_1(4) & \cdots & 1 & & -a_{12}(3) & \cdots & 0 & \\ \vdots & \vdots & \ddots & & & \vdots & & & \\ & & & & & -a_{12}(n_{12}) & 0 & & \\ -a_1(n_1) & 1 & & & & 0 & & & \\ 1 & & & & & 0 & & & \end{array}\right]$$

$$\boldsymbol{B} = \begin{bmatrix} \boldsymbol{r}_1 \\ \boldsymbol{r}_2 \\ \vdots \\ \boldsymbol{r}_{n_1-1} \\ \boldsymbol{r}_{n_1} \end{bmatrix}. \tag{4.62}$$

Similarly, the parameters $\theta_{21}$, $\theta_{22}$, and $\theta_{23}$ are expressed as

$$\begin{aligned}\boldsymbol{\theta}_{21} &= \left[\boldsymbol{e}_2^{\mathrm{T}}\boldsymbol{A}_2^{n_2-1}\boldsymbol{A}_{21}, \boldsymbol{e}_2^{\mathrm{T}}\boldsymbol{A}_2^{n_2}\right]^{\mathrm{T}} \\ &= [a_{21}(1), a_{21}(2), \ldots, a_{21}(n_{21}), 0, \ldots, 0, a_2(1), a_2(2), \ldots, a_2(n_2)]^{\mathrm{T}},\end{aligned}$$

$$\begin{aligned}\boldsymbol{\theta}_{22} &= \left[-\boldsymbol{e}_2^{\mathrm{T}}\boldsymbol{A}_2^{n_2}\boldsymbol{Q}_2 - \boldsymbol{l}_2 + \left[\boldsymbol{e}_2^{\mathrm{T}}\boldsymbol{A}_2^{n_2-1}\boldsymbol{F}_2, \boldsymbol{e}_2^{\mathrm{T}}\boldsymbol{A}_2^{n_2-2}\boldsymbol{F}_2, \cdots, \boldsymbol{e}_2^{\mathrm{T}}\boldsymbol{F}_2\right]\right]^{\mathrm{T}}, \\ &= \Big[-a_2(2)\boldsymbol{f}_{21} - a_2(3)\boldsymbol{f}_{22} - \cdots - a_2(n_2)\boldsymbol{f}_{2,(n_2-1)} + \boldsymbol{f}_{2,n_2} - a_{22}(2)\boldsymbol{f}_{11} \\ &\quad - a_{21}(3)\boldsymbol{f}_{12} - \cdots - a_{21}(n_{21})\boldsymbol{f}_{1,(n_{21}-1)}, - a_2(3)\boldsymbol{f}_{21} - a_2(4)\boldsymbol{f}_{22} - \cdots \\ &\quad - a_2(n_2)\boldsymbol{f}_{2,(n_2-2)} + \boldsymbol{f}_{2,(n_2-1)} - a_{21}(3)\boldsymbol{f}_{11} - a_{21}(4)\boldsymbol{f}_{12} - \cdots \\ &\quad - a_{21}(n_{21})\boldsymbol{f}_{1,(n_{21}-2)}, \ldots, - a_2(n_2)\boldsymbol{f}_{21} + \boldsymbol{f}_{22}, \boldsymbol{f}_{21}\Big]^{\mathrm{T}},\end{aligned} \tag{4.63}$$

$$\begin{aligned}\boldsymbol{\theta}_{23} &= \left[-\boldsymbol{e}_2^{\mathrm{T}}\boldsymbol{A}_2^{n_2}\boldsymbol{M}_2 - \boldsymbol{h}_2 + \left[\boldsymbol{e}_2^{\mathrm{T}}\boldsymbol{A}_2^{n_2-1}\boldsymbol{B}_2, \boldsymbol{e}_2^{\mathrm{T}}\boldsymbol{A}_2^{n_2-2}\boldsymbol{B}_2, \cdots, \boldsymbol{e}_2^{\mathrm{T}}\boldsymbol{B}_2\right]\right]^{\mathrm{T}} \\ &= \Big[-a_2(2)\boldsymbol{b}_{21} - a_2(3)\boldsymbol{b}_{22} - \cdots - a_2(n_2)\boldsymbol{b}_{2,(n_2-1)} + \boldsymbol{b}_{2,n_2} - a_{21}(2)\boldsymbol{b}_{11} \\ &\quad - a_{21}(3)\boldsymbol{b}_{12} - \cdots - a_{21}(n_{21})\boldsymbol{b}_{1,(n_{21}-1)}, - a_2(3)\boldsymbol{b}_{21} - a_2(4)\boldsymbol{b}_{22} \\ &\quad - \cdots - a_2(n_2)\boldsymbol{b}_{2,(n_2-2)} + \boldsymbol{b}_{2,(n_2-1)} - a_{21}(3)\boldsymbol{b}_{11} - a_{21}(4)\boldsymbol{b}_{12} - \cdots \\ &\quad - a_{21}(n_{21})\boldsymbol{b}_{1,(n_{21}-2)}, \ldots, - a_2(n_2)\boldsymbol{b}_{21} + \boldsymbol{b}_{22}, \boldsymbol{b}_{21}\Big]^{\mathrm{T}}.\end{aligned} \tag{4.64}$$

Let $\theta_{22}^{\mathrm{T}} =: [h_1, h_2, \ldots, h_{n_2}] \in R^{1 \times (2n_2)}$.
Using Eq. (4.63), the following equation is established.

$$\left[\begin{array}{cccc|ccccc} -a_{21}(2) & \cdots & -a_{21}(n_{21}) & 0 & -a_2(2) & -a_2(3) & \cdots & -a_2(n_2) & 1 \\ -a_{21}(3) & \cdots & 0 & & -a_2(3) & a_2(4) & \cdots & 1 & \\ \vdots & & & & \vdots & \cdot & & & \\ -a_{21}(n_{21}) & 0 & & & \cdot & \cdot & & & \\ 0 & & & & -a_2(n_2) & 1 & & & \\ 0 & & & & 1 & & & & \end{array}\right]$$

$$\boldsymbol{F} = \begin{bmatrix} \boldsymbol{h}_1 \\ \boldsymbol{h}_2 \\ \vdots \\ \boldsymbol{h}_{n_2-1} \\ \boldsymbol{h}_{n_2} \end{bmatrix}. \tag{4.65}$$

Let $\theta_{23}^{\mathrm{T}} =: \left[s_1, s_2, \ldots, s_{n_2}\right] \in R^{1 \times (nn_2)}$. Combining with Eq. (4.64), we have

$$\left[\begin{array}{cccc|ccccc} -a_{21}(2) & \cdots & -a_{21}(n_{21}) & 0 & -a_2(2) & -a_2(3) & \cdots & -a_2(n_2) & 1 \\ -a_{21}(3) & \cdots & 0 & & -a_2(3) & -a_2(4) & \cdots & 1 & \\ \vdots & & & & \vdots & . & & & \\ -a_{21}(n_{21}) & 0 & & & . & . & & & \\ 0 & & & & -a_2(n_2) & 1 & & & \\ 0 & & & & 1 & & & & \end{array}\right]$$

$$\boldsymbol{B} = \begin{bmatrix} \boldsymbol{s}_1 \\ \boldsymbol{s}_2 \\ \vdots \\ \boldsymbol{s}_{n_2-1} \\ \boldsymbol{s}_{n_2} \end{bmatrix}. \tag{4.66}$$

Combining Eqs. (4.61) and (4.65) gives

$$\left[\begin{array}{cccc|cccc} -a_1(2) & \cdots & -a_1(n_1) & 1 & -a_{12}(2) & \cdots & -a_{12}(n_{12}) & \mathbf{0} \\ -a_1(3) & \cdots & 1 & & -a_{12}(3) & \cdots & 0 & \\ . & & & & \vdots & & & \\ . & & & & -a_{12}(n_{12}) & 0 & & \\ -a_1(n_1) & & & & 0 & & & \\ 1 & & & & 0 & & & \\ \hline -a_{21}(2) & \cdots & -a_{21}(n_{21}) & 0 & -a_2(2) & \cdots & -a_2(n_2) & 1 \\ -a_{21}(3) & \cdots & 0 & & -a_2(3) & \cdots & 1 & \\ \vdots & & & & . & & & \\ -a_{21}(n_{21}) & 0 & & & . & & & \\ 0 & & & & -a_2(n_2) & & & \\ 0 & & & & 1 & & & \end{array}\right] \boldsymbol{F} = \begin{bmatrix} \boldsymbol{g}_1 \\ \boldsymbol{g}_2 \\ \vdots \\ \boldsymbol{g}_{n_1} \\ \boldsymbol{h}_1 \\ \boldsymbol{h}_2 \\ \vdots \\ \boldsymbol{h}_{n_2} \end{bmatrix}. \tag{4.67}$$

Similarly, combining Eqs. (4.62) and (4.66), we have

$$\left[\begin{array}{cccc|cccc} -a_1(2) & \cdots & -a_1(n_1) & 1 & -a_{12}(2) & \cdots & -a_{12}(n_{12}) & \mathbf{0} \\ -a_1(3) & \cdots & 1 & & -a_{12}(3) & \cdots & 0 & \\ . & & & & . & & & \\ . & & & & -a_{12}(n_{12}) & 0 & & \\ -a_1(n_1) & & & & 0 & & & \\ 1 & & & & 0 & & & \\ \hline -a_{21}(2) & \cdots & -a_{21}(n_{21}) & 0 & -a_2(2) & \cdots & -a_2(n_2) & 1 \\ -a_{21}(3) & \cdots & 0 & & -a_2(3) & \cdots & 1 & \\ . & & & & . & & & \\ -a_{21}(n_{21}) & 0 & & & . & & & \\ 0 & & & & -a_2(n_2) & & & \\ 0 & & & & 1 & & & \end{array}\right] \boldsymbol{B} = \begin{bmatrix} \boldsymbol{r}_1 \\ \boldsymbol{r}_2 \\ \vdots \\ \boldsymbol{r}_{n_1} \\ \boldsymbol{s}_1 \\ \boldsymbol{s}_2 \\ \vdots \\ \boldsymbol{s}_{n_2} \end{bmatrix}. \tag{4.68}$$

Define the large matrix:

$$
\boldsymbol{W} := \left[\begin{array}{cccc|cccc}
-a_1(2) & \cdots & -a_1(n_1) & 1 & -a_{12}(2) & \cdots & -a_{12}(n_{12}) & \mathbf{0} \\
-a_1(3) & \cdots & 1 & & -a_{12}(3) & \cdots & 0 & \\
\vdots & & & & \vdots & & & \\
\vdots & & & & -a_{12}(n_{12}) & 0 & & \\
-a_1(n_1) & & & & 0 & & & \\
1 & & & & 0 & & & \\
\hline
-a_{21}(2) & \cdots & -a_{21}(n_{21}) & 0 & -a_2(2) & \cdots & -a_2(n_2) & 1 \\
-a_{21}(3) & \cdots & 0 & & -a_2(3) & \cdots & 1 & \\
\vdots & & & & \vdots & & & \\
-a_{21}(n_{21}) & 0 & & & \vdots & & & \\
0 & & & & -a_2(n_2) & & &
\end{array}\right],
$$

$$
\boldsymbol{K} := [\boldsymbol{g}_1^{\mathrm{T}}, \boldsymbol{g}_2^{\mathrm{T}}, \ldots, \boldsymbol{g}_{n_1}^{\mathrm{T}}, \boldsymbol{h}_1^{\mathrm{T}}, \boldsymbol{h}_2^{\mathrm{T}}, \ldots, \boldsymbol{h}_{n_2}^{\mathrm{T}}]^{\mathrm{T}},
$$

$$
\boldsymbol{J} := [\boldsymbol{r}_1^{\mathrm{T}}, \boldsymbol{r}_2^{\mathrm{T}}, \ldots, \boldsymbol{r}_{n_1}^{\mathrm{T}}, \boldsymbol{s}_1^{\mathrm{T}}, \boldsymbol{s}_2^{\mathrm{T}}, \ldots, \boldsymbol{s}_{n_2}^{\mathrm{T}}]^{\mathrm{T}}.
$$

Eqs. (4.62) and (4.66) can be expressed as $WF = K$, $WB = J$. Applying the estimates $\hat{a}_i$, $\hat{F}(t)$ and $\hat{B}(t)$ to set up: $\hat{F}(t) = \hat{W}^{-1}(t)\hat{K}(t)$, $\hat{B}(t) = \hat{W}^{-1}(t)\hat{J}(t)$. Replacing $t-n$ in (4.33) and (4.34) to $t$ yields

$$
\boldsymbol{x}_i(t-n_i) = \boldsymbol{Y}_i(t) - \boldsymbol{M}_i\boldsymbol{x}(t-d) - \boldsymbol{Q}_i\boldsymbol{U}_i(t) - \boldsymbol{V}_i(t), \quad i = 1, 2.
$$

Using $\hat{M}_1, \hat{M}_2, \hat{Q}_1, \hat{Q}_2, \hat{V}_1$, and $\hat{V}_2$ to replace $M_1, M_2, Q_1, Q_2, V_1$, and $V_2$ in the above equations, we get the estimates of state vectors:

$$
\hat{\boldsymbol{x}}_i(t-n_i) = \boldsymbol{Y}_i(t) - \hat{\boldsymbol{M}}_i(t)\hat{\boldsymbol{x}}(t-d) - \hat{\boldsymbol{Q}}_i(t)\boldsymbol{U}_i(t) - \hat{\boldsymbol{V}}_i(t), \quad i = 1, 2.
$$

Under the known $\hat{\theta}_1(t)$ and $\hat{\theta}_2(t)$, according to the least squares principle, the state estimation algorithm of the two-input two-output systems with time delay is summarized as follows:

$$
\hat{\boldsymbol{x}}_i(t-n_i) = \boldsymbol{Y}_i(t) - \hat{\boldsymbol{M}}_i(t)\hat{\boldsymbol{x}}(t-d) - \hat{\boldsymbol{Q}}_i(t)\boldsymbol{U}_i(t) - \hat{\boldsymbol{V}}_i(t), \tag{4.69}
$$

$$
\boldsymbol{Y}_i(t) = \left[y_i(t-n_i), y_i(t-n_i+1), \ldots, y_i(t-1)\right]^{\mathrm{T}}, \tag{4.70}
$$

$$
\boldsymbol{U}_i(t) = \left[\boldsymbol{u}^{\mathrm{T}}(t-n_i), \boldsymbol{u}^{\mathrm{T}}(t-n_i+1), \ldots, \boldsymbol{u}^{\mathrm{T}}(t-1)\right]^{\mathrm{T}}, \tag{4.71}
$$

$$
\hat{\boldsymbol{V}}_i(t) = [\hat{v}_i(t-n_i), \hat{v}_i(t-n_i+1), \ldots, \hat{v}_i(t-1)]^{\mathrm{T}}, \tag{4.72}
$$

$$
\hat{\boldsymbol{X}}(t-d+n_i) = \left[\hat{\boldsymbol{x}}^{\mathrm{T}}(t-d), \hat{\boldsymbol{x}}^{\mathrm{T}}(t-d+1), \ldots, \hat{\boldsymbol{x}}^{\mathrm{T}}(t-d+n_i-1)\right]^{\mathrm{T}}, \tag{4.73}
$$

$$\hat{\boldsymbol{M}}_i(t) = \begin{bmatrix} 0 & \cdots & 0 & 0 \\ \hat{\boldsymbol{b}}_{i1}(t) & \cdots & 0 & 0 \\ \hat{\boldsymbol{b}}_{i2}(t) & \ddots & \vdots & \vdots \\ \vdots & \ddots & 0 & 0 \\ \hat{\boldsymbol{b}}_{i,n_i-1}(t) & \cdots & \hat{\boldsymbol{b}}_{i1}(t) & 0 \end{bmatrix}, \tag{4.74}$$

$$\hat{\boldsymbol{Q}}_i(t) = \begin{bmatrix} 0 & \cdots & 0 & 0 \\ \hat{\boldsymbol{f}}_{i1}(t) & \cdots & 0 & 0 \\ \hat{\boldsymbol{f}}_{i2}(t) & \ddots & \vdots & \vdots \\ \vdots & \ddots & 0 & 0 \\ \hat{\boldsymbol{f}}_{i,n_i-1}(t) & \cdots & \hat{\boldsymbol{f}}_{i1}(t) & 0 \end{bmatrix}, \quad i = 1, 2, \tag{4.75}$$

$$\hat{\boldsymbol{F}}(t) = \hat{\boldsymbol{W}}^{-1}(t)\hat{\boldsymbol{K}}(t), \tag{4.76}$$

$$\hat{\boldsymbol{B}}(t) = \hat{\boldsymbol{W}}^{-1}(t)\hat{\boldsymbol{J}}(t), \tag{4.77}$$

$$\hat{\boldsymbol{\theta}}_1(t) = \begin{bmatrix} \hat{\boldsymbol{\theta}}_{11}(t) \\ \hat{\boldsymbol{\theta}}_{12}(t) \\ \hat{\boldsymbol{\theta}}_{13}(t) \end{bmatrix}, \qquad \hat{\boldsymbol{\theta}}_1(t) = \begin{bmatrix} \hat{\boldsymbol{\theta}}_{21}(t) \\ \hat{\boldsymbol{\theta}}_{22}(t) \\ \hat{\boldsymbol{\theta}}_{23}(t) \end{bmatrix}, \tag{4.78}$$

$$\hat{\boldsymbol{\theta}}_{11}(t) = [\hat{a}_1(1),\ \hat{a}_1(2), \ldots, \hat{a}_1(n_1), \hat{a}_{12}(1), \hat{a}_{12}(2), \ldots, \hat{a}_{12}(n_{12}), 0, \ldots, 0]^{\mathrm{T}}, \tag{4.79}$$

$$\hat{\boldsymbol{\theta}}_{12}(t) = \left[\hat{\boldsymbol{g}}_1^{\mathrm{T}}(t),\ \hat{\boldsymbol{g}}_2^{\mathrm{T}}(t), \ldots, \hat{\boldsymbol{g}}_{n_1}^{\mathrm{T}}(t)\right]^{\mathrm{T}}, \tag{4.80}$$

$$\hat{\boldsymbol{\theta}}_{13}(t) = \left[\hat{\boldsymbol{r}}_1^{\mathrm{T}}(t),\ \hat{\boldsymbol{r}}_2^{\mathrm{T}}(t), \ldots, \hat{\boldsymbol{r}}_{n_1}^{\mathrm{T}}(t)\right]^{\mathrm{T}}, \tag{4.81}$$

$$\hat{\boldsymbol{\theta}}_{21}(t) = [\hat{a}_{21}(1),\ \hat{a}_{21}(2), \ldots, \hat{a}_{21}(n_{21}), 0, \ldots, 0, \hat{a}_2(1),\ \hat{a}_2(2), \ldots, \hat{a}_2(n_2)]^{\mathrm{T}}, \tag{4.82}$$

$$\hat{\boldsymbol{\theta}}_{22}(t) = \left[\hat{\boldsymbol{h}}_1^{\mathrm{T}}(t),\ \hat{\boldsymbol{h}}_2^{\mathrm{T}}(t), \ldots, \hat{\boldsymbol{h}}_{n_2}^{\mathrm{T}}(t),\right]^{\mathrm{T}}, \tag{4.83}$$

$$\hat{\boldsymbol{\theta}}_{23}(t) = \left[\hat{\boldsymbol{s}}_1^{\mathrm{T}}(t),\ \hat{\boldsymbol{s}}_2^{\mathrm{T}}(t), \ldots, \hat{\boldsymbol{s}}_{n_2}^{\mathrm{T}}(t),\right]^{\mathrm{T}}, \tag{4.84}$$

$$\hat{\boldsymbol{W}}(t)=\left[\begin{array}{cccc|cccc}
-\hat{a}_1(2) & \cdots & -\hat{a}_1(n_1) & 1 & -\hat{a}_{12}(2) & \cdots & -\hat{a}_{12}(n_{12}) & \mathbf{0} \\
-\hat{a}_1(3) & \cdots & 1 & & -\hat{a}_{12}(3) & \cdots & 0 & \\
\vdots & & & & \vdots & & & \\
 & & & & -\hat{a}_{12}(n_{12}) & 0 & & \\
-\hat{a}_1(n_1) & & & & 0 & & & \\
1 & & & & 0 & & & \\
\hline
-\hat{a}_{21}(2) & \cdots & -\hat{a}_{21}(n_{21}) & 0 & -\hat{a}_2(2) & \cdots & -\hat{a}_2(n_2) & 1 \\
-\hat{a}_{21}(3) & \cdots & 0 & & -\hat{a}_2(3) & \cdots & 1 & \\
\vdots & & & & \vdots & & & \\
-\hat{a}_{21}(n_{21}) & 0 & & & & & & \\
0 & & & & -\hat{a}_2(n_2) & & &
\end{array}\right], \tag{4.85}$$

$$\hat{\boldsymbol{K}}(t)=\left[\hat{\boldsymbol{g}}_1^{\mathrm{T}}(t),\ \hat{\boldsymbol{g}}_2^{\mathrm{T}}(t),\ldots,\hat{\boldsymbol{g}}_{n_1}^{\mathrm{T}}(t),\ \hat{\boldsymbol{h}}_1^{\mathrm{T}}(t),\ldots,\hat{\boldsymbol{h}}_{n_2}^{\mathrm{T}}(t)\right]^{\mathrm{T}}, \tag{4.86}$$

$$\hat{\boldsymbol{J}}(t)=\left[\hat{\boldsymbol{r}}_1^{\mathrm{T}}(t),\ \hat{\boldsymbol{r}}_2^{\mathrm{T}}(t),\ldots,\hat{\boldsymbol{r}}_{n_1}^{\mathrm{T}}(t),\hat{\boldsymbol{s}}_1^{\mathrm{T}}(t),\hat{\boldsymbol{s}}_2^{\mathrm{T}}(t),\ldots,\hat{\boldsymbol{s}}_{n_2}^{\mathrm{T}}(t)\right]^{\mathrm{T}}. \tag{4.87}$$

The main steps of the proposed algorithm are summarized as follows.

1. Input and output: $u(t), y(t)$.

   Form the information vector $\hat{\varphi}_1(t)$ and $\hat{\varphi}_2(t)$, and further get the output $Y_1(t)$, $Y_2(t)$ and the input $U_1(t)$, $U_2(t)$.
2. Initialization

   Let $t=1$, and set the initial values $\hat{\theta}_1(0)=1_n/p_0, \hat{\theta}_2(0)=1_n/p_0$, the initial covariance matrix $P_1(0)=p_0I, P_2(0)=p_0I, p_0=10^6, u(t)=0$, $y(t)=0$ and $\hat{v}(t)=1/p_0$ for $t\leq 0$.
3. Compute the parameter estimates

   Calculate the gain vector $L_1(t), L_2(t)$ and the covariance matrix $P_1(t), P_2(t)$.

   Update the parameter estimate $\hat{\theta}_1(t)$ and $\hat{\theta}_2(t)$.

   Compute $\hat{v}_1(t)$, $\hat{v}_2(t)$, and form $\hat{V}_1(t)$, $\hat{V}_2(t)$.

   If $||\hat{\theta}_1(t)-\hat{\theta}_1(t-1)||\leq\varepsilon, ||\hat{\theta}_2(t)-\hat{\theta}_2(t-1)||\leq\varepsilon, \varepsilon>0$, then terminate the procedure and obtain the estimates $\hat{\theta}_1(t)$ and $\hat{\theta}_2(t)$; otherwise, increase $t$ by 1 and go to Step 1.
4. Compute the states

   Compute $\hat{a}_i(t)$, $\hat{b}_i(t)$,$\hat{f}_i(t)$, and form $\hat{M}_i(t)$, $\hat{Q}_i(t)$.

   Compute $\hat{x}_i(t-n_i)$.

**Remark 4**: According to the state equation at different time $t$, the state vector is represented by measurable input and output variables, and the identification model of the system is derived. Then, the single-input single-output model algorithm is generalized, and its corresponding residual-based augmented least squares algorithm is derived. The estimated parameters are used to identify system status. The proposed algorithm is computationally intensive and highly accurate.

### 4.2.4 Example

Considering the two-input two-output state space system with two-step state delay, the parameters are

$$\boldsymbol{A}=\begin{bmatrix} 0 & 1 & 0 & 0 \\ 0.32 & 0.49 & 0.55 & 0.01 \\ 0 & 0 & 0 & 1 \\ -0.40 & -0.86 & -0.60 & -0.65 \end{bmatrix},$$

$$\boldsymbol{B}=\begin{bmatrix} 0.20 & -0.25 & 0.01 & 0.01 \\ 0.10 & 0.10 & 0.23 & 0.14 \\ 0.30 & 0.01 & 0.20 & 0.10 \\ 0.10 & 0.50 & 0.10 & 0.20 \end{bmatrix}, \quad F=\begin{bmatrix} 0.1 & 5 \\ 0.5 & 1 \\ -1 & 0 \\ 1 & 1 \end{bmatrix}, \quad C=\begin{bmatrix} 1 & 0 & 0 & 0 \\ 0 & 0 & 1 & 0 \end{bmatrix}.$$

In simulation, the input $\{u(t)\}$ is generated from uniform distribution and is taken as an uncorrelated persistent excitation signal sequence with zero mean and unit variance and $\{v(t)\}$ as a white noise sequence is generated from Gaussian distribution with zero mean and variances $\sigma^2 = 0.10^2$ and $\sigma^2 = 0.50^2$. Apply the estimation algorithm in (4.43)–(4.47) to estimate the parameter vector, using the estimated parameters and the state estimation algorithm in (4.69)–(4.87) to estimate the state vector. Simulation results for different noise variances are shown in Tables 4.5 and 4.6, the parameter estimation errors $\delta$ versus $t$ are shown in Figs. 4.5 and 4.6, where $\delta := ||\hat{\theta}(t) - \theta||/||\theta||$, and the state estimates and estimation errors versus $t$ are shown in Figs. 4.7–4.10 (solid line: state true value $x(t)$, dot line: state estimated value $\hat{x}(t)$).

From Tables 4.5 and 4.6 and Figs. 4.5–4.10, we can draw the following conclusions.

- The parameter estimation errors become smaller (in general) with the increasing of $t$.
- In the case of the same zero mean variance, the parameter estimation accuracy improves as the data length $t$ increases.
- The data converges faster when the noise variance is lower.
- The state estimates are close to their true values with $t$ increasing.

**Table 4.5** The parameter estimates and errors ($\sigma^2 = 0.10^2$).

| $t$ | 100 | 200 | 500 | 1000 | 2000 | 3000 |
|---|---|---|---|---|---|---|
| $\boldsymbol{\theta}_{11}(1) = 0.32000$ | 0.33323 | 0.33133 | 0.32130 | 0.32189 | 0.31938 | 0.32011 |
| $\boldsymbol{\theta}_{11}(2) = 0.49000$ | 0.55353 | 0.55688 | 0.56101 | 0.53621 | 0.52862 | 0.51793 |
| $\boldsymbol{\theta}_{11}(3) = 0.55000$ | 0.47000 | 0.47997 | 0.53305 | 0.53626 | 0.54508 | 0.54398 |
| $\boldsymbol{\theta}_{11}(4) = 0.01000$ | −0.10662 | −0.09320 | −0.04877 | −0.03429 | −0.01833 | −0.01297 |
| $\boldsymbol{\theta}_{12}(1) = 0.46100$ | 0.33646 | 0.34914 | 0.39761 | 0.41544 | 0.43096 | 0.43595 |
| $\boldsymbol{\theta}_{12}(2) = -1.45000$ | −1.78539 | −1.80924 | −1.81610 | −1.68662 | −1.64324 | −1.58898 |
| $\boldsymbol{\theta}_{12}(3) = 0.10000$ | 0.09481 | 0.09715 | 0.09261 | 0.09561 | 0.09820 | 0.09904 |
| $\boldsymbol{\theta}_{12}(4) = 5.00000$ | 5.01411 | 5.01028 | 5.00554 | 5.00143 | 5.00036 | 5.00066 |
| $\boldsymbol{\theta}_{13}(1) = -0.00100$ | −0.03347 | −0.02524 | −0.02297 | −0.01680 | −0.01204 | −0.00967 |
| $\boldsymbol{\theta}_{13}(2) = 0.22240$ | 0.27315 | 0.26559 | 0.23827 | 0.23503 | 0.22993 | 0.229,54 |
| $\boldsymbol{\theta}_{13}(3) = 0.22310$ | 0.24304 | 0.23747 | 0.22903 | 0.22507 | 0.22503 | 0.22468 |
| $\boldsymbol{\theta}_{13}(4) = 0.13410$ | 0.05497 | 0.07462 | 0.10181 | 0.11434 | 0.12335 | 0.12490 |
| $\boldsymbol{\theta}_{13}(5) = 0.20000$ | 0.11593 | 0.12433 | 0.14428 | 0.16035 | 0.17142 | 0.17731 |
| $\boldsymbol{\theta}_{13}(6) = -0.25000$ | −0.28450 | −0.29954 | −0.27371 | −0.26484 | −0.26212 | −0.25808 |
| $\boldsymbol{\theta}_{13}(7) = 0.01000$ | −0.00505 | −0.02683 | −0.01839 | −0.01375 | −0.00950 | −0.00370 |
| $\boldsymbol{\theta}_{13}(8) = 0.01000$ | −0.00178 | −0.00225 | −0.01584 | −0.01028 | −0.00560 | 0.00020 |
| $\delta_1(\%)$ | 7.79630 | 7.94004 | 7.38806 | 4.83661 | 3.88061 | 2.81674 |

| $t$ | **100** | **200** | **500** | **1000** | **2000** | **3000** |
|---|---|---|---|---|---|---|
| $\boldsymbol{\theta}_{11}(1) = -0.40000$ | −0.37980 | −0.41822 | −0.39763 | −0.39283 | −0.39233 | −0.39148 |
| $\boldsymbol{\theta}_{11}(2) = -0.86000$ | −0.80657 | −0.68193 | −0.78550 | −0.82997 | −0.84960 | −0.86341 |
| $\boldsymbol{\theta}_{11}(3) = -0.60000$ | −0.88774 | −0.67550 | −0.69762 | −0.69267 | −0.67007 | −0.66176 |
| $\boldsymbol{\theta}_{11}(4) = -0.65000$ | −0.89629 | −0.76787 | −0.75996 | −0.73791 | −0.71229 | −0.69996 |
| $\boldsymbol{\theta}_{12}(1) = 0.43600$ | 0.21078 | 0.31174 | 0.32703 | 0.34737 | 0.37277 | 0.38543 |
| $\boldsymbol{\theta}_{12}(2) = 5.30000$ | 5.01916 | 4.39057 | 4.91191 | 5.14152 | 5.24458 | 5.31549 |
| $\boldsymbol{\theta}_{12}(3) = -1.00000$ | −0.99814 | −0.99074 | −0.99320 | −0.99918 | −0.99897 | −0.99938 |
| $\boldsymbol{\theta}_{12}(4) = 0.00000$ | 0.00723 | −0.02473 | −0.01129 | −0.00755 | −0.00353 | −0.00085 |
| $\boldsymbol{\theta}_{13}(1) = 0.46700$ | 0.33712 | 0.31694 | 0.35822 | 0.39296 | 0.41551 | 0.42825 |
| $\boldsymbol{\theta}_{13}(2) = 0.29150$ | 0.36207 | 0.27650 | 0.28655 | 0.29273 | 0.29063 | 0.29260 |
| $\boldsymbol{\theta}_{13}(3) = 0.23860$ | 0.15888 | 0.11444 | 0.14511 | 0.17147 | 0.18971 | 0.20111 |
| $\boldsymbol{\theta}_{13}(4) = 0.27360$ | 0.17982 | 0.23558 | 0.23143 | 0.23078 | 0.24020 | 0.24426 |
| $\boldsymbol{\theta}_{13}(5) = 0.30000$ | 0.13147 | 0.14948 | 0.19789 | 0.23163 | 0.25606 | 0.26937 |
| $\boldsymbol{\theta}_{13}(6) = 0.01000$ | 0.09225 | 0.12605 | 0.11173 | 0.08131 | 0.06317 | 0.05272 |
| $\boldsymbol{\theta}_{13}(7) = 0.20000$ | 0.05266 | 0.04479 | 0.10487 | 0.13806 | 0.16192 | 0.17608 |
| $\boldsymbol{\theta}_{13}(8) = 0.10000$ | 0.13306 | 0.01138 | 0.05407 | 0.07719 | 0.08686 | 0.09461 |
| $\delta_2(\%)$ | 10.85096 | 17.82920 | 8.79871 | 4.91667 | 3.01024 | 2.27523 |

**Table 4.6** The parameter estimates and errors ($\sigma^2 = 0.50^2$).

| $t$ | 100 | 200 | 500 | 1000 | 2000 | 3000 |
|---|---|---|---|---|---|---|
| $\boldsymbol{\theta}_{11}(1) = 0.32000$ | 0.35824 | 0.35174 | 0.32679 | 0.33053 | 0.32039 | 0.32089 |
| $\boldsymbol{\theta}_{11}(2) = 0.49000$ | 0.56783 | 0.62383 | 0.63961 | 0.57088 | 0.57876 | 0.55134 |
| $\boldsymbol{\theta}_{11}(3) = 0.55000$ | 0.37275 | 0.41366 | 0.50710 | 0.49560 | 0.52760 | 0.53498 |
| $\boldsymbol{\theta}_{11}(4) = 0.01000$ | −0.23034 | −0.18695 | −0.11321 | −0.09020 | −0.05915 | −0.03801 |
| $\boldsymbol{\theta}_{12}(1) = 0.46100$ | 0.15183 | 0.18228 | 0.30506 | 0.35143 | 0.37912 | 0.39998 |
| $\boldsymbol{\theta}_{12}(2) = -1.45000$ | −1.89288 | −2.23143 | −2.26346 | −1.88623 | −1.89789 | −1.75371 |
| $\boldsymbol{\theta}_{12}(3) = 0.10000$ | 0.13195 | 0.14960 | 0.10786 | 0.10041 | 0.10083 | 0.10245 |
| $\boldsymbol{\theta}_{12}(4) = 5.00000$ | 5.06358 | 5.02870 | 5.00778 | 4.99859 | 4.99795 | 5.00106 |
| $\boldsymbol{\theta}_{13}(1) = -0.00100$ | −0.01936 | −0.04388 | −0.04573 | −0.02174 | −0.01536 | −0.00829 |
| $\boldsymbol{\theta}_{13}(2) = 0.22240$ | 0.39880 | 0.34557 | 0.28545 | 0.27757 | 0.25987 | 0.25313 |
| $\boldsymbol{\theta}_{13}(3) = 0.22310$ | 0.29878 | 0.25992 | 0.24702 | 0.24213 | 0.24361 | 0.24201 |
| $\boldsymbol{\theta}_{13}(4) = 0.13410$ | 0.04245 | 0.04720 | 0.09592 | 0.11043 | 0.12364 | 0.12610 |
| $\boldsymbol{\theta}_{13}(5) = 0.20000$ | 0.05028 | 0.05727 | 0.08788 | 0.12666 | 0.13952 | 0.15643 |
| $\boldsymbol{\theta}_{13}(6) = -0.25000$ | −0.40911 | −0.38944 | −0.32104 | −0.31158 | −0.30945 | −0.29278 |
| $\boldsymbol{\theta}_{13}(7) = 0.01000$ | −0.02743 | −0.06997 | −0.05712 | −0.04923 | −0.04426 | −0.02234 |
| $\boldsymbol{\theta}_{13}(8) = 0.01000$ | 0.02134 | −0.00932 | −0.02996 | −0.01028 | −0.01329 | −0.00056 |
| $\delta$i(%) | 13.16158 | 17.26615 | 16.39064 | 9.21619 | 9.09722 | 6.20602 |

| $t$ | 100 | 200 | 500 | 1000 | 2000 | 3000 |
|---|---|---|---|---|---|---|
| $\boldsymbol{\theta}_{11}(1) = -0.40000$ | −0.45308 | −0.44816 | −0.44920 | −0.44329 | −0.43244 | −0.41958 |
| $\boldsymbol{\theta}_{11}(2) = -0.86000$ | −0.52383 | −0.46546 | −0.50568 | −0.59307 | −0.66158 | −0.73180 |
| $\boldsymbol{\theta}_{11}(3) = -0.60000$ | −0.64117 | −0.66040 | −0.67489 | −0.66275 | −0.63280 | −0.63739 |
| $\boldsymbol{\theta}_{11}(4) = -0.65000$ | −0.79551 | −0.86462 | −0.84811 | −0.80126 | −0.75338 | −0.73133 |
| $\boldsymbol{\theta}_{12}(1) = 0.43600$ | 0.28401 | 0.21731 | 0.26339 | 0.28629 | 0.32767 | 0.35017 |
| $\boldsymbol{\theta}_{12}(2) = 5.30000$ | 3.56959 | 3.22069 | 3.44162 | 3.91283 | 4.29220 | 4.65195 |
| $\boldsymbol{\theta}_{12}(3) = -1.00000$ | −1.04865 | −0.99878 | −0.98843 | −0.99868 | −0.99514 | −0.99810 |
| $\boldsymbol{\theta}_{12}(4) = 0.00000$ | −0.06058 | −0.02795 | −0.02279 | −0.01566 | −0.00584 | 0.00137 |
| $\boldsymbol{\theta}_{13}(1) = 0.46700$ | 0.20094 | 0.21362 | 0.24797 | 0.29672 | 0.31582 | 0.33734 |
| $\boldsymbol{\theta}_{13}(2) = 0.29150$ | 0.14114 | 0.22112 | 0.26196 | 0.26052 | 0.24329 | 0.24893 |
| $\boldsymbol{\theta}_{13}(3) = 0.23860$ | 0.02026 | 0.04474 | 0.08310 | 0.09669 | 0.10066 | 0.11759 |
| $\boldsymbol{\theta}_{13}(4) = 0.27360$ | 0.21213 | 0.19479 | 0.23864 | 0.24027 | 0.24118 | 0.23474 |
| $\boldsymbol{\theta}_{13}(5) = 0.30000$ | 0.06717 | 0.02043 | 0.04688 | 0.10141 | 0.15251 | 0.19085 |
| $\boldsymbol{\theta}_{13}(6) = 0.01000$ | 0.20522 | 0.13866 | 0.13498 | 0.13232 | 0.14350 | 0.14225 |
| $\boldsymbol{\theta}_{13}(7) = 0.20000$ | −0.04823 | −0.09095 | −0.08206 | −0.03265 | 0.03308 | 0.08847 |
| $\boldsymbol{\theta}_{13}(8) = 0.10000$ | −0.14198 | −0.13627 | −0.08112 | −0.04237 | −0.03059 | −0.00158 |
| $\delta_2(\%)$ | 33.40628 | 39.52957 | 35.27017 | 26.57254 | 19.58267 | 13.07450 |

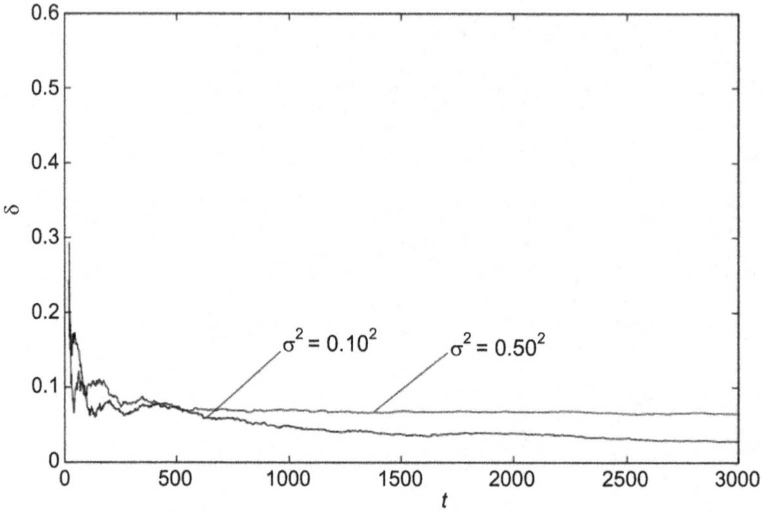


**Figure 4.5** The parameter estimation errors $\delta_1$ versus $t$.

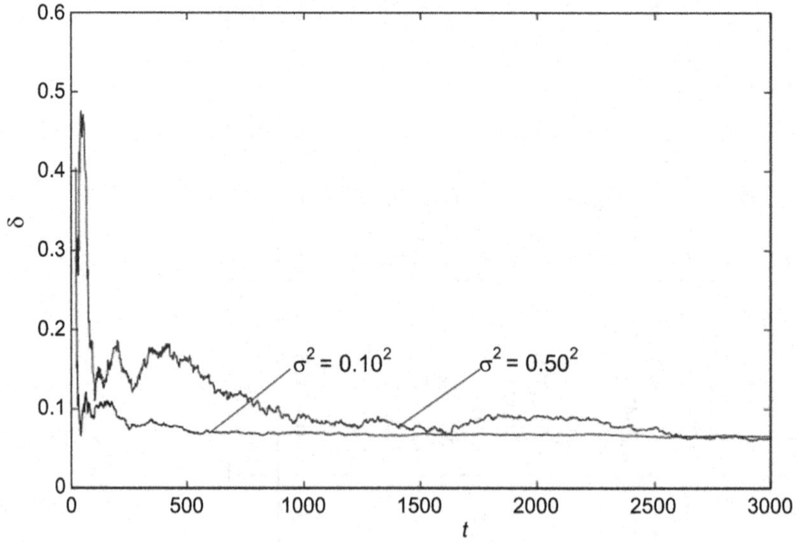


**Figure 4.6** The parameter estimation errors $\delta_2$ versus $t$.

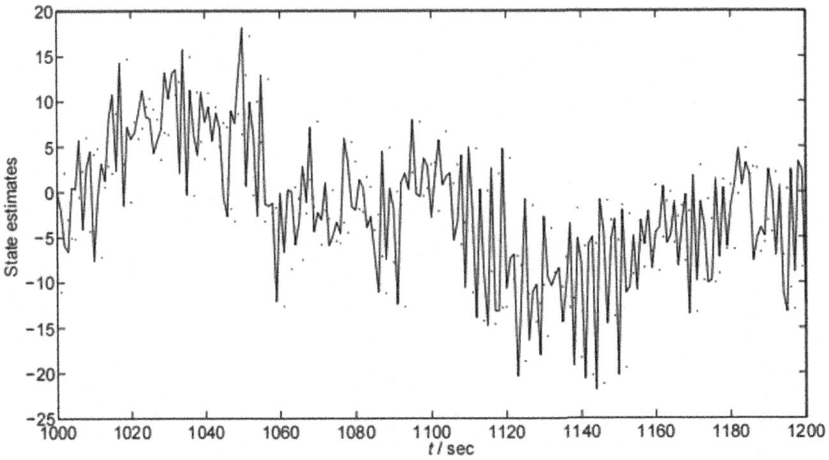


Solid line: the true $x_1(t)$, dots: the estimated $\hat{x}_1(t)$

**Figure 4.7** The state and state estimate $\hat{x}_1(t)$ versus $t$.

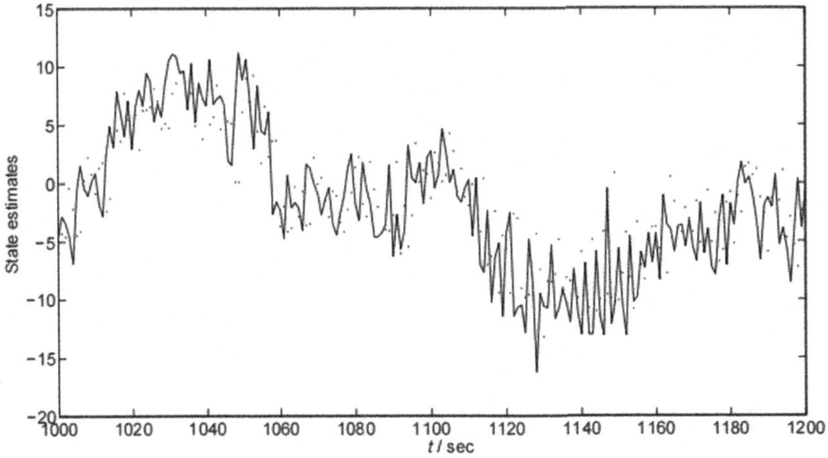


Solid line: the true $x_2(t)$, dots: the estimated $\hat{x}_2(t)$

**Figure 4.8** The state and state estimate $\hat{x}_2(t)$ versus $t$.

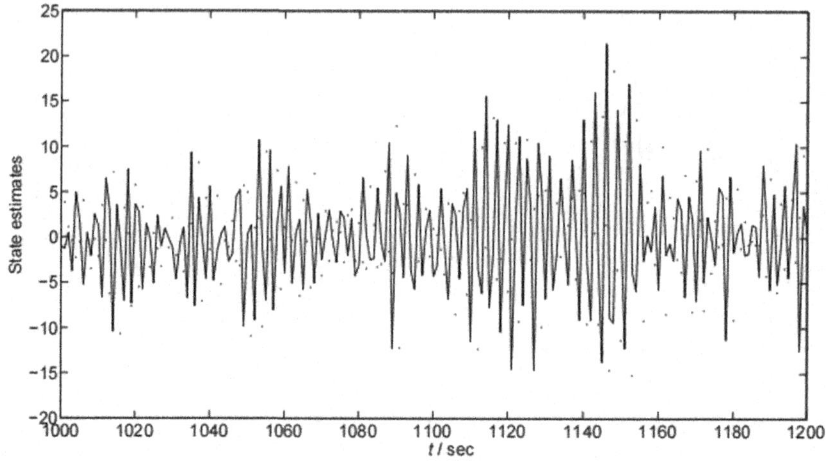


Solid line: the true $x_3(t)$, dots: the estimated $\hat{x}_3(t)$

**Figure 4.9** The state and state estimate $\hat{x}_3(t)$ versus $t$.

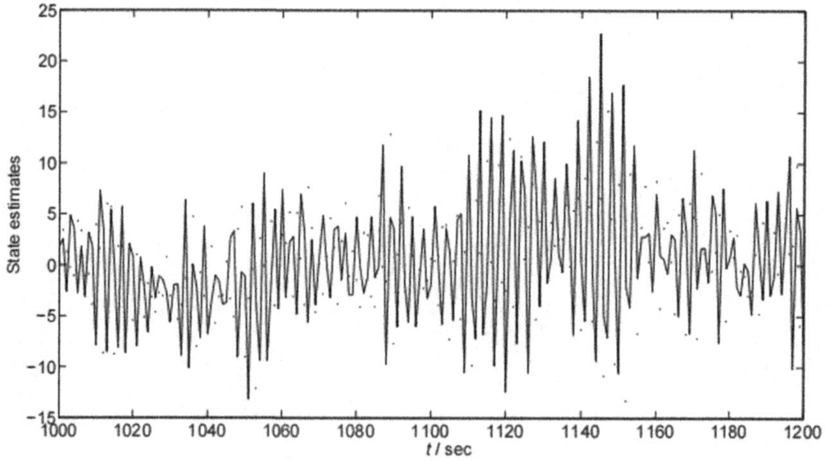


Solid line: the true $x_4(t)$, dots: the estimated $\hat{x}_4(t)$

**Figure 4.10** The state and state estimate $\hat{x}_4(t)$ versus $t$.

## 4.2.5 Conclusions

The basic algorithm derivation principle of this chapter is similar to the corresponding multiinput single-output model, but the number of parameters of this model is large, the dimension is complex, and there is coupling. When calculating the system state according to the hierarchical identification principle, it is necessary to combine the identification of two subsystems. The parameters make the identification more difficult, and because the recursive algorithm calculates the inverse of the matrix in the calculation process, the calculation amount is relatively large, which affects the identification accuracy. This chapter starts with a bivariate model with few dimensions to study the recursive least squares algorithm based on residuals. If the effect is good, it can be extended to the multivariate/nonlinear model, which can identify multivariate/nonlinear models with colored noise interference in the actual system to achieve accuracy optimal identification parameters.

## References

Chen, J., Huang, B., Ding, F., Gu, Y., 2018. Variational Bayesian approach for ARX systems with missing observations and varying time-delays. Automatica 94, 194–204.

Ding, F., Xu, L., Meng, D.D., Jin, X.B., Alsaedi, A., Hayat, T., 2020. Gradient estimation algorithms for the parameter identification of bilinear systems using the auxiliary model. Journal of Computational and Applied Mathematics 369.

Gan, M., Li, H.X., Peng, H., 2015. A variable projection approach for efficient estimation of RBF-ARX model. IEEE Transactions on Cybernetics 45, 462–471.

Sui, T.J., Marelli, D.E., Fu, M.Y., Lu, R.Q., 2018. Accuracy analysis for distributed weighted least-squares estimation in finite steps and loopy networks. Automatica 97, 82–91.

CHAPTER 5

# Nonlinear time-delay system identification

## 5.1 Parameter estimation for a Hammerstein state-space system with time delay

Almost all physical systems are nonlinear to a certain extent, and it is normal to describe a system using nonlinear models. A typical nonlinear system is the Hammerstein nonlinear system which has a static nonlinear block followed by a linear dynamic block (Ibrir, 2018). Plenty of identification methods have been proposed for Hammerstein systems, such as the auxiliary model identification algorithm, the maximum likelihood stochastic gradient estimation algorithm, the gradient-based estimation algorithm, the newton iteration method, and the iteration identification method.

For the identification of Hammerstein systems, there exist some estimation methods: the stochastic gradient method and the iterative identification algorithm (Ghosal and Rao, 2019). The stochastic gradient methods are a class of important stochastic approximation methods owing to their computational convenience and have been widely applied in many areas, including adaptive control and optimization. The iterative identification algorithms make sufficient use of all the measured information, so the accuracy of parameter estimation can be improved.

The identification of the time-delay system is significant for system control and system analysis. The effects of time delays on the function of control systems have received considerable attention from many researchers in industrial processes. The main contributions of this paper are as follows.

1. To derive a gradient-based identification algorithm for the Hammerstein state time-delay systems by using the gradient search method.
2. To present a new-type gradient-based iterative algorithm and a least squares-based iterative algorithm for a Hammerstein state time-delay system with colored noise using the iterative identification principle.
3. To analyze and compare the performances of the proposed algorithms using a numerical example, including the convergence rates and the estimation errors of the algorithms for finite measurement data.

*State Space Systems With Time-Delays Analysis, Identification, and Applications*
DOI: https://doi.org/10.1016/B978-0-323-91768-1.00008-3

### 5.1.1 The system description and input–output representation

Let us introduce some notation. "$A = :X$" or "$X: = A$" stands for "$A$ is defined as $X$"; the symbol $I(I_n)$ stands for an identity matrix of an appropriate size ($n \times n$); $z$ represents a unit forward shift operator: $zx(t) = x(t+1)$ and $z^{-1}x(t) = x(t-1)$; the superscript T denotes the matrix/vector transpose; $\hat{\theta}(t)$ denotes the estimate of $\theta$ at time $t$; and $1_n$ represents an $n \times 1$ vector whose elements are all unity.

Consider the following Hammerstein state time-delay system with colored noise depicted in Fig. 5.1,

$$\boldsymbol{x}(t+1) = \boldsymbol{A}\boldsymbol{x}(t) + \boldsymbol{B}\boldsymbol{x}(t-1) + \boldsymbol{g}\bar{u}(t), \tag{5.1}$$

$$\bar{u}(t) = f(u(t)), \tag{5.2}$$

$$y(t) = \boldsymbol{c}\boldsymbol{x}(t) + w(t), \tag{5.3}$$

$$w(t) = \sum_{s=1}^{n_d} d_s v(t-s) + v(t), \tag{5.4}$$

$$\boldsymbol{A} := \begin{bmatrix} 0 & 1 & 0 & \cdots & 0 \\ 0 & 0 & 1 & \ddots & \vdots \\ \vdots & \vdots & \ddots & \ddots & 0 \\ 0 & 0 & \cdots & 0 & 1 \\ a_n & a_{n-1} & a_{n-2} & \cdots & a_1 \end{bmatrix} \in \mathbb{R}^{n \times n}, \boldsymbol{g} := \begin{bmatrix} g_1 \\ g_2 \\ \vdots \\ g_n \end{bmatrix} \in \mathbb{R}^n,$$

$$\boldsymbol{B} := \begin{bmatrix} \boldsymbol{b}_1 \\ \boldsymbol{b}_2 \\ \vdots \\ \boldsymbol{b}_n \end{bmatrix} \in \mathbb{R}^{n \times n}, \boldsymbol{b}_i \in \mathbb{R}^{1 \times n}, \boldsymbol{d} := \begin{bmatrix} d_1 \\ d_2 \\ \vdots \\ d_{n_d} \end{bmatrix} \in \mathbb{R}^{n_d},$$

$$\boldsymbol{c} := \left[1, 0, \ldots, 0\right] \in \mathbb{R}^{1 \times n},$$

where $x(t) = [x_1(t), x_2(t), \ldots, x_n(t)]^{\mathrm{T}} \in R^n$ is the state vector, $u(t) \in R$ is the system input, $y(t) \in R$ is the system output, $\bar{u}(t) \in R$ is the unobserved output of the nonlinear block, $v(t) \in R$ is stochastic white noise with zero mean, and $A$, $B$, $g$, $c$, and $d$ are the system parameter matrices/vectors.

The nonlinear block is a nonlinear function:

$$\bar{u}(t) = f(u(t)) = \sum_{l=1}^{m} p_l(t) f_l(u(t)) = \boldsymbol{f}(u(t))\boldsymbol{p}, \tag{5.5}$$

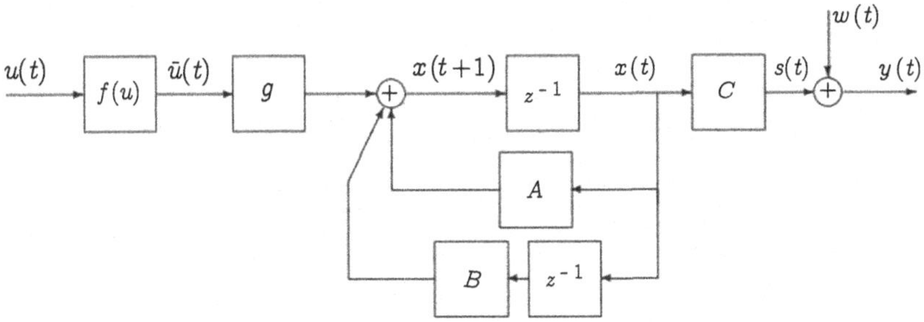


**Figure 5.1** The input nonlinear state time-delay system.

where the known basis $f := (f_1, f_2, \ldots, f_m)$ of $\overline{u} = f(u)$ are known functions of the system input $u(t)$, the unknown parameters $p := (p_1, p_2, \ldots, p_m)$ are the coefficients of the nonlinear functions, and $m$ is the number of the parameter for the nonlinear block.

The goal of this chapter is to develop iterative algorithms for estimating the unknown parameters $A$, $B$, $g$, $p$, and $d$ of the system in (5.1)–(5.3) by using the available input–output measurement data $\{u(t), y(t)\}$ and to evaluate the accuracy of the parameter estimates by simulations on computers.

Without loss of generality, assume that $u(t) = 0$ and $y(t) = 0$ for $t \leq 0$, and $n_1 := n^2 + mn + n_d$. The following transforms the Hammerstein state time-delay model in (5.1)–(5.4) into an input–output representation and gives its identification model.

From (5.1), we have

$$x_i(t+1) = x_{i+1}(t) + \boldsymbol{b}_i\boldsymbol{x}(t-1) + g_i\overline{u}(t), i = 1, 2, \ldots, n-1, \tag{5.6}$$

$$x_n(t+1) = a_n x_1(t) + a_{n-1}x_2(t) + \cdots + a_i x_n(t) + \boldsymbol{b}_n\boldsymbol{x}(t-1) + g_n\overline{u}(t). \tag{5.7}$$

Let $a := [a_n, a_{n-1}, \ldots, a_1]^{\mathrm{T}} \in R^n$. Using the property of the shift operator $z$, multiplying Eq. (5.6) by $z^{-i}$ and Eq. (5.7) by $z^{-n}$ gives

$$\begin{aligned} &x_i(t-i+1) = x_{i+1}(t-i) + \boldsymbol{b}_i\boldsymbol{x}(t-i-1) + g_i\overline{u}(t-i), i = 1, 2, \ldots, n-1, \\ &x_n(t-n+1) = \boldsymbol{ax}(t-n) + \boldsymbol{b}_n\boldsymbol{x}(t-n-1) + g_n\overline{u}(t-n). \end{aligned}$$

Adding all expressions gives

$$\begin{aligned} x_1(t) &= \boldsymbol{ax}(t-n) + \boldsymbol{b}_1\boldsymbol{x}(t-2) + \boldsymbol{b}_2\boldsymbol{x}(t-3) + \cdots + \boldsymbol{b}_{n-1}\boldsymbol{x}(t-n) + \boldsymbol{b}_n\boldsymbol{x}(t-n-1) \\ &\quad + g_1\overline{u}(t-1) + g_2\overline{u}(t-2) + \cdots + g_n\overline{u}(t-n) \\ &= \boldsymbol{ax}(t-n) + \sum_{i=1}^{n}\boldsymbol{b}_i\boldsymbol{x}(t-i-1) + \sum_{j=1}^{n} g_j \sum_{l=1}^{m} p_l f_l(u(t-j)). \end{aligned} \tag{5.8}$$

From (5.3) and (5.8), we have

$$y(t) = \boldsymbol{a}\boldsymbol{x}(t-n) + \sum_{i=1}^{n} \boldsymbol{b}_i \boldsymbol{x}(t-i-1) + \sum_{j=1}^{n} g_j \sum_{l=1}^{m} p_l f_l(u(t-j))$$
$$+ \sum_{s=1}^{n_d} d_s v(t-s) + v(t).$$

Let

$$\boldsymbol{\varphi}(t) := \begin{bmatrix} \boldsymbol{\varphi}_1(t) \\ \boldsymbol{\varphi}_2(t) \\ \boldsymbol{\varphi}_3(t) \end{bmatrix} \in \mathbb{R}^{n_1},$$
$$\boldsymbol{\varphi}_1(t) := \left[\boldsymbol{x}^{\mathrm{T}}(t-2), \boldsymbol{x}^{\mathrm{T}}(t-3), \ldots, \boldsymbol{x}^{\mathrm{T}}(t-n), \boldsymbol{x}^{\mathrm{T}}(t-n-1)\right]^{\mathrm{T}} \in \mathbb{R}^{n^2},$$
$$\boldsymbol{\varphi}_2(t) := \left[\boldsymbol{\psi}_1^{\mathrm{T}}(t), \boldsymbol{\psi}_2^{\mathrm{T}}(t), \ldots, \boldsymbol{\psi}_m^{\mathrm{T}}(t)\right]^{\mathrm{T}} \in \mathbb{R}^{mn},$$
$$\boldsymbol{\psi}_l(t) := \left[f_l(u(t-1)), f_l(u(t-2)), \ldots, f_l(u(t-n))\right]^{\mathrm{T}} \in \mathbb{R}^{n}, l = 1, 2 \ldots, m,$$
$$\boldsymbol{\varphi}_3(t) := [v(t-1), v(t-2), \ldots, v(t-n_d)]^{\mathrm{T}} \in \mathbb{R}^{n_d},$$
$$\boldsymbol{\theta} := \left[\boldsymbol{b}_1, \boldsymbol{b}_2, \ldots, \boldsymbol{a}+\boldsymbol{b}_{n-1}, \boldsymbol{b}_n, p_1\boldsymbol{g}^{\mathrm{T}}, p_2\boldsymbol{g}^{\mathrm{T}}, \ldots, p_m\boldsymbol{g}^{\mathrm{T}}, \boldsymbol{d}^{\mathrm{T}}\right]^{\mathrm{T}} \in \mathbb{R}^{n_1}.$$

Thus, we can obtain the identification model of Hammerstein state time-delay system with colored noise in (5.1)–(5.4),

$$y(t) = \boldsymbol{\varphi}^{\mathrm{T}}(t)\boldsymbol{\theta} + v(t). \tag{5.9}$$

### 5.1.2 The stochastic gradient algorithm

In order to show the advantages of the iterative identification methods proposed in the next section, the following simply discusses the comparable stochastic gradient algorithm for Hammerstein state time-delay models.

Let E denote an expectation operator and $||X||^2 := \mathrm{tr}[XX^{\mathrm{T}}]$ be the norm of the matrix $X$. Since $v(t)$ is a white noise, forming a quadratic cost function,

$$J_1(\boldsymbol{\theta}) = \mathrm{E}\left[\left\| y(t) - \boldsymbol{\varphi}^{\mathrm{T}}(t)\boldsymbol{\theta}\right\|^2\right],$$

and minimizing $J_1(\theta)$ lead to the following stochastic gradient algorithm of estimating $\theta$,

$$\hat{\boldsymbol{\theta}}(t) = \hat{\boldsymbol{\theta}}(t-1) + \frac{\boldsymbol{\varphi}(t)}{r(t)}\left[y(t) - \boldsymbol{\varphi}^{\mathrm{T}}(t)\hat{\boldsymbol{\theta}}(t-1)\right], \tag{5.10}$$

$$r(t) = r(t-1) + \left\|\varphi(t)\right\|^2, r(0) = 1. \tag{5.11}$$

Note that the information vector $\varphi(t)$ on the right-hand side contains the unknown state terms $x(t-i)(i = 2, 3, \ldots, n+1)$ and the unknown noise terms $v(t-s)(s = 1, 2, \ldots, n_d)$, and the algorithm in (5.10) and (5.11) is impossible to realize. The solution here is to replace the unknown variables $x(t-i)$ and $v(t-s)$ with their corresponding estimates $\hat{x}(t-i)$ and $\hat{v}(t-s)$ and to define

$$\hat{\varphi}(t) = \left[\hat{\boldsymbol{x}}^{\mathrm{T}}(t-2), \hat{\boldsymbol{x}}^{\mathrm{T}}(t-3), \ldots, \hat{\boldsymbol{x}}^{\mathrm{T}}(t-n), \hat{\boldsymbol{x}}^{\mathrm{T}}(t-n-1),\right.$$

$$\left.\varphi_2^{\mathrm{T}}(t), \hat{v}(t-1), \hat{v}(t-2), \ldots, \hat{v}(t-n_d)\right]^{\mathrm{T}}. \tag{5.12}$$

From (5.9), we have $v(t) = y(t) - \varphi^{\mathrm{T}}(t)\theta$. Replacing $\varphi(t)$ and $\theta$ with $\hat{\varphi}(t)$ and $\hat{\theta}(t)$, the estimated residual can be computed through

$$\hat{v}(t) = y(t) - \hat{\varphi}^{\mathrm{T}}(t)\hat{\boldsymbol{\theta}}(t). \tag{5.13}$$

Replacing $A$, $B$, $g$, $x(t)$, and $\bar{u}(t)$ in (5.1) with their estimates $\hat{A}(t)$, $\hat{B}(t)$, $\hat{g}(t)$, $\hat{x}(t)$, and $\hat{\bar{u}}(t)$, we have

$$\hat{\boldsymbol{x}}(t+1) = \hat{\boldsymbol{A}}(t)\hat{\boldsymbol{x}}(t) + \hat{\boldsymbol{B}}(t)\hat{\boldsymbol{x}}(t-1) + \hat{\boldsymbol{g}}(t)\hat{\bar{u}}(t). \tag{5.14}$$

Replacing $p_l$ in (5.5) with its estimate $\hat{p}_l(t)$, the estimate $\hat{\bar{u}}(t)$ can be computed through

$$\hat{\bar{u}}(t) = \sum_{l=1}^{m} \hat{p}_l(t) f_l(u(t)). \tag{5.15}$$

Replacing $\varphi(t)$ in (5.10)–(5.11) with $\hat{\varphi}(t)$, we can summarize the stochastic gradient (SG) identification algorithm for estimating the parameter vector $\theta$ as

$$\hat{\boldsymbol{\theta}}(t) = \hat{\boldsymbol{\theta}}(t-) + \frac{\hat{\varphi}(t)}{r(t)}\left[y(t) - \hat{\varphi}^{\mathrm{T}}(t)\hat{\boldsymbol{\theta}}(t-1)\right], \tag{5.16}$$

$$r(t) = r(t-1) + \left\|\hat{\varphi}(t)\right\|^2, r(0) = 1, \tag{5.17}$$

$$\hat{\boldsymbol{\theta}}(t) = \left[\hat{\boldsymbol{b}}_1(t), \hat{\boldsymbol{b}}_2(t), \ldots, \widehat{\boldsymbol{a}+\boldsymbol{b}_{n-1}}(t), \hat{\boldsymbol{b}}_n(t), \widehat{p_1\boldsymbol{g}}^{\mathrm{T}}(t), \widehat{p_2\boldsymbol{g}}^{\mathrm{T}}(t), \ldots, \widehat{p_m\boldsymbol{g}}^{\mathrm{T}}(t), \hat{\boldsymbol{d}}^{\mathrm{T}}(t)\right]^{\mathrm{T}}, \tag{5.18}$$

$$\hat{\boldsymbol{A}}(t) = \begin{bmatrix} 0 & 1 & 0 & \cdots & 0 \\ 0 & 0 & 1 & \ddots & \vdots \\ \vdots & \vdots & \ddots & \ddots & 0 \\ 0 & 0 & \cdots & 0 & 1 \\ \hat{a}_n(t) & \hat{a}_{n-1}(t) & \hat{a}_{n-2}(t) & \cdots & \hat{a}_1(t) \end{bmatrix}, \tag{5.19}$$

$$\hat{\boldsymbol{B}}(t) = \left[\hat{\boldsymbol{b}}_1(t), \hat{\boldsymbol{b}}_2(t), \ldots, \hat{\boldsymbol{b}}_n(t)\right]^{\mathrm{T}}, \tag{5.20}$$

$$\hat{\boldsymbol{g}}(t) = \left[\hat{g}_1(t), \hat{g}_2(t), \ldots, \hat{g}_n(t)\right]^{\mathrm{T}}, \tag{5.21}$$

$$\hat{\boldsymbol{p}}(t) = \left[\hat{p}_1(t), \hat{p}_2(t), \ldots, \hat{p}_m(t)\right]^{\mathrm{T}}. \tag{5.22}$$

The parameters $\hat{A}(t)$, $\hat{B}(t)$, $\hat{g}(t)$, and $\hat{p}(t)$ are the coefficients of equations in (5.14) and (5.15). The SG algorithm also includes the expressions in (5.12)–(5.13).

To initialize this algorithm, the initial value $\hat{\theta}(0)$ is generally taken to be zero or a small real vector, for example, $\hat{\theta}(0) = 1_{n_0}/p_0$ with $1_{n_0}$ being an $n_0$-dimensional column vector whose elements are all 1 and with $p_0$ being normally a large positive number (e.g., $p_0 = 10^6$).

### 5.1.3 The iterative algorithms

The stochastic gradient algorithm is an on-line algorithm which is necessary for many control problems, and the iterative algorithm is an off-line algorithm that makes use of all the observed data and has limitations for many control problems. Although the SG algorithm can generate the parameter estimates of the Hammerstein state time-delay models, its convergence rate is relatively slow. To improve the parameter estimation accuracy, we investigate the iterative identification algorithms for the Hammerstein state time-delay systems.

#### *5.1.3.1 The gradient-based iterative algorithm*

Define a quadratic criterion function,

$$J_2(\boldsymbol{\theta}) := \left\| \boldsymbol{Y}(L) - \boldsymbol{\Phi}(L)\boldsymbol{\theta} \right\|^2. \tag{5.23}$$

Let $k = 1, 2, 3, \ldots$ be an iteration variable and $\hat{\theta}_k$ be the iterative estimate of $\theta$.

Minimizing $J_2(\theta)$ and using the negative gradient search lead to the iterative algorithm of computing $\hat{\theta}_k$:

$$\hat{\boldsymbol{\theta}}_k = \hat{\boldsymbol{\theta}}_{k-1} - \frac{\mu_k}{2}\mathrm{grad}\left[J_2(\hat{\boldsymbol{\theta}}_{k-1})\right] = \hat{\boldsymbol{\theta}}_{k-1} + \mu_k\boldsymbol{\Phi}^{\mathrm{T}}(L)\left[\boldsymbol{Y}(L) - \boldsymbol{\Phi}(L)\hat{\boldsymbol{\theta}}_{k-1}\right], \tag{5.24}$$

where $\mu_k$ is the iterative step size or convergence factor to be given later. Here, a difficulty arises because $\Phi(L)$ in (5.24) contains the unknown state vectors $x(t-i)(i=2,3,\ldots,n+1)$ and unknown noise variables $v(t-s)(s=1,2,\ldots,n_d)$, so it is impossible to compute the iterative solution $\hat{\theta}_k$ of $\theta$ using (5.24). In order to solve this difficulty, the approach here is based on the hierarchical identification principle. Let $\hat{x}_k(t)$ be the estimate of $x(t)$ and $\hat{v}_k(t)$ be the estimate of $v(t)$ at iteration $k$, and $\hat{\varphi}_k(t)$ denotes the information vector obtained by replacing $x(t-i)$ and $v(t-s)$ with $\hat{x}_{k-1}(t-i)$ and $\hat{v}_{k-1}(t-s)$, defining

$$\hat{\varphi}_k(t) := \begin{bmatrix} \hat{\varphi}_{1,k}(t) \\ \varphi_2(t) \\ \hat{\varphi}_{3,k}(t) \end{bmatrix} \in \mathbb{R}^{n_1}, \tag{5.25}$$

$$\hat{\varphi}_{1,k}(t) := \left[\hat{\boldsymbol{x}}_{k-1}^{\mathrm{T}}(t-2), \hat{\boldsymbol{x}}_{k-1}^{\mathrm{T}}(t-3), \ldots, \hat{\boldsymbol{x}}_{k-1}^{\mathrm{T}}(t-n), \hat{\boldsymbol{x}}_{k-1}^{\mathrm{T}}(t-n-1)\right]^{\mathrm{T}} \in \mathbb{R}^{n^2}, \tag{5.26}$$

$$\varphi_2(t) := \left[\boldsymbol{\psi}_1^{\mathrm{T}}(t), \boldsymbol{\psi}_2^{\mathrm{T}}(t), \ldots, \boldsymbol{\psi}_m^{\mathrm{T}}(t)\right]^{\mathrm{T}} \in \mathbb{R}^{mn}, \tag{5.27}$$

$$\boldsymbol{\psi}_l(t) := \left[f_l(u(t-1)), f_l(u(t-2)), \ldots, f_l(u(t-n))\right]^{\mathrm{T}}, l = 1, 2, \ldots, m, \tag{5.28}$$

$$\hat{\varphi}_{3,k}(t) := \left[\hat{v}_{k-1}(t-1), \hat{v}_{k-1}(t-2), \ldots, \hat{v}_{k-1}(t-n_d)^{\mathrm{T}}\right] \in \mathbb{R}^{n_d}, \tag{5.29}$$

$$\hat{v}_k(t) := y(t) - \hat{\varphi}_k^{\mathrm{T}}(t)\hat{\boldsymbol{\theta}}_k, t = 1, 2, \ldots, L. \tag{5.30}$$

Let $\hat{A}_k$, $\hat{B}_k$, $\hat{g}_k$, and $\hat{\overline{u}}_k(t)$ be the estimate of $A$, $B$, $g$, and $\overline{u}(t)$ at iteration $k$, and we can compute the estimate $\hat{x}_k(t+1)$ through

$$\hat{\boldsymbol{x}}_k(t+1) = \hat{\boldsymbol{A}}_k\hat{\boldsymbol{x}}_{k-1}(t) + \hat{\boldsymbol{B}}_k\hat{\boldsymbol{x}}_{k-1}(t-1) + \hat{\boldsymbol{g}}_k\hat{\overline{u}}_k(t). \tag{5.31}$$

Replacing $p_l$ in (5.26) with its estimate $\hat{p}_{l,k}$, the estimate $\hat{\bar{u}}_k(t)$ can be computed through

$$\hat{\bar{u}}_k(t) = \sum_{l=1}^{m} \hat{p}_{l,k} f_l(u(t)). \tag{5.32}$$

Similarly, from (5.9), we can compute the estimate $\hat{v}_k(t)$ through $\hat{v}_k(t) = y(t) - \hat{\varphi}_k^{\mathrm{T}}(t)\hat{\theta}_k$. Define

$$\hat{\boldsymbol{\Phi}}_k(L) := \left[\hat{\varphi}_k(L), \hat{\varphi}_k(L-1), \ldots, \hat{\varphi}_k(1)\right]^{\mathrm{T}} \in \mathbb{R}^{L \times n_1}. \tag{5.33}$$

Replacing $\Phi(L)$ in (5.24) with $\hat{\Phi}_k(L)$ yields the gradient-based iterative (GI) identification algorithm for Hammerstein state time-delay systems and can be summarized as

$$\hat{\boldsymbol{\theta}}_k = \hat{\boldsymbol{\theta}}_{k-1} + \mu_k \hat{\boldsymbol{\Phi}}_k^{\mathrm{T}}(L)\left[\mathbf{Y}(L) - \hat{\boldsymbol{\Phi}}_k(L)\hat{\boldsymbol{\theta}}_{k-1}\right], k = 1, 2, 3, \ldots \tag{5.34}$$

$$\mathbf{Y}(L) = [y(L), y(L-1), \ldots, y(1)]^{\mathrm{T}}, \tag{5.35}$$

$$0 < \mu_k \leqslant 2\lambda_{\max}^{-1}\left[\hat{\boldsymbol{\Phi}}_k^{\mathrm{T}}(L)\hat{\boldsymbol{\Phi}}_k(L)\right], \tag{5.36}$$

$$\hat{\boldsymbol{\theta}}_k = \left[\hat{\boldsymbol{b}}_{1,k}, \hat{\boldsymbol{b}}_{2,k}, \ldots, \widehat{\boldsymbol{a}+\boldsymbol{b}_{n-1k}}, \hat{\boldsymbol{b}}_{n,k}, \widehat{p_1\boldsymbol{g}_k^{\mathrm{T}}}, \widehat{p_2\boldsymbol{g}_k^{\mathrm{T}}}, \ldots \widehat{p_m\boldsymbol{g}_k^{\mathrm{T}}}, \hat{\boldsymbol{d}}_k^{\mathrm{T}}\right]^{\mathrm{T}}, \tag{5.37}$$

$$\hat{\boldsymbol{A}}_k = \begin{bmatrix} 0 & 1 & 0 & \cdots & 0 \\ 0 & 0 & 1 & \ddots & \vdots \\ \vdots & \vdots & \ddots & \ddots & 0 \\ 0 & 0 & \cdots & 0 & 1 \\ \hat{a}_{n,k} & \hat{a}_{n-1,k} & \hat{a}_{n-2,k} & \cdots & \hat{a}_{1,k} \end{bmatrix}, \tag{5.38}$$

$$\hat{\boldsymbol{B}}_k = \left[\hat{\boldsymbol{b}}_{1,k}, \hat{\boldsymbol{b}}_{2,k}, \ldots, \hat{\boldsymbol{b}}_{n,k}\right]^{\mathrm{T}}, \tag{5.39}$$

$$\hat{\boldsymbol{g}}_k = \left[\hat{g}_{1,k}, \hat{g}_{2,k}, \ldots, \hat{g}_{n,k}\right]^{\mathrm{T}}, \tag{5.40}$$

$$\hat{\boldsymbol{p}}_k = \left[\hat{p}_{1,k}, \hat{p}_{2,k}, \ldots, \hat{p}_{m,k}\right]^{\mathrm{T}}. \tag{5.41}$$

The GI algorithm also includes the expressions in (5.25)–(5.33).

To summarize, we list the steps involved in the gradient-based iterative identification algorithm to compute $\hat{\theta}_k$ as $k$ increases.

1. Collect the input–output data $\{u(t), y(t): t = 0, 1, 2, \ldots, L\}$, and form $Y(L)$ using (5.35).
2. To initialize, let $k = 1$, $\hat{\theta}_0 = \mathbf{1}_{n_0}/p_0$, $\hat{v}_0(t) = 1/p_0$, $p_0 = 10^6$, and $\hat{x}_0(t) =$ a random vector.
3. Form $\hat{\varphi}_{1,k}(t)$ using (5.26), $\psi_l(t)$ using (5.28), $\varphi_2(t)$ using (5.27), $\hat{\varphi}_{3,k}(t)$ using (5.29), $\hat{\varphi}_k(t)$ using (5.25), and $\hat{\Phi}_k(L)$ using (5.33), respectively.
4. Choose a large $\mu_k$ satisfying (5.36), and update the estimate $\hat{\theta}_k$ using (5.34).
5. Compute $\hat{v}_k(t)$ using (5.30).
6. Form $\hat{A}_k$ using (5.38), $\hat{B}_k$ using (5.39), $\hat{g}_k$ using (5.40), and $\hat{p}_k$ using (5.41), respectively.
7. Compute the state estimation vector $\hat{x}_k(t+1)$ using (5.31) and $\hat{\bar{u}}_k(t)$ using (5.32).
8. If $||\hat{\theta}_k - \hat{\theta}_{k-1}|| \leq \varepsilon$ for some pre-set small $\varepsilon$, obtain the parameter estimate $\hat{\theta}_k$; otherwise, increase $k$ by 1 and go to step 3.

The computational burden of the GI algorithm is shown in Table 5.1.

#### 5.1.3.2 The least squares-based iterative algorithm

Another iterative algorithm is the least squares-based iterative algorithm for the Hammerstein state time-delay systems to be discussed in the following.

Provided that $\varphi(t)$ is persistently exciting, minimizing $J_2(\theta)$ in (5.23) gives

$$\hat{\boldsymbol{\theta}} = \left[\boldsymbol{\Phi}^{\mathrm{T}}(L)\boldsymbol{\Phi}(L)\right]^{-1}\boldsymbol{\Phi}^{\mathrm{T}}(L)\mathbf{Y}(L). \tag{5.42}$$

However, a similar difficulty arises in that $\Phi(L)$ in (5.42) contains the unknown inner variables $x(t-i)$ and $v(t-s)$ and so the estimate of $\theta$ is also impossible to compute by (5.42). A similar way to that of deriving the GI algorithm can result in the following least squares-based iterative (LSI) identification algorithm for estimating $\theta$ of the Hammerstein state time-delay systems:

$$\hat{\boldsymbol{\theta}}_k = \left[\hat{\boldsymbol{\Phi}}_k^{\mathrm{T}}(L)\hat{\boldsymbol{\Phi}}_k(L)\right]^{-1}\hat{\boldsymbol{\Phi}}_k^{\mathrm{T}}(L)\mathbf{Y}(L), k = 1, 2, 3, \ldots \tag{5.43}$$

$$\mathbf{Y}(L) = [y(L), y(L-1), \ldots, y(1)]^{\mathrm{T}}, \tag{5.44}$$

**Table 5.1** The computational burden of the GI algorithm.

| Variables | Computational sequences | Number of multiplications | Number of additions |
|---|---|---|---|
| $\hat{\boldsymbol{\theta}}_k$ | $\alpha_1 := \mathbf{Y}(L) - \hat{\boldsymbol{\Phi}}_k(L)\hat{\boldsymbol{\theta}}_{k-1} \in \mathbb{R}^L$ | $n_1 L$ | $n_1 L$ |
| | $\beta_1 := \hat{\boldsymbol{\Phi}}_k^{\mathrm{T}}(L)\left[\mathbf{Y}(L) - \hat{\boldsymbol{\Phi}}_k(L)\hat{\boldsymbol{\theta}}_{k-1}\right] \in \mathbb{R}^{n_1}$ | $n_1 L$ | $n_1(\mathrm{L} - 1)$ |
| | $\hat{\boldsymbol{\theta}}_k = \hat{\boldsymbol{\theta}}_{k-1} + \mu_k \beta_1 \in \mathbb{R}^{n_1}$ | $n_1$ | $n_1$ |
| $\hat{v}_k(t)$ | $\hat{v}_k(t) = y(t) - \hat{\varphi}_k^{\mathrm{T}}(t)\hat{\boldsymbol{\theta}}_k \in \mathbb{R}$ | $n_1 L$ | $n_1 L$ |
| $\hat{\boldsymbol{x}}_k(t+1)$ | $\boldsymbol{p}_1 := \hat{\boldsymbol{A}}_k \hat{\boldsymbol{x}}_{k-1}(t) \in \mathbb{R}^n$ | $n^2 L$ | $n(n-1)L$ |
| | $\boldsymbol{q}_1 := \hat{\boldsymbol{B}}_k \hat{\boldsymbol{x}}_{k-1}(t-1) \in \mathbb{R}^n$ | $n^2 L$ | $n(n-1)L$ |
| | $\boldsymbol{s}_1 := \hat{\boldsymbol{g}}_k \hat{\bar{u}}_k(t) \in \mathbb{R}^n$ | $nL$ | |
| | $\hat{\boldsymbol{x}}_k(t+1) = \boldsymbol{p}_1 + \boldsymbol{q}_1 + \boldsymbol{s}_1 \in \mathbb{R}^n$ | | $2nL$ |
| $\hat{\bar{u}}_k(t)$ | $\hat{\bar{u}}_k(t) = \sum_{l=1}^{m} \hat{p}_{l,k} f_l(u(t)) \in \mathbb{R}^n$ | $mL$ | $(m-1)L$ |
| | Sum | $(2n^2 + n + 3n_1 + m)L + n_1$ | $(2n^2 + 3n_1 + m - 1)L$ |
| | Total flops | $4n^2 + n + 6n_1 + 2m - 1)L + n_1$ [178011] | |

$$\hat{\boldsymbol{\Phi}}_k(L) = \left[\hat{\varphi}_k(L), \hat{\varphi}_k(L-1), \ldots, \hat{\varphi}_k(1)\right]^{\mathrm{T}}, \tag{5.45}$$

$$\hat{\varphi}_k(t) = \left[\hat{\varphi}_{1,k}^{\mathrm{T}}(t), \hat{\varphi}_2^{\mathrm{T}}(t), \hat{\varphi}_{3,k}^{\mathrm{T}}(t)\right]^{\mathrm{T}}, \tag{5.46}$$

$$\hat{\varphi}_{1,k}(t) = \left[\hat{\boldsymbol{x}}_{k-1}^{\mathrm{T}}(t-2), \hat{\boldsymbol{x}}_{k-1}^{\mathrm{T}}(t-3), \ldots, \hat{\boldsymbol{x}}_{k-1}^{\mathrm{T}}(t-n), \hat{\boldsymbol{x}}_{k-1}^{\mathrm{T}}(t-n-1)\right]^{\mathrm{T}}, \tag{5.47}$$

$$\varphi_2(t) = \left[\psi_1^{\mathrm{T}}(t), \psi_2^{\mathrm{T}}(t), \ldots, \psi_m^{\mathrm{T}}(t)\right]^{\mathrm{T}}, \tag{5.48}$$

$$\psi_l(t) = \left[f_l(u(t-1)), f_l(u(t-2)), \ldots, f_l(u(t-n))\right]^{\mathrm{T}}, l = 1, 2, \ldots, m, \tag{5.49}$$

$$\hat{\varphi}_{3,k}(t) = [\hat{v}_{k-1}(t-1), \hat{v}_{k-1}(t-2), \ldots, \hat{v}_{k-1}(t-n_d)]^{\mathrm{T}}, \tag{5.50}$$

$$\hat{v}_k(t) = y(t) - \hat{\varphi}_k^{\mathrm{T}}(t)\hat{\boldsymbol{\theta}}_k, t = 1, 2, \ldots, L, \tag{5.51}$$

$$\hat{\boldsymbol{x}}_k(t+1) = \hat{\boldsymbol{A}}_k\hat{\boldsymbol{x}}_{k-1}(t) + \hat{\boldsymbol{B}}_k\hat{\boldsymbol{x}}_{k-1}(t-1) + \hat{\boldsymbol{g}}_k\hat{\bar{u}}_k(t), \tag{5.52}$$

$$\hat{\boldsymbol{A}}_k = \begin{bmatrix} 0 & 1 & 0 & \cdots & 0 \\ 0 & 0 & 1 & \ddots & \vdots \\ \vdots & \vdots & \ddots & \ddots & 0 \\ 0 & 0 & \cdots & 0 & 1 \\ \hat{a}_{n,k} & \hat{a}_{n-1,k} & \hat{a}_{n-2,k} & \cdots & \hat{a}_{1,k} \end{bmatrix}, \tag{5.53}$$

$$\hat{\boldsymbol{B}}_k = \left[\hat{\boldsymbol{b}}_{1,k}, \hat{\boldsymbol{b}}_{2,k}, \ldots, \hat{\boldsymbol{b}}_{n,k}\right]^{\mathrm{T}}, \tag{5.54}$$

$$\hat{\boldsymbol{g}}_k = \left[\hat{g}_{1,k}, \hat{g}_{2,k}, \ldots, \hat{g}_{n,k}\right]^{\mathrm{T}}, \tag{5.55}$$

$$\hat{\boldsymbol{p}}_k = \left[\hat{p}_{1,k}, \hat{p}_{2,k}, \ldots, \hat{p}_{m,k}\right]^{\mathrm{T}}. \tag{5.56}$$

Like the GI algorithm, the LSI algorithm also makes full use of all the measured input–output data $\{u(i), y(i): i = 0, 1, 2, \ldots, t, \ldots, L\}$ at each iteration $k = 1, 2, 3, \ldots$ and thus has higher parameter estimation accuracy.

The steps of computing the parameter estimation vector $\hat{\theta}_k$ using the LSI algorithm are listed in the following,

1. Collect the input–output data $\{u(t), y(t): t=0, 1, 2, \ldots, L\}$, and form $Y(L)$ using (5.44).
2. To initialize, let $k=1$, $\hat{x}_0(t)=$ random number for all $t$.
3. Form $\hat{\varphi}_{1,k}(t)$ using (5.47), $\psi_l(t)$ using (5.49), $\varphi_2(t)$ using (5.48), $\hat{\varphi}_{3,k}(t)$ using (5.50), $\hat{\varphi}_k(t)$ using (5.46), and $\hat{\Phi}_k(L)$ using (5.45), respectively.
4. Update the estimate $\hat{\theta}_k$ using (5.43).
5. Compute $\hat{v}_k(t)$ using (5.51).
6. Form $\hat{A}_k$ using (5.53), $\hat{B}_k$ using (5.54), $\hat{g}_k$ using (5.55), and $\hat{p}_k$ using (5.56), respectively.
7. Compute the state estimation vector $\hat{x}_k(t+1)$ using (5.52).
8. For some preset small $\varepsilon$, if $||\hat{\theta}_k - \hat{\theta}_{k-1}|| \leq \varepsilon$, then terminate the procedure and obtain the iterative times k and estimate $\hat{\theta}_k$; otherwise, increase $k$ by 1 and go to step 3.

The computational burden of the LSI algorithm is shown in Table 5.2, where the numbers in the brackets denote the computation loads for a system with $n=2$, $n_d=1$, $m=3$, and $L=2000$ at each step. Since $n_1=n^2+mn+n_d$, the GI algorithm has smaller computational efforts than the LSI algorithm

## 5.1.4 Example

Consider the following Hammerstein state time-delay system:

$$
\begin{aligned}
\bar{u}(t) &= f(u(t)) = p_1 f_1(u(t)) + p_2 f_2(u(t)) + p_3 f_3(u(t)) \\
&= u(t) + 0.50u^2(t) + 0.10u^3(t),
\end{aligned}
$$

$$
\boldsymbol{x}(t+1) = \begin{bmatrix} 0 & 1 \\ -0.04 & 0.20 \end{bmatrix} \boldsymbol{x}(t) + \begin{bmatrix} 0.24 & -0.08 \\ 0.16 & 0.12 \end{bmatrix} \boldsymbol{x}(t-1) + \begin{bmatrix} 0.50 \\ 1.00 \end{bmatrix} \bar{u}(t),
$$

$$
y(t) = [1, 0]\boldsymbol{x}(t) + 0.20v(t-1) + v(t).
$$

The parameter vector to be identified is

$$
\begin{aligned}
\boldsymbol{\theta} &= [\theta_1, \theta_2, \theta_3, \theta_4, \theta_5, \theta_6, \theta_7, \theta_8, \theta_9, \theta_{10}, \theta_{11}]^{\mathrm{T}} \\
&= \left[a_2 + b_{11}, a_1 + b_{12}, b_{21}, b_{22}, g_1, g_2, p_2 g_1, p_2, g_2, p_3 g_1, p_3 g_2, d_1\right]^{\mathrm{T}} \\
&= [0.20, 0.12, 0.16, 0.12, 0.50, 1.00, 0.25, 0.50, 0.05, 0.10, 0.20]^{\mathrm{T}}.
\end{aligned}
$$

In simulation, the input $\{u(t)\}$ is taken as an uncorrelated stochastic signal sequence with zero mean and unit variance and $\{v(t)\}$ as a white noise sequence with zero mean and variance $\sigma^2=0.10^2$ and $\sigma^2=0.50^2$,

**Table 5.2** The computational burden of the LSI algorithm.

| Variables | Computational sequences | Number of multiplications | Number of additions |
|---|---|---|---|
| $\hat{\boldsymbol{\theta}}_k$ | $\alpha_2 := \hat{\boldsymbol{\Phi}}_k^{\mathrm{T}}(L)\boldsymbol{Y}(L) \in \mathbb{R}^{n_1}$ | $n_1 L$ | $n_1(L-1)$ |
| | $\boldsymbol{H}_2 := \hat{\boldsymbol{\Phi}}_k^{\mathrm{T}}(L)\hat{\boldsymbol{\Phi}}_k(L) \in \mathbb{R}^{n_1 \times n_1}$ | $n_1^2 L$ | $n_1^2(L-1)$ |
| | $\boldsymbol{H}_2^{'} := \boldsymbol{H}_2^{-1} \in \mathbb{R}^{n_1 \times n_1}$ | $n_1^3$ | $n_1^3 - n_1^2$ |
| | $\hat{\boldsymbol{\theta}}_k = \boldsymbol{H}_2^{'}\boldsymbol{\alpha}_2 \in \mathbb{R}^{n_1}$ | $n_1^2$ | $n_1(n_1-1)$ |
| $\hat{v}_k(t)$ | $\hat{v}_k(t) = y(t) - \hat{\varphi}_k^{\mathrm{T}}(t)\hat{\boldsymbol{\theta}}_k \in \mathbb{R}$ | $n_1 L$ | $n_1 L$ |
| $\hat{\boldsymbol{x}}_k(t+1)$ | $\boldsymbol{p}_2 := \hat{\boldsymbol{A}}_k\hat{\boldsymbol{x}}_{k-1}(t) \in \mathbb{R}^n$ | $n^2 L$ | $n(n-1)L$ |
| | $\boldsymbol{q}_2 := \hat{\boldsymbol{B}}_k\hat{\boldsymbol{x}}_{k-1}(t-1) \in \mathbb{R}^n$ | $n^2 L$ | $n(n-1)L$ |
| | $\boldsymbol{s}_2 := \hat{\boldsymbol{g}}_k\hat{\bar{u}}_k(t) \in \mathbb{R}^n$ | $nL$ | |
| | $\hat{\boldsymbol{x}}_k(t+1) = \boldsymbol{p}_2 + \boldsymbol{q}_2 + \boldsymbol{s}_2 \in \mathbb{R}^n$ | | $2nL$ |
| | Sum | $(2n^2 + n + n_1^2 + 2n_1)L + n_1^3 + n_1^2$ | $(2n^2 + n_1^2 + 2n_1)L + n_1^3 - n_1^2 - 2n_1$ |
| | Total flops | $4n^2 + n + 2n_1^2 + 4_{n-1})L + 2n_1^3 - 2n_1$ [610640] | |

respectively. Apply the GI algorithm and LSI algorithm to estimate the parameters of this example system, the parameter estimates and their errors are shown in Tables 5.3–5.6 with the data length $L=2000$, and the parameter estimation error $\delta := ||\hat{\theta}_k - \theta||/||\theta||$ versus $k$ is shown in Figs. 5.2–5.3. When the noise variance is $\sigma^2 = 0.10^2$ and $\sigma^2 = 0.50^2$, the noise-to-signal ratio of the system is $\delta_{ns} = 7.88\%$ and $\delta_{ns} = 39.38\%$, respectively, where $\delta_{ns}$ is the noise-to-signal ratio defined by the square root of the ratio of the variances of $w(t)$ and $s(t)$, that is

$$\delta_{\text{ns}} = \sqrt{\frac{\text{var}[w(t)]}{\text{var}[s(t)]}} \times 100\% = \frac{\sigma_w}{\sigma_s} \times 100\%,$$

$$w(t) = 0.20v(t-1) + v(t),$$

$$s(t) = G(z)\overline{u}(t) = \frac{\beta(z)}{\alpha(z)}\overline{u}(t),$$

$$\begin{aligned}\alpha(z) &= 1 + \alpha_1 z^{-1} + \alpha_2 z^{-2} - \alpha_3 z^{-3}\alpha_4 z^{-4}\\ &= 1 - 0.2z^{-1} - 0.32z^{-2} - 0.1152z^{-3} + 0.0416z^{-4},\\ \beta(z) &= \beta_1 z^{-1} + \beta_2 z^{-2} + \beta_3 z^{-3}\\ &= 0.5z^{-1} + 0.9z^{-2} - 0.14z^{-3}.\end{aligned}$$

From Tables 5.3–5.6 and Figs. 5.2–5.3, we can draw the following conclusions: (1) as the noise-to-signal ratio decreases, the parameter estimation errors given by the GI algorithm and LSI algorithm become small, and the parameter estimates are close to their true values slowly—see Tables 5.3–5.6; (2) the parameter estimation accuracy of the LSI algorithm is higher than that of the GI algorithm—see Figs. 5.2–5.3; (3) the LSI algorithm has fast convergence rates and needs several iterations to converge to their true values—see Tables 5.5–5.6.

### 5.1.5 Conclusions

This chapter presents the iterative parameter estimation algorithms for nonlinear systems based on the state time-delay model with colored noise. The simulation results show that the proposed algorithm has fast convergence rates and can generate highly accurate parameter estimates after iterations. The proposed techniques can be extended to identify Wiener nonlinear models, Hammerstein–Wiener systems, and nonuniformly sampled multirate systems.

**Table 5.3** The GI parameter estimates and errors ($\sigma^2 = 0.10^2, \delta_{ns} = 7.88\%$).

| $k$ | 1 | 2 | 5 | 10 | 20 | 50 | 100 | 200 | 500 |
|---|---|---|---|---|---|---|---|---|---|
| $\theta_1 = 0.20$ | 0.24839 | 0.23948 | 0.21918 | 0.20254 | 0.19289 | 0.19316 | 0.19336 | 0.19297 | 0.19165 |
| $\theta_2 = 0.12$ | 0.15177 | 0.14912 | 0.14007 | 0.13297 | 0.12558 | 0.11870 | 0.11918 | 0.12111 | 0.12308 |
| $\theta_3 = 0.16$ | 0.23826 | 0.23384 | 0.22123 | 0.20300 | 0.17963 | 0.16173 | 0.16036 | 0.16013 | 0.15986 |
| $\theta_4 = 0.12$ | 0.17190 | 0.16273 | 0.14316 | 0.12607 | 0.11704 | 0.12106 | 0.12350 | 0.12522 | 0.12736 |
| $\theta_5 = 0.50$ | 0.04101 | 0.06986 | 0.11783 | 0.14996 | 0.18484 | 0.26366 | 0.35225 | 0.43923 | 0.48784 |
| $\theta_6 = 1.00$ | 0.08424 | 0.14359 | 0.23626 | 0.29634 | 0.36316 | 0.51745 | 0.69400 | 0.87702 | 0.99402 |
| $\theta_7 = 0.25$ | 0.12399 | 0.14506 | 0.19121 | 0.23248 | 0.25570 | 0.24860 | 0.24537 | 0.24644 | 0.24730 |
| $\theta_8 = 0.50$ | 0.14033 | 0.17668 | 0.26251 | 0.35536 | 0.44487 | 0.50114 | 0.50507 | 0.50282 | 0.50130 |
| $\theta_9 = 0.05$ | 0.07580 | 0.12503 | 0.19457 | 0.21516 | 0.20417 | 0.16546 | 0.12180 | 0.07895 | 0.05494 |
| $\theta_{10} = 0.10$ | 0.15632 | 0.25949 | 0.39533 | 0.43218 | 0.41005 | 0.33557 | 0.25011 | 0.16154 | 0.10494 |
| $\theta_{11} = 0.20$ | 0.00100 | 0.02948 | 0.06326 | 0.08436 | 0.10623 | 0.14470 | 0.17309 | 0.18797 | 0.19361 |
| $\delta$ (%) | 85.51773 | 80.35799 | 73.40554 | 68.06531 | 61.05053 | 45.93240 | 29.02240 | 11.77127 | 1.55905 |

**Table 5.4** The GI parameter estimates and errors ($\sigma^2 = 0.50^2, \delta_{ns} = 39.38\%$).

| $k$ | 1 | 2 | 5 | 10 | 20 | 50 | 100 | 200 | 500 |
|---|---|---|---|---|---|---|---|---|---|
| $\theta_1 = 0.20$ | 0.24584 | 0.23693 | 0.21635 | 0.19948 | 0.18935 | 0.18808 | 0.18592 | 0.18106 | 0.16865 |
| $\theta_2 = 0.12$ | 0.14996 | 0.14711 | 0.13684 | 0.12863 | 0.12074 | 0.11473 | 0.11712 | 0.12203 | 0.13013 |
| $\theta_3 = 0.16$ | 0.23618 | 0.23196 | 0.22003 | 0.20264 | 0.17980 | 0.16202 | 0.16061 | 0.16009 | 0.15949 |
| $\theta_4 = 0.12$ | 0.17040 | 0.16148 | 0.14270 | 0.12659 | 0.11844 | 0.12370 | 0.12818 | 0.13417 | 0.14699 |
| $\theta_5 = 0.50$ | 0.04025 | 0.06848 | 0.11503 | 0.14530 | 0.17701 | 0.24822 | 0.32830 | 0.40672 | 0.45008 |
| $\theta_6 = 1.00$ | 0.08545 | 0.14559 | 0.23926 | 0.29987 | 0.36691 | 0.52099 | 0.69804 | 0.88358 | 1.00275 |
| $\theta_7 = 0.25$ | 0.12243 | 0.14296 | 0.18750 | 0.22669 | 0.24751 | 0.23808 | 0.23416 | 0.23528 | 0.23640 |
| $\theta_8 = 0.50$ | 0.13947 | 0.17594 | 0.26200 | 0.35543 | 0.44644 | 0.50503 | 0.50927 | 0.50707 | 0.50584 |
| $\theta_9 = 0.05$ | 0.07510 | 0.12396 | 0.19319 | 0.21413 | 0.20437 | 0.16900 | 0.12932 | 0.09063 | 0.06919 |
| $\theta_{10} = 0.10$ | 0.15865 | 0.26331 | 0.40086 | 0.43836 | 0.41639 | 0.34200 | 0.25632 | 0.16657 | 0.10894 |
| $\theta_{11} = 0.20$ | 0.00050 | 0.03257 | 0.07346 | 0.10290 | 0.13617 | 0.18781 | 0.21170 | 0.21125 | 0.20713 |
| $\delta$ (%) | 85.48754 | 80.28680 | 73.32903 | 67.99166 | 61.02128 | 46.27633 | 29.83243 | 13.09033 | 5.42554 |

**Table 5.5** The LSI parameter estimates and errors ($\sigma^2 = 0.10^2$, $\delta_{ns} = 7.88\%$).

| $k$ | 1 | 2 | 3 | 4 | 5 | 10 |
|---|---|---|---|---|---|---|
| $\theta_1 = 0.20$ | 0.20097 | 0.20120 | 0.20124 | 0.20125 | 0.20125 | 0.20125 |
| $\theta_2 = 0.12$ | 0.11788 | 0.11773 | 0.11773 | 0.11773 | 0.11773 | 0.11773 |
| $\theta_3 = 0.16$ | 0.15776 | 0.15770 | 0.15770 | 0.15771 | 0.15771 | 0.15771 |
| $\theta_4 = 0.12$ | 0.12060 | 0.12034 | 0.12031 | 0.12030 | 0.12030 | 0.12030 |
| $\theta_5 = 0.50$ | 0.50034 | 0.49916 | 0.49927 | 0.49928 | 0.49928 | 0.49928 |
| $\theta_6 = 1.00$ | 0.99914 | 0.99919 | 0.99916 | 0.99918 | 0.99918 | 0.99918 |
| $\theta_7 = 0.25$ | 0.24948 | 0.24995 | 0.24981 | 0.24981 | 0.24981 | 0.24981 |
| $\theta_8 = 0.50$ | 0.50104 | 0.50082 | 0.50090 | 0.50088 | 0.50087 | 0.50087 |
| $\theta_9 = 0.05$ | 0.04931 | 0.04995 | 0.04993 | 0.04993 | 0.04993 | 0.04993 |
| $\theta_{10} = 0.10$ | 0.10049 | 0.10051 | 0.10052 | 0.10052 | 0.10051 | 0.10051 |
| $\theta_{11} = 0.20$ | 0.00203 | 0.18370 | 0.18879 | 0.18875 | 0.18875 | 0.18875 |
| $\delta$ (%) | 15.14432 | 1.27949 | 0.90521 | 0.90796 | 0.90796 | 0.90797 |

**Table 5.6** The LSI parameter estimates and errors ($\sigma^2 = 0.50^2$, $\delta_{ns} = 39.38\%$).

| $k$ | 1 | 2 | 3 | 4 | 5 | 10 |
|---|---|---|---|---|---|---|
| $\theta_1 = 0.20$ | 0.20485 | 0.20602 | 0.20621 | 0.20626 | 0.20626 | 0.20626 |
| $\theta_2 = 0.12$ | 0.10939 | 0.10864 | 0.10864 | 0.10863 | 0.10863 | 0.10863 |
| $\theta_3 = 0.16$ | 0.14880 | 0.14851 | 0.14852 | 0.14854 | 0.14854 | 0.14854 |
| $\theta_4 = 0.12$ | 0.12301 | 0.12169 | 0.12155 | 0.12152 | 0.12151 | 0.12151 |
| $\theta_5 = 0.50$ | 0.50171 | 0.49584 | 0.49636 | 0.49638 | 0.49638 | 0.49638 |
| $\theta_6 = 1.00$ | 0.99569 | 0.99597 | 0.99581 | 0.99592 | 0.99592 | 0.99592 |
| $\theta_7 = 0.25$ | 0.24742 | 0.24974 | 0.24903 | 0.24903 | 0.24903 | 0.24903 |
| $\theta_8 = 0.50$ | 0.50519 | 0.50408 | 0.50449 | 0.50438 | 0.50437 | 0.50437 |
| $\theta_9 = 0.05$ | 0.04657 | 0.04975 | 0.04965 | 0.04964 | 0.04965 | 0.04965 |
| $\theta_{10} = 0.10$ | 0.10244 | 0.10254 | 0.10260 | 0.10258 | 0.10257 | 0.10257 |
| $\theta_{11} = 0.20$ | 0.01016 | 0.18420 | 0.18879 | 0.18875 | 0.18875 | 0.18875 |
| $\delta$ (%) | 14.58907 | 1.88368 | 1.68655 | 1.68494 | 1.68473 | 1.68481 |

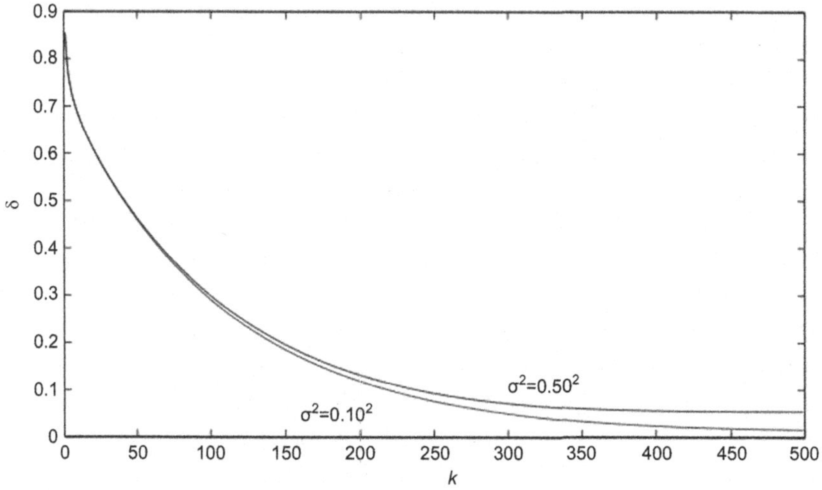


**Figure 5.2** The GI estimation errors $\delta$ versus $k$ with $\sigma^2 = 0.10^2$ and $\sigma^2 = 0.50^2$.

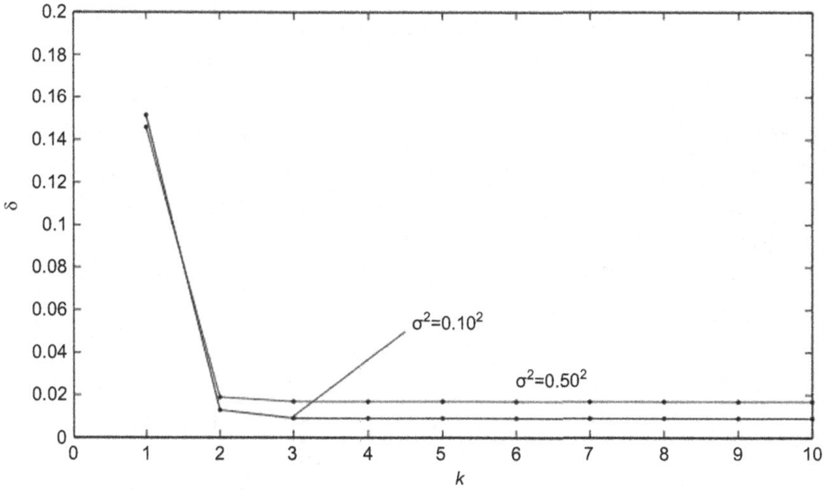


**Figure 5.3** The LSI estimation errors $\delta$ versus $k$ with $\sigma^2 = 0.10^2$ and $\sigma^2 = 0.50^2$.

## 5.2 The bias compensation-based parameter and state estimation for a nonlinear system

The actual part of the control system in process industry is always nonlinear to some extent, for example, as the amplifier transistor magnifying glass, because of their components have a linear working range, once beyond the scope of amplifier saturation phenomenon will appear that saturation nonlinear characteristics. The nonlinear characteristic has a great influence on the design of system controller. To eliminate these effects, the design of a better robust controller requires that all parameters of the nonlinear system be known in advance. Therefore, the research for nonlinear system identification has been a hot topic over the past decades (Beikzadeh et al., 2018).

The identification of state-space systems has received considerable attention in the past few decades. For state-space systems, there are many identification methods, such as subspace identification algorithm, recursive least squares algorithm, and Kalman filter algorithm. Most existing work assumes that the system is linear and that the system has no time delay. However, due to hardware limitations, the system is nonlinear and may have time delays. It is well known that the lack of time delay reduces the accuracy of the estimation and even cannot be identified by the system. Therefore, how to identify a system with a time delay has attracted widespread attention.

Time-delay systems often exist in engineering practice. For example, in the communication network, due to the network congestion, the signals which are transmitted over a communication channel often have time delays. In chemical processes, some variables such as chemical component concentrations are often measured through laboratory analysis, which will introduce time delays. In a communications network, because of network congestion, signals sent through communication channels usually have a time delay. In chemistry, most of the key variables (such as chemical composition concentration) are usually measured by online analyzers or laboratory analysis, which cause time delays. Recently, there are many recognition algorithms for time-delay systems.

The main contributions of the paper are as follows.

1. This communique derives the state and parameter identification of the state-space model with colored noise and time delay and presents a bias compensation-based identification algorithm for jointly estimating the system parameters and states based on the bias compensation.
2. The convergence of the proposed algorithm is analyzed using the stochastic process theory.

## 5.2.1 The system description and identification model

Consider the following nonlinear state-space system with time delay,

$$\boldsymbol{x}(t+1) = \boldsymbol{A}\boldsymbol{x}(t) + \boldsymbol{B}\boldsymbol{x}(t-1) + \boldsymbol{g}\bar{u}(t), \tag{5.57}$$

$$\overline{u}(t) = f(u(t)), \tag{5.58}$$

$$y(t) = cx(t) + w(t), \tag{5.59}$$

$$w(t) = \sum_{s=1}^{n_d} d_s v(t-s) + v(t). \tag{5.60}$$

$$A := \begin{bmatrix} 0 & 1 & \cdots & 0 \\ 0 & 0 & \ddots & \vdots \\ \vdots & \vdots & \ddots & 0 \\ 0 & 0 & 0 & 1 \\ -a_n & -a_{n-1} & \cdots & -a_1 \end{bmatrix} \in \mathbb{R}^{n\times n},$$

$$B := \left[b_1^{\mathrm{T}}, b_2^{\mathrm{T}}, \ldots, b_n^{\mathrm{T}}\right]^{\mathrm{T}} \in \mathbb{R}^{n\times n}, b_i \in \mathbb{R}^{1\times n},$$

$$f := \left[f1, f2, \ldots, f_n\right]^{\mathrm{T}} \in \mathbb{R}^n,$$

$$c := \begin{bmatrix} 1, & 0, & \ldots & 0 \end{bmatrix} \in \mathbb{R}^{1\times n}.$$

where $x(t) \in R^n$ is the state vector, $u(t) \in R$ is the system input, $y(t) \in R$ is the system output, $v(t) \in R$ is a random noise with zero mean, and $A \in R^{n\times n}$, $B \in R^{n\times n}$, $g \in R^n$, and $c \in R^{1\times n}$ are the system parameter matrices/vectors. Assume that $(c, A)$ is observable and $u(t) = 0$, $y(t) = 0$, and $v(t) = 0$ for $t \leq 0$.

The external disturbance $w(t)$ can be fitted by a moving average process, an autoregressive process or an autoregressive moving average process. Without loss of generality, we consider $w(t)$ as a moving average noise process:

$$w(t) := \left(1 + w_1 z^{-1} + w_2 z^{-2} + \cdots + w_{n_d} z^{-n_d}\right) v(t). \tag{5.61}$$

Since $w(t)$ is a colored noise, the parameter estimation of the conventional least square algorithm is biased. Here, the principle of deviation compensation is used to study the identification problem of such systems. The basic idea is to add a correction term on the basis of the least square estimation to obtain the unbiased estimation. The nonlinear block is a nonlinear function:

$$\overline{u}(t) = f(u(t)) = \sum_{l=1}^{m} p_l(t) f_l(u(t)) = \boldsymbol{f}(u(t))\boldsymbol{p}, \tag{5.62}$$

where the known basis $f := (f_1, f_2, \ldots, f_m)$ are known functions of the system input $u(t)$, the unknown parameters $p := (p_1, p_2, \ldots, p_m)$ are the coefficients of the nonlinear functions, and $m$ is the number of the parameters of the nonlinear block. The system in (5.57)–(5.60) is an observability canonical form, and its transformation matrix T is a nonsingular matrix, that is,

$$\boldsymbol{T} := \left[c^{\mathrm{T}}, (c\boldsymbol{A})^{\mathrm{T}}, \ldots, (c\boldsymbol{A}^{n-1})^{\mathrm{T}}\right]^{\mathrm{T}} = \boldsymbol{I}_n. \tag{5.63}$$

Due to the unknown state variable in (5.57), it needs to derive a new identity expression that only involves the available input–output data $\{u(t), y(t)\}$. The following transforms the state-space model with time delay in (5.57)–(5.60) into an input–output representation and gives its identification model.

From (5.57)–(5.60), we have

$$\begin{aligned} y(t+i) = {} & \boldsymbol{c}\boldsymbol{A}^{i}\boldsymbol{x}(t) + \boldsymbol{c}\boldsymbol{A}^{i-1}\boldsymbol{B}\boldsymbol{x}(t-1) + \boldsymbol{c}\boldsymbol{A}^{i-2}\boldsymbol{B}\boldsymbol{x}(t) \\ & + \cdots + \boldsymbol{c}\boldsymbol{B}\boldsymbol{x}(t-1+i-1) + \boldsymbol{c}\boldsymbol{A}^{i-1}\boldsymbol{g}\boldsymbol{f}(u(t))p_1 \\ & + \boldsymbol{c}\boldsymbol{A}^{i-2}\boldsymbol{g}\boldsymbol{f}(u(t+1))p_2 + \cdots \\ & + \boldsymbol{c}\boldsymbol{g}\boldsymbol{f}(u(t+i-1))p_m + w(t+i), \end{aligned} \tag{5.64}$$

$$\begin{aligned} y(t+n) = {} & \boldsymbol{c}\boldsymbol{A}^{n}\boldsymbol{x}(t) + \boldsymbol{c}\boldsymbol{A}^{n-1}\boldsymbol{B}\boldsymbol{x}(t-1) + \boldsymbol{c}\boldsymbol{A}^{n-2}\boldsymbol{B}\boldsymbol{x}(t) \\ & + \cdots + \boldsymbol{c}\boldsymbol{B}\boldsymbol{x}(t-2+n) + \boldsymbol{c}\boldsymbol{A}^{n-1}\boldsymbol{g}\boldsymbol{f}(u(t))p_1 \\ & + \boldsymbol{c}\boldsymbol{A}^{n-2}\boldsymbol{g}\boldsymbol{f}(u(t+1))p_2 + \cdots \\ & + \boldsymbol{c}\boldsymbol{g}\boldsymbol{f}(u(t+n-1))p_m + w(t+n). \end{aligned} \tag{5.65}$$

Define some vectors/matrices,

$$\begin{aligned} & \phi_y(t) := [y(t), y(t+1), \ldots, y(t+n-1)]^{\mathrm{T}} \in \mathbb{R}^n, \\ & \phi_u(t) := \left[\boldsymbol{f}(u(t)), \boldsymbol{f}(u(t+1)), \ldots, \boldsymbol{f}(u(t+n-1))\right]^{\mathrm{T}} \in \mathbb{R}^n, \\ & \mathbf{X}(t-1+n) := \left[\boldsymbol{x}^{\mathrm{T}}(t-1), \boldsymbol{x}^{\mathrm{T}}(t), \ldots, \boldsymbol{x}^{\mathrm{T}}(t-2+n)\right]^{\mathrm{T}} \in \mathbb{R}^{n^2}, \\ & \phi_w(t) := [w(t), w(t+1), \ldots, w(t+n-1)]^{\mathrm{T}} \in \mathbb{R}^n, \end{aligned}$$

$$\boldsymbol{M} := \begin{bmatrix} 0 & \cdots & 0 & 0 \\ \boldsymbol{cB} & \cdots & 0 & 0 \\ \boldsymbol{cAB} & \ddots & \vdots & \vdots \\ \vdots & \ddots & 0 & 0 \\ \boldsymbol{cA}^{n-2}\boldsymbol{B} & \cdots & \boldsymbol{cB} & 0 \end{bmatrix} \in \mathbb{R}^{n \times n^2},$$

$$\mathbf{Q} := \begin{bmatrix} 0 & \cdots & 0 & 0 \\ \boldsymbol{c}\boldsymbol{g}p_m & \cdots & 0 & 0 \\ \boldsymbol{cAg}p_{m-1} & \ddots & \vdots & \vdots \\ \vdots & \ddots & 0 & 0 \\ \boldsymbol{cA}^{n-2}\boldsymbol{g}p_2 & \cdots & \boldsymbol{cg}p_m & 0 \end{bmatrix} \in \mathbb{R}^{n \times n}.$$

From Eqs. (5.64) and (5.65) and the above definitions, we have

$$\begin{aligned} \phi_y(t+n) &= \boldsymbol{Tx}(t) + \boldsymbol{MX}(t-1+n) + \mathbf{Q}\phi_u(t+n) \\ &\quad + \phi_w(t+n) \\ &= \boldsymbol{x}(t) + \boldsymbol{MX}(t-1+n) + \mathbf{Q}\phi_u(t+n) \\ &\quad + \phi_w(t+n), \end{aligned}$$

or

$$\boldsymbol{x}(t) = \phi_u(t+n) - \boldsymbol{MX}(t-1+n) - \mathbf{Q}\phi_u(t+n) - \phi_w(t+n). \tag{5.66}$$

Define the information vectors $\varphi_s(t)$, $\varphi_n(t)$ and the parameter vector $\theta$ as

$$\varphi_s(t) := \begin{bmatrix} \phi_y(t) \\ \boldsymbol{X}(t-1) \\ \phi_u(t) \end{bmatrix} \in \mathbb{R}^{2n+n^2},$$

$$\varphi_n(t) := \begin{bmatrix} -\phi_w(t) \\ 0 \\ 0 \end{bmatrix} \in \mathbb{R}^{2n+n^2},$$

$$\boldsymbol{\theta} := \begin{bmatrix} \boldsymbol{\theta}_a \\ \boldsymbol{\theta}_b \\ \boldsymbol{\theta}_c \end{bmatrix} \in \mathbb{R}^{2n+n^2},$$

$$\boldsymbol{\theta}_a := [\boldsymbol{cA}^n]^{\mathrm{T}} \in \mathbb{R}^n,$$

$$\boldsymbol{\theta}_b := \left[-\boldsymbol{cA}^n\boldsymbol{M} + [\boldsymbol{cA}^{n-1}\boldsymbol{B}, \ldots, \boldsymbol{cB}]\right]^{\mathrm{T}} \in \mathbb{R}^{n^2},$$

$$\boldsymbol{\theta}_c := \left[-\boldsymbol{cA}^n\mathbf{Q} + [\boldsymbol{cA}^{n-1}\boldsymbol{g}p_1, \ldots, \boldsymbol{cg}p_m]\right]^{\mathrm{T}} \in \mathbb{R}^n.$$

Inserting (5.66) into (5.65) yields

$$\begin{aligned}
y(t+n) &= \boldsymbol{c}\boldsymbol{A}^n[\phi_y(t+n) - \boldsymbol{M}\boldsymbol{X}(t-1+n) \\
&\quad - \mathbf{Q}\phi_u(t+n) - \phi_w(t+n)] \\
&\quad + \boldsymbol{c}\boldsymbol{A}^{n-1}\mathbf{B}\boldsymbol{x}(t-1) + \boldsymbol{c}\boldsymbol{A}^{n-2}\mathbf{B}\boldsymbol{x}(t) + \ldots \\
&\quad + \boldsymbol{c}\mathbf{B}\boldsymbol{x}(t-2+n) + \boldsymbol{c}\boldsymbol{A}^{n-1}\boldsymbol{g}p_1 f(u(t)) \\
&\quad + \boldsymbol{c}\boldsymbol{A}^{n-2}\boldsymbol{g}p_2 f(u(t+1)) + \ldots \\
&\quad + \boldsymbol{c}\boldsymbol{g}f(u(t+n-1))p_m + w(t+n) \\
&= \boldsymbol{c}\boldsymbol{A}^n[\phi_y(t+n) - \boldsymbol{M}\boldsymbol{X}(t-1+n) \\
&\quad - \mathbf{Q}\phi_u(t+n) - \phi_w(t+n)] \\
&\quad + \left[\boldsymbol{c}\boldsymbol{A}^{n-1}\mathbf{B}, \ldots, \boldsymbol{c}\mathbf{B}\right]\begin{bmatrix} \boldsymbol{x}(t-1) \\ \boldsymbol{x}(t) \\ \vdots \\ \boldsymbol{x}(t-2+n \end{bmatrix} \\
&\quad + \left[\boldsymbol{c}\boldsymbol{A}^{n-1}\boldsymbol{g}p_1, \boldsymbol{c}\boldsymbol{A}^{n-2}\boldsymbol{g}p_2, \ldots, \boldsymbol{c}\boldsymbol{g}p_m\right] \\
&\quad \times \begin{bmatrix} f(u(t)) \\ f(u(t+1)) \\ \vdots \\ f(u(t+n-1)) \end{bmatrix} + w(t+n) \\
&= \boldsymbol{c}\boldsymbol{A}^n\phi_y(t+n) + [-\boldsymbol{c}\boldsymbol{A}^n\boldsymbol{M} + [\boldsymbol{c}\boldsymbol{A}^{n-1}\mathbf{B}, \\
&\quad \boldsymbol{c}\boldsymbol{A}^{n-2}\mathbf{B}, \ldots, \boldsymbol{c}\mathbf{B}]]\boldsymbol{X}(t-1+n) \\
&\quad + \left[-\boldsymbol{c}\boldsymbol{A}^n\mathbf{Q} + \left[\boldsymbol{c}\boldsymbol{A}^{n-1}\boldsymbol{g}p_1, \ldots, \boldsymbol{c}\boldsymbol{g}p_m\right]\right] \\
&\quad \times \phi_u(t+n) - \boldsymbol{c}\boldsymbol{A}^n\phi_w(t+n) + w(t+n) \\
&= \phi_y^{\mathrm{T}}(t+n)\boldsymbol{\theta}_a + \boldsymbol{X}^{\mathrm{T}}(t-1+n)\boldsymbol{\theta}_b \\
&\quad + \phi_u^{\mathrm{T}}(t+n)\boldsymbol{\theta}_c - \phi_w^{\mathrm{T}}(t+n)\boldsymbol{\theta}_a + w(t+n) \\
&= \varphi_s^{\mathrm{T}}(t+n)\boldsymbol{\theta} + \varphi_n^{\mathrm{T}}(t+n)\boldsymbol{\theta} + w(t+n).
\end{aligned} \tag{5.67}$$

Replacing $t+n$ in Eq. (5.67) with $t$ gives:

$$\begin{aligned}
y(t) &= \boldsymbol{\varphi}_s^{\mathrm{T}}(t)\boldsymbol{\theta} + \boldsymbol{\varphi}_n^{\mathrm{T}}(t)\boldsymbol{\theta} + w(t) \\
&=: \boldsymbol{\varphi}_s^{\mathrm{T}}(t)\boldsymbol{\theta} + e(t).
\end{aligned} \tag{5.68}$$

The above expression is the system identification model in least squares form of the state-space system in Eqs. (5.57)–(5.60), where the information vector $\phi_y(t)$ and $\phi_u(t)$ are composed of the known data. Define the parameter vector $\theta_v$ and the information vector $\varphi_v(t)$ as:

$$\boldsymbol{\theta}_v := \left[d_1, d_2, \ldots, d_{n_d}\right] \in \mathbb{R}^{n_d},$$
$$\boldsymbol{\varphi}_v(t) := [v(t-1), v(t-2)\ldots, v(t-n_d)]^{\mathrm{T}} \in \mathbb{R}^{n_d}.$$

From Eq. (5.59), we have:

$$w(t) = \boldsymbol{\varphi}_v^{\mathrm{T}}(t)\boldsymbol{\theta}_v + v(t). \tag{5.69}$$

Based on the system model in Eq. (5.68) and the noise model in Eq. (5.69), the identification algorithm based on bias compensation is derived below.

**Remark 5.1**: For the identification model in (5.57)−(5.60), the input−output data $\{u(t), y(t)\}$ are available. That is, only $y(t)$ and $f(u(t))$ are available, but $x(t)$ and $w(t)$ are unavailable and unknown.

**Remark 5.2**: According to our knowledge, parameter estimation of nonlinear state-space systems has been studied for many years. Although a least squares algorithm has been developed for state-space system with colored noise, the state variable is eliminated, so the calculation is large, but the state estimation is not considered. In addition, less work has focused on theoretical analysis of the performance of the proposed algorithm. This inspired us to explore new and effective methods for state and parameter combination estimation and provided theoretical analysis to prove the convergence of the proposed algorithm.

**Remark 5.3**: The difficulty of identification is that the information vectors $\varphi_s(t)$ and $\varphi_n(t)$ contain the unknown inner variable $X(t)$ and $\phi_w(t)$, respectively. The solution here is to use the measured data $u(t)$ and $y(t)$ and to replace the unknown $x(t-i)$ and $w(t)$ in the identification algorithm with the estimated value $\hat{x}(t-i)$ and $\hat{w}(t)$.

## 5.2.2 The parameter estimation algorithm

Using the least squares search and minimizing the cost function give the least squares estimate of the parameter vector $\theta$:

$$\hat{\boldsymbol{\theta}}_{LS}(t) = \left[\sum_{i=1}^{t} \varphi_s(i)\varphi_s^{\mathrm{T}}(i)\right]^{-1} \sum_{i=1}^{t} \varphi_s(i)y(i). \tag{5.70}$$

Using Eqs. (5.68) and (5.70) gives

$$\left[\sum_{i=1}^{t}\varphi_s(i)\varphi_s^{\mathrm{T}}(i)\right]\left[\hat{\theta}_{LS}(t)-\theta\right]$$
$$=\sum_{i=1}^{t}\varphi_s(i)y(i)-\sum_{i=1}^{t}\varphi_s(i)\varphi_s^{\mathrm{T}}(i)\theta$$
$$=\sum_{i=1}^{t}\varphi_s(i)\left[y(i)-\varphi_s^{\mathrm{T}}(i)\theta\right]$$
$$=\sum_{i=1}^{t}\varphi_s(i)\left[\varphi_s^{\mathrm{T}}(i)\theta+w(i)\right].$$

Dividing by $t$ and taking limits on both sides give:

$$\lim_{t\to\infty}\frac{1}{t}\left[\left[\sum_{i=1}^{t}\varphi_s(i)\varphi_s^{\mathrm{T}}(i)\right]\left[\hat{\theta}_{LS}(t)-\theta\right]\right]$$
$$=\lim_{t\to\infty}\left[\frac{1}{t}\sum_{i=1}^{t}\varphi_s(i)\varphi_n^{\mathrm{T}}(i)\right]\theta$$
$$+\lim_{t\to\infty}\frac{1}{t}\sum_{i=1}^{t}\varphi_s(i)w(i). \tag{5.71}$$

Note that $w(t)$ in Eq. (5.71) is the moving average noise and $v(t)$ is white noise with zero mean and variance $\delta$ and is independent of the inputs.

From Eq. (5.61), we have:

$$H(0):=\mathrm{E}\left[w^2(t)\right]$$
$$=\left[1+\theta_v^{\mathrm{T}}\theta_v\right]\delta,$$
$$H(i):=\mathrm{E}[w(t-1)w(t)]$$
$$:=\theta_v^{\mathrm{T}}(i{:}n_d)\left[\theta_v(1{:}^{1}_{n_d}-i)\right]\delta, i=1,2,\ldots,n_d,$$

where $H(i)$ is the autocorrelation function of the noise $w(t)$ and $H(i)=0$ when $i>n_d$. Define the autocorrelation function vectors $h$ and the autocorrelation function matrices $H$, $\Lambda$ as:

$$\boldsymbol{h}:=[H(n),H(n-1),\ldots,H(1),0,0,\ldots,0]\in\mathbb{R}^{2n}$$
$$\boldsymbol{H}:=\mathrm{diag}[\Lambda,0]\in\mathbb{R}^{(2n)\times(2n)},$$

$$\mathbf{\Lambda} := \begin{bmatrix} H(0) & H(1) & \dots & H(n-1) \\ H(1) & H(0) & \dots & H(n-2) \\ \vdots & \vdots & \ddots & \vdots \\ H(n-1) & H(n-2) & \dots & H(0) \end{bmatrix} \in \mathbb{R}^{n\times n}.$$

Eq. (5.70) can be rewritten as:

$$\lim_{t\to\infty} \hat{\boldsymbol{\theta}}_{\mathrm{LS}}(t) = \boldsymbol{\theta} - t\boldsymbol{P}(t)(H\vartheta - h).$$

The above equation shows that the least square estimation of model parameters is biased. If compensation $tP(t)(H\vartheta - h)$ is introduced into the least square estimation $\hat{\theta}_{LS}(t)$, $\hat{\theta}_C(t) = \hat{\theta}_{LS}(t) + tP(t)[\hat{H}(t)\hat{\theta}_C(t) - \hat{h}(t)]$ is the unbiased estimation of $\theta$. This is the basic idea of the deviation-compensated least square method. Let us write it recursively as

$$\hat{\boldsymbol{\theta}}_{\mathrm{C}}(t) = \hat{\boldsymbol{\theta}}_{\mathrm{LS}}(t) + t\boldsymbol{P}(t)\left[\hat{\boldsymbol{H}}(t)\hat{\boldsymbol{\theta}}_{\mathrm{C}}(t) - \hat{\boldsymbol{h}}(t)\right],$$

where $\hat{\theta}_C(t)$ and $\hat{\theta}_{LS}(t)$ are the bias compensation least squares estimation and the least squares estimation at time $t$ of parameter $\theta$, respectively. $\hat{\theta}_C(t)$ is related to the estimates of $H$ and $h$ (i.e., $\delta$ and $\theta_v$).

The following gives the estimates based on the interactive identification. Define the covariance matrix:

$$\boldsymbol{P}(t) = \left[\sum_{i=1}^{t} \boldsymbol{\varphi}_s(i)\boldsymbol{\varphi}_s^{\mathrm{T}}(i)\right]^{-1}.$$

The question turns into how to estimate $\hat{\delta}(t)$. Define

$$\varepsilon_{\mathrm{LS}}(i) = y(i) - \boldsymbol{\varphi}_s^{\mathrm{T}}(i)\hat{\boldsymbol{\theta}}_{\mathrm{LS}}(t), i = 1, 2, \dots, t.$$

Using Eq. (5.68) and the following

$$\sum_{i=1}^{t} \varepsilon_{\mathrm{LS}}(i)\boldsymbol{\varphi}_s^{\mathrm{T}}(i) = \mathbf{0},$$

we have

$$\begin{aligned} \sum_{i=1}^{t} \varepsilon_{\mathrm{LS}}^2(i) &= \sum_{i=1}^{t} \varepsilon_{\mathrm{LS}}(i)\left[y(i) - \boldsymbol{\varphi}_s^{\mathrm{T}}(i)\hat{\boldsymbol{\theta}}_{\mathrm{LS}}(t)\right] \\ &= \sum_{i=1}^{t} \boldsymbol{\varphi}_s^{\mathrm{T}}(i)\left[\boldsymbol{\theta} - \hat{\boldsymbol{\theta}}_{\mathrm{LS}}(t)\right]\left[\boldsymbol{\varphi}_n^{\mathrm{T}}(i)\boldsymbol{\theta} + w(i)\right] \\ &\quad + \sum_{i=1}^{t} \left[\boldsymbol{\varphi}_n^{\mathrm{T}}(i)\boldsymbol{\theta} + w(i)\right]^2. \end{aligned}$$

Divide both sides of the above equation by $t$ to take the limit, and use the white noise characteristics of $v(t)$ to obtain

$$\begin{aligned}\lim_{t\to\infty}\frac{1}{t}\sum_{i=1}^{t}\varepsilon_{\mathrm{LS}}^{2}(i)=&\lim_{t\to\infty}\frac{1}{t}\sum_{i=1}^{t}\boldsymbol{\varphi}_{s}^{\mathrm{T}}(i)\left[\boldsymbol{\theta}-\hat{\boldsymbol{\theta}}_{\mathrm{LS}}(t)\right]\\&\times\left[\boldsymbol{\varphi}_{\mathrm{n}}^{\mathrm{T}}(i)\boldsymbol{\theta}+w(i)\right]\\&+\lim_{t\to\infty}\frac{1}{t}\sum_{i=1}^{t}\left[\boldsymbol{\varphi}_{\mathrm{n}}^{\mathrm{T}}(i)\boldsymbol{\theta}+w(i)\right]^{2}\\&=\boldsymbol{\theta}^{\mathrm{T}}\boldsymbol{h}\hat{\boldsymbol{\theta}}_{\mathrm{LS}}(t)-\boldsymbol{h}^{\mathrm{T}}\left[\boldsymbol{\theta}+\hat{\boldsymbol{\theta}}_{\mathrm{LS}}(t)\right]\\&+\delta\left[1+\boldsymbol{\theta}_{v}^{\mathrm{T}}\boldsymbol{\theta}_{v}\right].\end{aligned}$$

In conclusion, the bias compensation-based recursive least squares algorithm is as follows,

$$\hat{\boldsymbol{\theta}}_{\mathrm{C}}(t)=\hat{\boldsymbol{\theta}}_{\mathrm{LS}}(t)+t\boldsymbol{P}(t)\left[\hat{\boldsymbol{H}}(t)\hat{\boldsymbol{\theta}}_{\mathrm{C}}(t)-\hat{\boldsymbol{h}}(t)\right],$$

$$\hat{\boldsymbol{\theta}}_{\mathrm{LS}}(t)\hat{\boldsymbol{\theta}}_{\mathrm{LS}}(t-1)+\boldsymbol{L}(t)\left[y(t)-\boldsymbol{\varphi}_{s}^{\mathrm{T}}(t)\hat{\boldsymbol{\theta}}_{\mathrm{LS}}(t-1)\right],$$

$$\boldsymbol{L}(t)=\frac{\boldsymbol{P}(t-1)\boldsymbol{\varphi}_{s}(t)}{1+\boldsymbol{\varphi}_{s}^{\mathrm{T}}(t)\boldsymbol{P}(t-1)\boldsymbol{\varphi}_{s}(t)},$$

$$\boldsymbol{P}(t)=\left[\boldsymbol{I}-\boldsymbol{L}(t)\boldsymbol{\varphi}_{s}^{\mathrm{T}}(t)\right]\boldsymbol{P}(t-1),\ P(0)=p_{0}\boldsymbol{I},$$

$$\hat{\delta}(t)=\frac{\frac{1}{t}J(t)}{\chi},$$

$$\begin{aligned}\chi=&\hat{\boldsymbol{\theta}}_{C}^{\mathrm{T}}(t)\hat{\boldsymbol{R}}(t)\hat{\boldsymbol{\theta}}_{\mathrm{LS}}(t)-\hat{\boldsymbol{h}}^{\mathrm{T}}(t)\left[\hat{\boldsymbol{\theta}}_{\mathrm{C}}(t)+\hat{\boldsymbol{\theta}}_{\mathrm{LS}}(t)\right]/\delta\\&+1+\hat{\boldsymbol{\theta}}_{v}^{\mathrm{T}}(t)\hat{\boldsymbol{\theta}}_{v}(t),\end{aligned}$$

$$J(t)=J(t-1)+\frac{\left[y(t)-\boldsymbol{\varphi}_{s}^{\mathrm{T}}(t)\hat{\boldsymbol{\theta}}_{\mathrm{LS}}(t-1)\right]^{2}}{1+\boldsymbol{\varphi}_{s}^{\mathrm{T}}(t)\boldsymbol{P}(t-1)\boldsymbol{\varphi}_{s}(t)},$$

$$\hat{\boldsymbol{h}}=\left[\hat{H}(n),\hat{H}(n-1),\ldots,\hat{H}(1),0,0,\ldots,0\right]\in\mathbb{R}^{2n},$$

$$\hat{\boldsymbol{R}}(t) = \begin{bmatrix} 1+\hat{\boldsymbol{\theta}}_v^{\mathrm{T}}(t)\hat{\boldsymbol{\theta}}_v(t) & \cdots & \dfrac{\hat{H}(n-1)}{\hat{\delta}(t-1)} & 0 \\ \dfrac{\hat{H}(1)}{\hat{\delta}(t-1)} & \cdots & \dfrac{\hat{H}(n-2)}{\hat{\delta}(t-1)} & 0 \\ \vdots & \ddots & \vdots & \vdots \\ \dfrac{\hat{H}(n-1)}{\hat{\delta}(t-1)} & \cdots & 1+\hat{\boldsymbol{\theta}}_v^{\mathrm{T}}(t)\hat{\boldsymbol{\theta}}_v(t) & 0 \\ 0 & \cdots & 0 & 0 \end{bmatrix},$$

$$\frac{\hat{H}(i)}{\hat{\delta}(t-1)} = \hat{\boldsymbol{\theta}}_v^{\mathrm{T}}(i{:}n_d)\left[\hat{\boldsymbol{\theta}}_v\,(1{:}n_d^{1} - i)\right], i = 1, 2, \ldots, n_d.$$

The above equation involves the estimate $\hat{\theta}_v(t)$ of the noise parameter vector $\theta_v$, which can be computed by the noise model. Using the least squares principle, the estimate $\hat{\theta}_v(t)$ of $\theta_v(t)$ is as follows,

$$\begin{aligned} \hat{\boldsymbol{\theta}}_v(t) &= \hat{\boldsymbol{\theta}}_v(t-1) \\ &\quad + \boldsymbol{P}_v(t)\hat{\boldsymbol{\varphi}}_v(t)\left[\hat{w}(t) - \boldsymbol{\varphi}_v^{\mathrm{T}}(t)\hat{\boldsymbol{\theta}}_v(t-1)\right], \\ \boldsymbol{P}_v^{-1}(t) &= \boldsymbol{P}_v^{-1}(t-1) + \hat{\boldsymbol{\varphi}}_v(t)\boldsymbol{\varphi}_v^{\mathrm{T}}(t), \\ \hat{\boldsymbol{\varphi}}_v(t) &= [\hat{v}(t-1), \hat{v}(t-2), \ldots, \hat{v}(t-n_d)]^{\mathrm{T}}, \\ \hat{w}(t) &= y(t) - \boldsymbol{\varphi}_s^{\mathrm{T}}(t)\hat{\boldsymbol{\theta}}_C(t) - \boldsymbol{\varphi}_s^{\mathrm{T}}(t)\hat{\boldsymbol{\theta}}_C(t), \\ \hat{v}(t) &= \hat{w}(t) - \boldsymbol{\varphi}_v^{\mathrm{T}}(t)\hat{\boldsymbol{\theta}}_v(t). \end{aligned}$$

**Remark 5.4**: For the purpose of reducing the influence of the colored noise on parameter estimation, the interactive estimation process is introduced by using a compensation-based parameter estimation algorithm: compute the estimates $\hat{\theta}_v(t)$ and $\hat{\delta}(t)$ by the output $y(t)$ and the information vectors $\varphi_s(t)$, $\varphi_v(t)$; then update the unbiased estimate $\hat{\theta}_C(t)$ by these obtained variable estimates.

### 5.2.3 The state estimation algorithm

According to the principle of hierarchical identification, the first step is to recursively calculate the parameter estimation of the system based on the input–output data of the system, and the second step is to recursively calculate the state estimation of the system based on the input–output data and the obtained parameter estimation.

Since the relationship between the parameter vector $\theta$ and the matrices/vectors $A$, $B$, $g$, $p$, and $d$ has been established, here look for simple algebra to establish the relationship between the parameter vector $\theta$ and the parameters $A$, $B$, $g$, $p$, and $d$.

Post-multiplying Eq. (5.63) by $B$ gives

$$\boldsymbol{c}\boldsymbol{A}^{i-1}\boldsymbol{B} = \boldsymbol{b}_{i,}\ i = 1, 2, \ldots, n.$$

Then, the parameter matrix $M$ can be expressed as:

$$\boldsymbol{M} = \begin{bmatrix} \mathbf{0} & \cdots & \mathbf{0} & \mathbf{0} \\ \boldsymbol{b}_1 & \cdots & \mathbf{0} & \mathbf{0} \\ \boldsymbol{b}_2 & \ddots & \vdots & \vdots \\ \vdots & \ddots & \mathbf{0} & \mathbf{0} \\ \boldsymbol{b}_{n-1} & \cdots & \boldsymbol{b}_1 & \mathbf{0} \end{bmatrix}.$$

Post-multiplying Eq. (5.63) by $A$ and writing down the nth row:

$$\boldsymbol{c}\boldsymbol{A}^n = [-a_n, -a_{n-1}, \ldots, -a_1].$$

According to the definitions of $\theta_a$, $\theta_b$, and $\theta_c$, we have:

$$\boldsymbol{\theta}_a = [\boldsymbol{c}\boldsymbol{A}^n]^{\mathrm{T}} = [-a_n, -a_{n-1}, \ldots, -a_1]^{\mathrm{T}},$$

$$\begin{aligned} \boldsymbol{\theta}_b &= \left[-\boldsymbol{c}\boldsymbol{A}^n\boldsymbol{M} + \left[\boldsymbol{c}\boldsymbol{A}^{n-1}\boldsymbol{B}, cA^{n-2}\boldsymbol{B}, \ldots, cB\right]\right]^{\mathrm{T}} \\ &= \begin{bmatrix} \boldsymbol{b}_1^{\mathrm{T}} a_{n-1} + \boldsymbol{b}_2^{\mathrm{T}} a_{n-2} + \cdots + \boldsymbol{b}_{n-1}^{\mathrm{T}} a_1 + \boldsymbol{b}_n^{\mathrm{T}} \\ \boldsymbol{b}_1^{\mathrm{T}} a_{n-2} + \boldsymbol{b}_2^{\mathrm{T}} a_{n-3} + \cdots + \boldsymbol{b}_{n-2}^{\mathrm{T}} a_1 + \boldsymbol{b}_{n-1}^{\mathrm{T}} \\ \vdots \\ \boldsymbol{b}_1^{\mathrm{T}} a_1 + \boldsymbol{b}_2^{\mathrm{T}} \\ \boldsymbol{b}_1^{\mathrm{T}} \end{bmatrix} \\ &= \begin{bmatrix} \boldsymbol{b}_1^{\mathrm{T}} & \cdots & \boldsymbol{b}_{n-1}^{\mathrm{T}} & \boldsymbol{b}_n^{\mathrm{T}} \\ 0 & \cdots & \boldsymbol{b}_{n-2}^{\mathrm{T}} & \boldsymbol{b}_{n-1}^{\mathrm{T}} \\ \vdots & & \vdots & \vdots \\ 0 & \cdots & \boldsymbol{b}_1^{\mathrm{T}} & \boldsymbol{b}_2^{\mathrm{T}} \\ 0 & \cdots & 0 & \boldsymbol{b}_1^{\mathrm{T}} \end{bmatrix} \begin{bmatrix} a_{n-1} \\ a_{n-2} \\ \vdots \\ a_1 \\ 1 \end{bmatrix}, \end{aligned}$$

$$\begin{aligned} \boldsymbol{\theta}_c &= \left[-\boldsymbol{c}\boldsymbol{A}^n\boldsymbol{Q} + \left[\boldsymbol{c}\boldsymbol{A}^{n-1}gp_1, \boldsymbol{c}\boldsymbol{A}^{n-2}gp_2, \ldots, cgp_m\right]\right]^{\mathrm{T}} \\ &= \begin{bmatrix} g_1 p_m a_{n-1} + \ldots + g_{n-1} p_2 a_1 + g_n p_1 \\ g_1 p_m a_{n-2} + \ldots + g_{n-2} p_3 a_1 + g_{n-1} p_2 \\ \vdots \\ g_1 p_m a_1 + g_2 p_{m-1} \\ g_1 p_m \end{bmatrix} \end{aligned}$$

$$= \begin{bmatrix} a_{n-1} & \cdots & a_1 & 1 \\ a_{n-2} & \cdots & 1 & 0 \\ \vdots & \vdots & \vdots & \vdots \\ a_1 & 1 & 0 & 0 \\ 1 & 0 & 0 & 0 \end{bmatrix} \begin{bmatrix} g_1 p_m \\ g_2 p_{m-1} \\ \vdots \\ g_{n-1} p_2 \\ g_n p_1 \end{bmatrix}.$$

The following gives the state estimation algorithm:

$$\hat{\boldsymbol{x}}(t-n) = \phi_y(t) - \hat{\boldsymbol{M}}(t)\hat{\mathbf{X}}(t-1) - \hat{\mathbf{Q}}(t)\phi_u(t)\hat{\boldsymbol{p}}(t) - \hat{\phi}_w(t),$$

$$\phi_y(t) = [y(t), y(t+1), \ldots, y(t+n-1)]^{\mathrm{T}},$$

$$\phi_u(t) = \left[f(u(t)), f(u(t+1)), \ldots, f(u(t+n-1))\right]^{\mathrm{T}},$$

$$\hat{\mathbf{X}}(t-1) = \hat{\boldsymbol{x}}^{\mathrm{T}}(t-1-n), \hat{\boldsymbol{x}}^{\mathrm{T}}(t-n), \ldots, \hat{\boldsymbol{x}}^{\mathrm{T}}(t-2)]^{\mathrm{T}},$$

$$\hat{\phi}_w(t) = [\hat{w}(t), \hat{w}(t+1), \ldots, \hat{w}(t+n-1)]^{\mathrm{T}},$$

$$\hat{\boldsymbol{M}}(t) = \begin{bmatrix} 0 & 0 & \cdots & 0 & 0 \\ \hat{\boldsymbol{b}}_1(t) & 0 & \cdots & 0 & 0 \\ \hat{\boldsymbol{b}}_2(t) & \hat{\boldsymbol{b}}_1(t) & \ddots & \vdots & \vdots \\ \vdots & \vdots & \ddots & 0 & 0 \\ \hat{\boldsymbol{b}}_{n-1}(t) & \hat{\boldsymbol{b}}_{n-2}(t) & \cdots & \hat{\boldsymbol{b}}_1(t) & 0 \end{bmatrix},$$

$$\mathbf{Q}(t) = \begin{bmatrix} 0 & \cdots & 0 & 0 \\ \hat{g}_1(t)\hat{p}_m(t) & \cdots & 0 & 0 \\ \hat{g}_2(t)\hat{p}_{m-1}(t) & \ddots & \vdots & \vdots \\ \vdots & \ddots & 0 & 0 \\ \hat{g}_{n-1}(t)\hat{p}_2(t) & \cdots & \hat{g}_1(t)\hat{p}_m(t) & 0 \end{bmatrix}.$$

**Theorem 5.1**: For the system in (5.57)−(5.60), suppose that $\{v(t)\}$ is a white noise sequence with zero mean and variance $\sigma^2$, that is, $\mathrm{E}[v(t)] = 0$,

$\mathrm{E}[v^2(t)] = \sigma^2$, and that there exist constants $0 < \tau_i < \infty$ such that for large $t$, the following persistent excitation conditions hold,

$$\tau_1 \boldsymbol{I} \leqslant \frac{1}{t}\sum_{j=1}^{t} \hat{\phi}_y^{\mathrm{T}}(j)\hat{\phi}_y(j) \leqslant \tau_2 \boldsymbol{I}, \ a.s.,$$

$$\tau_3 \boldsymbol{I} \leqslant \frac{1}{t}\sum_{j=1}^{t} \hat{\boldsymbol{X}}^{\mathrm{T}}(j)\hat{\boldsymbol{X}}^{\mathrm{T}}(j) \leqslant \tau_4 \boldsymbol{I}, \ a.s.,$$

$$\tau_5 \boldsymbol{I} \leqslant \frac{1}{t}\sum_{j=1}^{t} \hat{\phi}_u^{\mathrm{T}}(j)\hat{\phi}_u(j) \leqslant \tau_6 \boldsymbol{I}, \ a.s.,$$

$$\tau_7 \boldsymbol{I} \leqslant \frac{1}{t}\sum_{j=1}^{t} \hat{\phi}_w^{\mathrm{T}}(j)\hat{\phi}_w(j) \leqslant \tau_8 \boldsymbol{I}, \ a.s..$$

The parameter estimation error given by the bias compensation least squares algorithm converges to zero.

### 5.2.4 Examples

**Example 1**: Consider the following input nonlinear state time-delay system:

$$\begin{aligned}
\boldsymbol{x}(t+1) &= \begin{bmatrix} 0 & 1 \\ -0.05 & 0.10 \end{bmatrix} \boldsymbol{x}(t) \\
&\quad + \begin{bmatrix} 0.24 & 0.11 \\ 0.19 & 0.10 \end{bmatrix} \boldsymbol{x}(t-1) + \begin{bmatrix} 0.50 \\ 1.00 \end{bmatrix} \overline{u}(t), \\
\overline{u}(t) &= f(u(t)) = p_1 f_1(u(t)) + p_2 f_2(u(t)) + p_3 f_3(u(t)) \\
&= u(t) + 5.00u^2(t) + 0.10u^3(t), \\
y(t) &= [1, 0]\boldsymbol{x}(t) + 0.10v(t-1) + v(t).
\end{aligned}$$

The parameter vector to be identified is

$$\begin{aligned}
\boldsymbol{\theta} &= [\theta_1, \theta_2, \theta_3, \theta_4, \theta_5, \theta_6, \theta_7 \theta_8, \theta_9 \theta_{10}, \theta_{11}]^{\mathrm{T}} \\
&= [0.19, 0.21, 0.19, 0.10, 0.50, 1.00, 2.50, 5.00, 0.05, 0.10, 0.10]^{\mathrm{T}}.
\end{aligned}$$

In simulation, the input $\{u(t)\}$ is taken as an uncorrelated stochastic signal sequence with zero mean and unit variance and $\{v(t)\}$ as a white noise sequence with zero mean and variance $\sigma^2 = 0.10^2$ and $\sigma^2 = 1.00^2$. Apply the compensation-based parameter estimation algorithm to estimate the parameters of this example system, the parameter estimates and their errors are shown in Table 5.7, the parameter estimation error versus $t$ is

**Table 5.7** The parameter estimates and errors.

| $\sigma^2$ | $t$ | $\theta_1$ | $\theta_2$ | $\theta_3$ | $\theta_4$ | $\theta_5$ | $\theta_6$ | $\theta_7$ | $\theta_8$ | $\theta_9$ | $\theta_{10}$ | $\theta_{11}$ | δ (%) |
|---|---|---|---|---|---|---|---|---|---|---|---|---|---|
| $0.10^2$ | 100 | 0.16820 | 0.23286 | 0.20131 | 0.10873 | 0.63414 | 0.84759 | 2.51580 | 4.85989 | −0.01254 | 0.13945 | 0.12257 | 4.88064 |
| | 200 | 0.15739 | 0.23406 | 0.20003 | 0.12220 | 0.56086 | 0.90025 | 2.50998 | 4.93119 | 0.02649 | 0.12156 | 0.11783 | 3.11804 |
| | 500 | 0.18939 | 0.21148 | 0.19012 | 0.10204 | 0.53059 | 0.93883 | 2.49241 | 4.97458 | 0.03688 | 0.11644 | 0.11346 | 2.20195 |
| | 1000 | 0.19120 | 0.21016 | 0.19007 | 0.09857 | 0.51011 | 0.97640 | 2.49812 | 4.98643 | 0.04669 | 0.10557 | 0.10793 | 1.82653 |
| | 2000 | 0.19308 | 0.20849 | 0.18968 | 0.09720 | 0.50217 | 0.98568 | 2.50119 | 4.99251 | 0.05051 | 0.10425 | 0.10381 | 1.77662 |
| | 3000 | 0.19262 | 0.20863 | 0.18979 | 0.09755 | 0.50079 | 0.99011 | 2.50039 | 4.99541 | 0.05089 | 0.10265 | 0.10386 | 1.76249 |
| $1.00^2$ | 100 | 0.09970 | 0.29621 | 0.21272 | 0.16476 | 0.60702 | 1.12974 | 2.51363 | 4.70900 | 0.04721 | -0.07817 | 0.11751 | 7.32544 |
| | 200 | 0.01317 | 0.34109 | 0.22453 | 0.23365 | 0.61154 | 0.84719 | 2.52238 | 4.86943 | 0.04799 | 0.09310 | 0.11231 | 6.33194 |
| | 500 | 0.16739 | 0.22915 | 0.18822 | 0.13397 | 0.63417 | 0.75695 | 2.40858 | 4.93747 | 0.00207 | 0.14538 | 0.11025 | 5.69091 |
| | 1000 | 0.19109 | 0.21569 | 0.18949 | 0.09499 | 0.52430 | 0.94984 | 2.47511 | 4.95838 | 0.05112 | 0.09468 | 0.10234 | 2.18209 |
| | 2000 | 0.21377 | 0.19814 | 0.18657 | 0.07773 | 0.48632 | 0.95326 | 2.50924 | 4.97018 | 0.07091 | 0.10961 | 0.10301 | 2.14980 |
| | 3000 | 0.21082 | 0.19866 | 0.18795 | 0.07985 | 0.48241 | 0.96714 | 2.50170 | 4.98449 | 0.07025 | 0.10362 | 0.10396 | 1.99755 |
| True values | | 0.19000 | 0.21000 | 0.19000 | 0.10000 | 0.50000 | 1.00000 | 2.50000 | 5.00000 | 0.05000 | 0.10000 | 0.10000 | |

shown in Fig. 5.4, and the state estimates $\hat{x}_1(t)$ and $\hat{x}_2(t)$ versus $t$ are shown in Figs. 5.5–5.6.

From Table 5.7 and Figs. 5.4–5.6, we can draw the following conclusions.

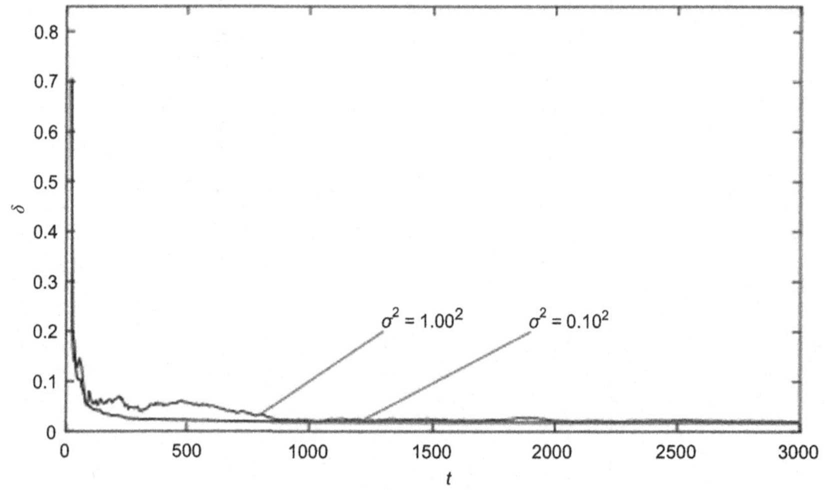


**Figure 5.4** The parameter estimation errors $\delta$ versus $t$ with $\sigma^2 = 0.10^2$ and $\sigma^2 = 1.00^2$.

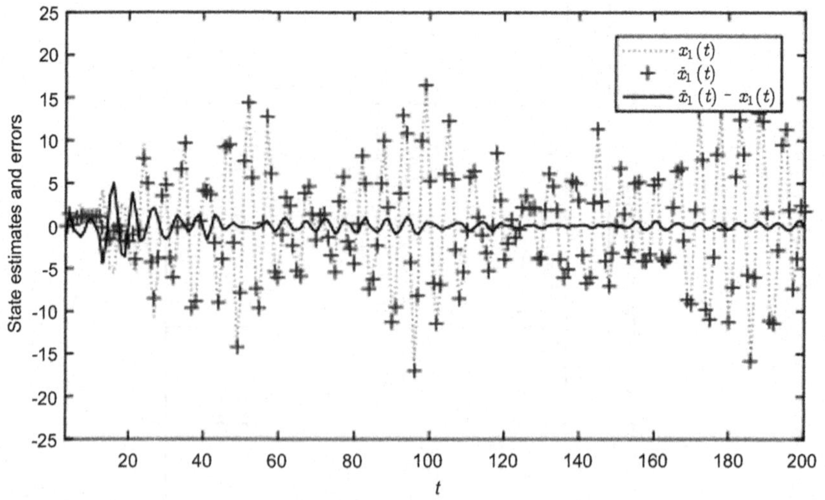


**Figure 5.5** The state and state estimate $\hat{x}_1(t)$ versus $t$.

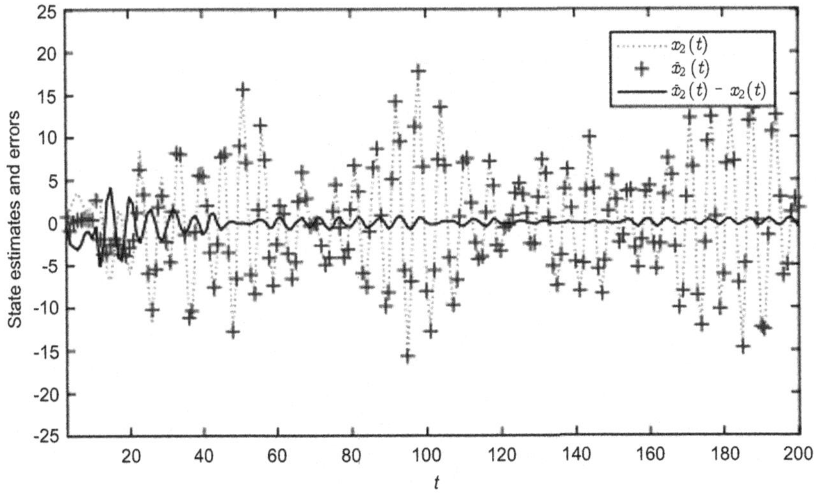


**Figure 5.6** The state and state estimate $\hat{x}_2(t)$ versus $t$.

1. The proposed algorithm is effective for estimating the parameters of the nonlinear state-space system with time delay. With the data length increasing, the parameter estimation errors become smaller and converge to zero.
2. A low noise variance leads to higher accuracy of parameter estimates.
3. It is clear that the proposed state observer can generate accurate state estimates because the state estimates are close to their true values as $t$ increases.

**Example 2**: An experiment is performed on a three-tank system which is shown in Fig. 5.7. In this experiment, a nonlinear function is designed for the input signal $f(u(t)) = u(t) + 5.00u^2(t) + 0.10u^3(t)$ to drive the pump to change the flow of the dollar inflow water. The water level in the middle tank is used as the system output signal. Therefore, consider the input function as a static nonlinear block. With the flow rate as input and the intermediate tank liquid level as the target output, a linear dynamic subsystem is constructed. The entire system is considered a Hammerstein model. The validation result of the identification model is given in Fig. 5.8, where the solid line is the measured output and the dashed line is the predicted output. It can be seen that the model output fits well with the measured output, indicating that the identification model can reflect the dynamic characteristics of the process and prove the effectiveness of the presented method.

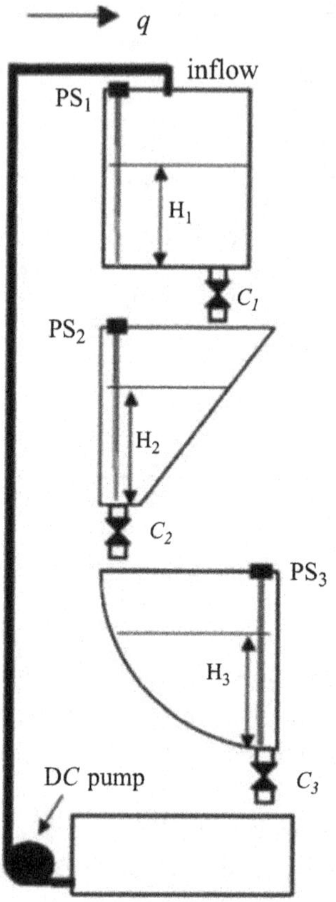


**Figure 5.7** Experimental apparatus.

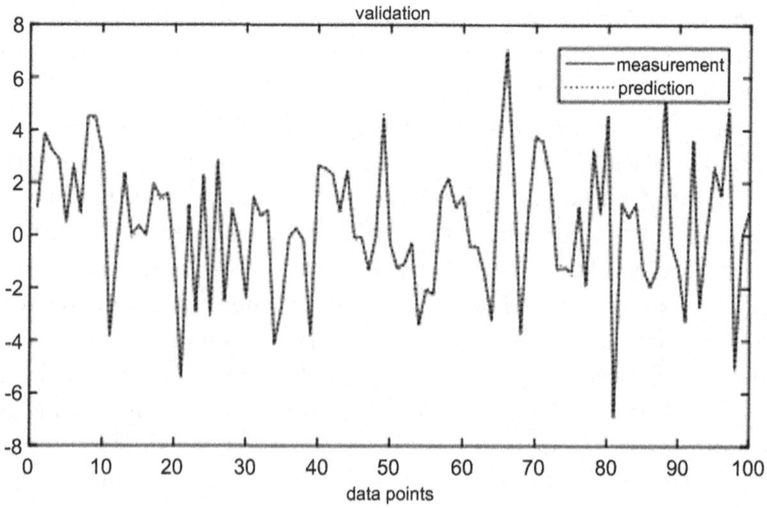


**Figure 5.8** The validation result.

## 5.3 Conclusion

This chapter discusses the bias compensation algorithm for estimating the parameters and states for nonlinear systems with colored noise based on the canonical state-space model with time delay. Compared with the recursive least squares methods, the advantage of this algorithm is that the parameter estimates are unbiased. A simulation example is given to show the effectiveness of the proposed algorithm. The convergence properties of the proposed algorithm can be analyzed by means of the stochastic process theory, and the proposed algorithm can be extended to other scalar or multivariable linear or nonlinear systems.

## Appendix

The proof of the theorem

The output equation in the identification model is as follows,

$$\begin{aligned} y(t) &= \phi_y^{\mathrm{T}}(t)\boldsymbol{\theta}_a + \mathbf{X}^{\mathrm{T}}(t-1)\boldsymbol{\theta}_b + \phi_u^{\mathrm{T}}(t)\boldsymbol{\theta}_c - \phi_w^{\mathrm{T}}(t)\boldsymbol{\theta}_a + w(t) \\ &= \left[\phi_y^{\mathrm{T}}(t) - \phi_w^{\mathrm{T}}(t)\right]\boldsymbol{\theta}_a + \mathbf{X}^{\mathrm{T}}(t-1)\boldsymbol{\theta}_b + \phi_u^{\mathrm{T}}(t)\boldsymbol{\theta}_c + w(t). \end{aligned}$$

Define the parameter estimation error vectors:

$$\begin{aligned} \tilde{\boldsymbol{\theta}}_a(t) &:= \hat{\boldsymbol{\theta}}_a(t) - \boldsymbol{\theta}_a, \\ \tilde{\boldsymbol{\theta}}_b(t) &:= \hat{\boldsymbol{\theta}}_b(t) - \boldsymbol{\theta}_b, \\ \tilde{\boldsymbol{\theta}}_c(t) &:= \hat{\boldsymbol{\theta}}_c(t) - \boldsymbol{\theta}_c, \\ \tilde{\boldsymbol{\theta}}_v(t) &:= \hat{\boldsymbol{\theta}}_v(t) - \boldsymbol{\theta}_v. \end{aligned}$$

Define some variables:

$$\begin{aligned} \tilde{y}(t) &:= \left[\hat{\phi}_y^{\mathrm{T}}(t) - \hat{\phi}_w^{\mathrm{T}}(t)\right]\tilde{\boldsymbol{\theta}}_a(t-1) + \hat{\mathbf{X}}^{\mathrm{T}}(t-1)\tilde{\boldsymbol{\theta}}_b(t-1) \\ &\quad + \hat{\phi}_u^{\mathrm{T}}(t)\tilde{\boldsymbol{\theta}}_c(t-1), \\ \tilde{w}(t) &:= \hat{\boldsymbol{\varphi}}_v^{\mathrm{T}}(t)\tilde{\boldsymbol{\theta}}_v(t-1), \\ \Upsilon_1(t) &:= \hat{y}(t) - y(t) + \left[\phi_y^{\mathrm{T}}(t) - \phi_w^{\mathrm{T}}(t) - \left[\hat{\phi}_y^{\mathrm{T}}(t) - \hat{\phi}_w^{\mathrm{T}}\right]\right]\boldsymbol{\theta}_a \\ &\quad + \left[\mathbf{X}^{\mathrm{T}}(t-1) - \hat{\mathbf{X}}^{\mathrm{T}}(t-1)\right]\boldsymbol{\theta}_b + \left[\phi_u^{\mathrm{T}}(t) - \hat{\phi}_y^{\mathrm{T}}(t)\right]\boldsymbol{\theta}_c, \\ \Upsilon_2(t) &:= \hat{w}(t) - w(t) + \left[\varphi_v^{\mathrm{T}}(t) - \varphi_v^{\mathrm{T}}(t)\right]\boldsymbol{\theta}_v, \\ e_v(t) &:= y(t) - \left[\phi_y^{\mathrm{T}}(t) - \phi_w^{\mathrm{T}}(t)\right]\hat{\boldsymbol{\theta}}_a(t-1) \\ &\quad - \hat{\mathbf{X}}^{\mathrm{T}}(t-1)\hat{\boldsymbol{\theta}}_v(t-1) \\ &\quad - \hat{\boldsymbol{\varphi}}_v^{\mathrm{T}}(t)\hat{\boldsymbol{\theta}}_v(t-1) \\ &= \hat{w}(t) - \hat{\boldsymbol{\varphi}}_v^{\mathrm{T}}(t)\hat{\boldsymbol{\theta}}_v(t-1). \end{aligned}$$

And, we have

$$\begin{aligned} e(t) &:= -\tilde{y}(t) + \Upsilon_1(t) + v(t), \\ e_v(t) &:= -\tilde{w}(t) + \Upsilon_2(t) + v(t). \end{aligned}$$

The parameter estimation error vectors in least squares form:

$$\begin{aligned} \tilde{\boldsymbol{\theta}}_a(t) &= \hat{\boldsymbol{\theta}}_a(t-1) + \boldsymbol{P}_a(t)\left[\phi_y^{\mathrm{T}}(t) - \phi_w^{\mathrm{T}}(t)\right] e(t), \\ \tilde{\boldsymbol{\theta}}_b(t) &= \hat{\boldsymbol{\theta}}_b(t-1) + \boldsymbol{P}_b(t)\hat{\boldsymbol{X}}^{\mathrm{T}}(t-1)e(t), \\ \tilde{\boldsymbol{\theta}}_c(t) &= \hat{\boldsymbol{\theta}}_c(t-1) + \boldsymbol{P}_c(t)\hat{\phi}_u^{\mathrm{T}}(t)e(t), \\ \tilde{\boldsymbol{\theta}}_v(t) &= \hat{\boldsymbol{\theta}}_v(t-1) + \boldsymbol{P}_v(t)\hat{\phi}_v^{\mathrm{T}}(t)e(t). \end{aligned}$$

Define the nonnegative functions:

$$\begin{aligned} \Omega_1(t) &:= \tilde{\boldsymbol{\theta}}_a^{\mathrm{T}}(t)\boldsymbol{P}_a^{-1}(t)\tilde{\boldsymbol{\theta}}_a(t), \\ \Omega_2(t) &:= \tilde{\boldsymbol{\theta}}_b^{\mathrm{T}}(t)\boldsymbol{P}_b^{-1}(t)\tilde{\boldsymbol{\theta}}_b(t), \\ \Omega_3(t) &:= \tilde{\boldsymbol{\theta}}_c^{\mathrm{T}}(t)\boldsymbol{P}_c^{-1}(t)\tilde{\boldsymbol{\theta}}_c(t), \\ \Omega_4(t) &:= \tilde{\boldsymbol{\theta}}_v^{\mathrm{T}}(t)\boldsymbol{P}_v^{-1}(t)\tilde{\boldsymbol{\theta}}_v(t), \\ \Omega_{123}(t) &:= \Omega_1(t) + \Omega_2(t) + \Omega_3(t), \\ \Omega(t) &:= \Omega_{123}(t) + \Omega_4(t), \\ \alpha_1(t) &:= \left[\hat{\phi}_y^{\mathrm{T}}(t) - \hat{\phi}_w^{\mathrm{T}}(t)\right]\boldsymbol{P}_a(t)\left[\hat{\phi}_y^{\mathrm{T}}(t) - \hat{\phi}_w^{\mathrm{T}}(t)\right]^{\mathrm{T}} \\ &\quad + \hat{\boldsymbol{X}}^{\mathrm{T}}(t-1)\boldsymbol{P}_b(t)\hat{\boldsymbol{X}}(t-1) + \hat{\phi}_u^{\mathrm{T}}(t)\boldsymbol{P}_c(t)\hat{\phi}_u(t), \\ \alpha_2(t) &:= \hat{\phi}_v^{\mathrm{T}}(t)\boldsymbol{P}_v(t)\hat{\phi}_v(t). \end{aligned}$$

Then, we have

$$\begin{aligned} \Omega_1(t) &= \tilde{\boldsymbol{\theta}}_a(t-1) + \boldsymbol{P}_a(t)\left[\hat{\phi}_y^{\mathrm{T}}(t) - \hat{\phi}_w^{\mathrm{T}}(t)\right]^{\mathrm{T}} e(t)^{\mathrm{T}}\boldsymbol{P}_a^{-1}(t) \\ &\quad \times \left\{\tilde{\boldsymbol{\theta}}_a(t-1) + \boldsymbol{P}_a(t)\left[\hat{\phi}_y^{\mathrm{T}}(t) - \hat{\phi}_w^{\mathrm{T}}(t)\right]^{\mathrm{T}} e(t)\right\} \\ &= \tilde{\boldsymbol{\theta}}_a^{\mathrm{T}}(t-1)\boldsymbol{P}_a^{-1}(t)\tilde{\boldsymbol{\theta}}_a(t-1) \\ &\quad + 2\tilde{\boldsymbol{\theta}}_a^{\mathrm{T}}(t-1)\left[\hat{\phi}_y^{\mathrm{T}}(t) - \hat{\phi}_w^{\mathrm{T}}(t)\right]^{\mathrm{T}} e(t) \\ &\quad + e^{\mathrm{T}}(t)\left[\hat{\phi}_y^{\mathrm{T}}(t) - \hat{\phi}_w^{\mathrm{T}}(t)\right]\boldsymbol{P}_a(t)\left[\hat{\phi}_y^{\mathrm{T}}(t) - \hat{\phi}_w^{\mathrm{T}}(t)\right]^{\mathrm{T}} e(t) \\ &= \Omega_1(t-1) + \tilde{\boldsymbol{\theta}}_a^{\mathrm{T}}(t-1)\left[\hat{\phi}_y^{\mathrm{T}}(t) - \hat{\phi}_w^{\mathrm{T}}(t)\right]^{\mathrm{T}} \\ &\quad \times \left[\hat{\phi}_y^{\mathrm{T}}(t) - \hat{\phi}_w^{\mathrm{T}}(t)\right] \\ &\quad + 2\tilde{\boldsymbol{\theta}}_a^{\mathrm{T}}(t-1)\left[\hat{\phi}_y^{\mathrm{T}}(t) - \hat{\phi}_w^{\mathrm{T}}(t)\right]^{\mathrm{T}} e(t) \\ &\quad + e^{\mathrm{T}}(t)\left[\hat{\phi}_y^{\mathrm{T}}(t) - \hat{\phi}_w^{\mathrm{T}}(t)\right]\boldsymbol{P}_a(t)\left[\hat{\phi}_y^{\mathrm{T}}(t) - \hat{\phi}_w^{\mathrm{T}}(t)\right]^{\mathrm{T}} e(t) \end{aligned}$$

$$\leqslant \Omega_1(t-1) + \left[\left[\hat{\phi}_y^{\mathrm{T}}(t) - \hat{\phi}_w^{\mathrm{T}}(t)\right]\tilde{\boldsymbol{\theta}}_a(t-1)\right]^2$$

$$+ 2\tilde{\boldsymbol{\theta}}_a^{\mathrm{T}}(t-1)\left[\hat{\phi}_y^{\mathrm{T}}(t) - \hat{\phi}_w^{\mathrm{T}}(t)\right]^{\mathrm{T}} e(t)$$

$$+ \left[\hat{\phi}_y^{\mathrm{T}}(t) - \hat{\phi}_w^{\mathrm{T}}(t)\right]\boldsymbol{P}_a(t)\left[\hat{\phi}_y^{\mathrm{T}}(t) - \hat{\phi}_w^{\mathrm{T}}(t)\right]^{\mathrm{T}} e^2(t).$$

Similarly,

$$\begin{aligned}
\Omega_2(t) &= \Omega_2(t-1) + \left[\hat{\mathbf{X}}^{\mathrm{T}}(t-1)\tilde{\boldsymbol{\theta}}_b(t-1)\right]^2 \\
&\quad + 2e(t)\hat{\mathbf{X}}^{\mathrm{T}}(t-1)\tilde{\boldsymbol{\theta}}_b(t-1) \\
&\quad + \hat{\mathbf{X}}^{\mathrm{T}}(t-1)\boldsymbol{P}_b(t)\hat{\mathbf{X}}(t-1)e^2(t), \\
\Omega_3(t) &= \Omega_3(t-1) + \left[\hat{\phi}_u^{\mathrm{T}}(t)\tilde{\boldsymbol{\theta}}_c(t-1)\right]^2 \\
&\quad + 2e(t)\hat{\phi}_u^{\mathrm{T}}(t)\tilde{\boldsymbol{\theta}}_c(t-1) \\
&\quad + \hat{\phi}_u^{\mathrm{T}}(t)\boldsymbol{P}_c(t)\hat{\phi}_u(t)e^2(t), \\
\Omega_4(t) &\leqslant \Omega_4(t-1) + \left[\hat{\phi}_v^{\mathrm{T}}(t)\tilde{\boldsymbol{\theta}}_{\boldsymbol{v}}(t-1)\right]^2 \\
&\quad + 2\tilde{\boldsymbol{\theta}}_{\boldsymbol{v}}^{\mathrm{T}}(t-1)\hat{\phi}_u(t)e_v(t) \\
&\quad + \hat{\phi}_{\boldsymbol{v}}^{\mathrm{T}}(t)\boldsymbol{P}_{\boldsymbol{v}}(t)\hat{\phi}_{\boldsymbol{v}}(t)e_{\boldsymbol{v}}^2(t) \\
&= \Omega_4(t-1) + \tilde{w}^2(t) + 2\tilde{w}(t)e_{\boldsymbol{v}}(t) + \alpha_2(t)e_{\boldsymbol{v}}^2(t) \\
&= \Omega_4(t-1) - \left[1 - \alpha_2(t)\tilde{w}^2(t)\right] \\
&\quad + 2[1-\alpha_2(t)]\tilde{w}(t)[\Upsilon_2(t) + v(t)] \\
&\quad \alpha_2(t)\left[v^2(t) + \Upsilon_2^2(t) + 2\Upsilon_2(t)v(t)\right], \\
\Omega_{123}(t) &\leqslant \Omega_{123}(t-1) + \tilde{y}^2(t) - 2\left[\hat{\phi}_y^{\mathrm{T}}(t) - \hat{\phi}_w^{\mathrm{T}}(t)\right]\tilde{\boldsymbol{\theta}}_a(t-1) \\
&\quad \times \hat{\mathbf{X}}^{\mathrm{T}}(t-1)\tilde{\boldsymbol{\theta}}_b(t-1)\hat{\phi}_u^{\mathrm{T}}(t)\tilde{\boldsymbol{\theta}}_c(t-1) \\
&\quad + 2\tilde{y}(t)e(t) + \alpha_1(t)e^2(t) \\
&= \Omega_{123}(t-1) - [1-\alpha_1(t)]\tilde{y}^2(t) - 2\left[\hat{\phi}_y^{\mathrm{T}}(t) - \hat{\phi}_w^{\mathrm{T}}(t)\right] \\
&\quad \times \tilde{\boldsymbol{\theta}}_{\boldsymbol{a}}(t-1)\hat{\mathbf{X}}^{\mathrm{T}}(t-1)\tilde{\boldsymbol{\theta}}_{\boldsymbol{b}}(t-1)\hat{\boldsymbol{\phi}}_{\boldsymbol{u}}^{\mathrm{T}}(t)\tilde{\boldsymbol{\theta}}_c(t-1) \\
&\quad + 2[1-\alpha_1(t)]\tilde{y}(t)[\boldsymbol{\Upsilon}_1(t) + v(t)] \\
&\quad \alpha_1(t)\left[v^2(t) + \boldsymbol{\Upsilon}_1^2(t) + 2\boldsymbol{\Upsilon}_1(t)v(t)\right], \\
\Omega(t) &\leqslant \Omega(t-1) - [1-\alpha_1(t)]\tilde{y}^2(t) - 2\left[\hat{\phi}_y^{\mathrm{T}}(t) - \hat{\phi}_{\boldsymbol{w}}^{\mathrm{T}}(t)\right] \\
&\quad \times \tilde{\boldsymbol{\theta}}_{\boldsymbol{a}}(t-1)\hat{\mathbf{X}}^{\mathrm{T}}(t-1)\tilde{\boldsymbol{\theta}}_{\boldsymbol{b}}(t-1)\hat{\boldsymbol{\phi}}_{\boldsymbol{u}}^{\mathrm{T}}(t)\tilde{\boldsymbol{\theta}}_c(t-1) \\
&\quad + 2[1-\alpha_1(t)]\tilde{y}(t)[\boldsymbol{\Upsilon}_1(t) + v(t)] \\
&\quad - [1-\alpha_2(t)]\tilde{w}^2(t) + 2[1-\alpha_2(t)] \\
&\quad \times \tilde{w}(t)[\boldsymbol{\Upsilon}_2(t) + v(t)] \\
&\quad + [\alpha_1(t) + \alpha_2(t)][2v^2(t) + \boldsymbol{\Upsilon}_1^2(t) + 2\boldsymbol{\Upsilon}_1(t)v(t) \\
&\quad + \boldsymbol{\Upsilon}_2^2(t) + 2\boldsymbol{\Upsilon}_2(t)v(t)]
\end{aligned}$$

$$
\begin{aligned}
&= \Omega(t-1) - [1-\alpha_1(t)]\tilde{y}^2(t) - 2\left[\hat{\phi}_y^{\mathrm{T}}(t) - \hat{\phi}_w^{\mathrm{T}}(t)\right] \\
&\quad \times \tilde{\boldsymbol{\theta}}_a(t-1)\hat{\mathbf{X}}^{\mathrm{T}}(t-1)\tilde{\boldsymbol{\theta}}_b(t-1)\hat{\boldsymbol{\phi}}_u^{\mathrm{T}}(t)\tilde{\boldsymbol{\theta}}_c(t-1) \\
&\quad + 2[1-\alpha_1(t)]\tilde{y}(t)\boldsymbol{\Upsilon}_1(t) - [1-\alpha_2(t)]\tilde{w}^2(t) \\
&\quad + 2[1-\alpha_2(t)]\tilde{w}(t)\boldsymbol{\Upsilon}_2(t) \\
&\quad + [\alpha_1(t)+\alpha_2(t)]\left[2v^2(t) + \boldsymbol{\Upsilon}_1^2(t) + \boldsymbol{\Upsilon}_2^2(t)\right] \\
&\quad + 2[1-\alpha_1(t)]\tilde{y}(t)v(t) + 2[1-\alpha_2(t)]\tilde{w}(t)v(t) \\
&\quad + [\alpha_1(t)+\alpha_2(t)][2\Upsilon_1(t)v(t) + 2\boldsymbol{\Upsilon}_2(t)v(t)].
\end{aligned}
$$

Let

$$
\begin{aligned}
\alpha_3(t) := {} & [1-\alpha_1(t)]\tilde{y}^2(t) + 2\left[\hat{\phi}_y^{\mathrm{T}}(t) - \hat{\phi}_w^{\mathrm{T}}(t)\right] \\
& \times \tilde{\boldsymbol{\theta}}_a(t-1)\hat{\mathbf{X}}^{\mathrm{T}}(t-1)\tilde{\boldsymbol{\theta}}_b(t-1)\hat{\phi}_u^{\mathrm{T}}(t)\tilde{\boldsymbol{\theta}}_c(t-1) \\
& - 2[1-\alpha_1(t)]\tilde{y}(t)\Upsilon_1(t) + [1-\alpha_2(t)]\tilde{w}^2(t) \\
& - 2[1-\alpha_2(t)]\tilde{w}(t)\Upsilon_2(t).
\end{aligned}
$$

For $\alpha_1(t)$, $\alpha_2(t)$, $\tilde{y}(t)$, $[\hat{\phi}_y^{\mathrm{T}}(t) - \hat{\phi}_w^{\mathrm{T}}(t)]\tilde{\theta}_a(t-1)\hat{X}^{\mathrm{T}}(t-1)\tilde{\theta}_b(t-1)$ $\hat{\phi}_u^{\mathrm{T}}(t)\tilde{\theta}_c(t-1)$, $\gamma_1(t)$, $\gamma_2(t)$, and $\tilde{w}(t)$ are uncorrelated with $v(t)$, taking the mathematical expectation yields

$$
\mathrm{E}[\Omega(t)] \le \mathrm{E}[\Omega(t-1)] - \mathrm{E}[\alpha_3(t)] + \mathrm{E}\left\{[\alpha_1(t)+\alpha_2(t)]\left[\Upsilon_1^2(t) + \Upsilon_2^2(t) + 2v^2(t)\right]\right\}.
$$

If $\alpha_3(t) \ge 0$ and $\gamma_i^2(t) \le \varepsilon$, $\varepsilon < \infty$, we have

$$
\begin{aligned}
\mathrm{E}[\Omega(t)] &\le \mathrm{E}[\Omega(t-1)] + \mathrm{E}\left\{[\alpha_1(t)+\alpha_2(t)]\left[\Upsilon_1^2(t) + \Upsilon_2^2(t) + 2v^2(t)\right]\right\} \\
&\le \mathrm{E}[\Omega(t-1)] + \mathrm{E}\left\{[\alpha_1(t)+\alpha_2(t)](2\sigma^2 + 2\varepsilon)\right\}.
\end{aligned}
$$

Otherwise, let $\hat{\theta}_a(t) = \hat{\theta}_a(t-1)$, $\hat{\theta}_b(t) = \hat{\theta}_b(t-1)$, $\hat{\theta}_c(t) = \hat{\theta}_c(t-1)$, and $\hat{\theta}_v(t) = \hat{\theta}_v(t-1)$. Then, we always have

$$
\mathrm{E}[\Omega(t)] \le \mathrm{E}[\Omega(t-1)] + \mathrm{E}\left\{[\alpha_1(t)+\alpha_2(t)](2\sigma^2 + 2\varepsilon)\right\}.
$$

## References

Beikzadeh, H., Liu, G.J., Marquez, H.J., 2018. Robust sensitive fault detection and estimation for single-rate and multirate nonlinear sampled-data systems. System Control Letter 119, 71–80.

Ghosal, M., Rao, V., 2019. Fusion of multirate measurements for nonlinear dynamic state estimation of the power systems. IEEE Transactions Smart Grid 10, 216–226.

Ibrir, S., 2018. Joint state and parameter estimation of non-linearly parameterized discrete-time nonlinear systems. Automatica 97, 226–233.

CHAPTER 6

# Uncertain state delay systems identification

## 6.1 State space model identification of multirate processes with uncertain time delay

Parameter estimation approaches have wide applications in many areas such as signal processing and system identification. Mathematical models are the foundation of system control. The state space model can be utilized to describe dynamic systems (Rafal et al., 2017). In industrial systems, many controlled objects need to be abstracted into state space models of the system (Bretas and Bretas, 2015). In general, a physical phenomenon with time delay (fixed or varying) accompanies with the controlled object. In practice, the irregularly sampled outputs are often available with random delays due to manual analysis in laboratory. Moreover, the associated sampling delays are uncertain since only the arrival time of the measurements is recorded, while the time instant that the samples are actually taken is unknown or not accurately recorded.

Traditional methods assume the delay is a fixed value and it can be treated as a parameter in the identification problem. So, the delay estimation may be solved by the recursive algorithms and the iterative algorithms along with the model parameters. However, the delay is usually associated with some process variable transmission (e.g., liquid flow rate). The fast flow rate results in small time delay while a slow flow rate results in a long time delay; thus, the varying delay is more reasonable at most situations.

The elements of a variety of measurement sampling functions and control signals operate at different sampling rates which lead to multirate (MR) systems. The theoretical research for such multirate sampled systems started in the 1950s. The application in the field is concerned with computer signal processing, network control system, and process control. Various methods have been proposed to model the MR systems and infer the unmeasurable or missing outputs (Zamani et al., 2016).

It develops a multirate model-based identification procedure of state space systems, taking into account uncertain random delays associated

*State Space Systems With Time-Delays Analysis, Identification, and Applications*
DOI: https://doi.org/10.1016/B978-0-323-91768-1.00021-6

with the irregularly sampled outputs. The expectation–maximization (EM) algorithm computes the maximum likelihood estimates of unknown parameters in probabilistic models involving latent variables. The EM algorithm is an iterative method that alternates between computing a conditional expectation and solving a maximization problem. We will in this work derive the EM algorithm and show that it provides a maximum likelihood estimate. There are two main motivations: to simultaneously estimate the discrete time delay and continuous states; to estimate the parameters of the state space model. To solve the above problems, we will use the EM algorithm. Under the EM framework, it will be shown that, instead of the point estimation of the scheduling variable, the complete probability distribution of the estimation of the scheduling variable is what one really needs for the estimation of system parameters. The states estimation is given by their expectation.

The main contributions of this paper are as follows.

- This paper proposes the identification of state space model with unknown time delay, which includes conventional nonuniformly sampled-data systems and multirate systems as special cases.
- By introducing two hidden variables, this paper presents the derived EM algorithm to estimate the unknown model parameters and the time delays simultaneously.
- By using a numerical example, this paper demonstrates the performances of the proposed algorithm, including the estimation errors of the EM algorithm for finite measurement data.

### 6.1.1 Problem statement

The mathematical formulation of state space model with time-varying time delay is as follows:

$$\boldsymbol{x}_{t+1} = \boldsymbol{A}\boldsymbol{x}_t + \boldsymbol{b}u_t + \boldsymbol{w}_t, \tag{6.1}$$

$$y_t = \boldsymbol{c}\boldsymbol{x}_{t-d_t} + v_t, \tag{6.2}$$

where $x_t$ is the unmeasurable state; $u_t, t = 1, 2, \ldots, T$ is the input and available at every sampling period $\Delta t$; $T$ represents the number of data points that have been collected; $y_t, t = T_1, T_2, \ldots, T_N$ is the irregularly sampled output and only available at time instant $t = T_i \cdot \Delta t$ with unknown time delay $d_{T_i} \cdot \Delta t$ (i.e., the delay can vary in each data sample); $w_t \in R^{n \times 1}$ is process noise and Gaussian distributed noise with zero mean and covariance matrix $P_0$, and the identically distributed Gaussian noise is the measurement noise $v_{T_i}$ with covariance matrix $R_0$. $A \in R^{n\times n}$, $b \in R^{n\times 1}$, and $c \in R^{1\times n}$ are the system parameter matrices/vectors to be estimated. The delay $d_{T_i}$ is a random integer that can follow any discrete distribution. In this article, it is assumed to be uniformly distributed between 0 and $q$, that is,

$$P(d_t = k) = \frac{1}{q+1}, k = 0, 1, \ldots, q.$$

## 6.1.2 Model identification using the EM algorithm

The mathematical form of the Q function of the EM algorithm for the state space system with time-varying time delay can be written as

$$\begin{aligned} Q(\Theta|\Theta^k) &= \mathrm{E}_{C_{mis}|C_{obs},\Theta^k}\{\log[p(C_{obs}, C_{mis}|\Theta)]\} \\ &= \int_{C_{mis}} \log[p(C_{obs}, C_{mis}|\Theta)]p(C_{mis}|C_{obs}, \Theta^k)\mathrm{d}C_{mis}, \end{aligned} \tag{6.3}$$

where $\Theta$ denotes the system parameters: $A$, $b$, and $c$, and $\Theta^k$ represents the parameter estimation results from the previous iteration which are used to compute the expectation of the complete data likelihood. The observed data set $C_{obs}$ are $\{y_{T_1}, \ldots, y_{T_N}\}$ and $\{u_1, \ldots, u_T\}$, while the hidden states $X = \{x_1, \ldots, x_T\}$ and the time delay $d_{T_i}$ can be viewed as the latent data $C_{mis}$.

The EM algorithm is a kind of important approach in statistics, which has been widely used in system identification, signal processing, machine learning, and computer vision. The EM algorithm can assure the nondecreasing function of Q in each iteration, so the iterative parameter estimation can always converge to a stable point with the increasing of iterations. Its convergence property was demonstrated by Wu under normal conditions.

Based on the distribution function, the expectation of the complete data with the expectation taking over the missing observation can be derived, which is known as the Q function. The E-step is to compute the Q function, and the M-step is to maximize the Q function with respect to $\Theta$ to obtain

$$\Theta^{k+1} = \arg\max_{\Theta} Q(\Theta|\Theta^k).$$

The E-step and M-step iterate until converges.

Consider the state space model described in (6.1) and (6.2). Let $p(C_{obs}, C_{mis}|\Theta)$ denote the complete likelihood function including both the hidden variables and observations. The $Q$ function is defined as the conditional expectation of the log-likelihood function $\log[p(y_{T_i:T_N}, x_{1:T}, d_{T_1:T_N}|\Theta)]$ over the missing data given $C_{obs}$ and $\Theta^k$,

$$\begin{aligned} Q(\Theta|\Theta^k) &= \mathrm{E}_{C_{mis}|C_{obs},\Theta^k}\{\log p(C_{obs}, C_{mis}|\Theta)\} \\ &= \mathrm{E}_{\boldsymbol{x}_{1:T},d_{T_1}:_{T_N}|C_{obs},\Theta^k}\{\log p(y_{T_1:T_N}, \boldsymbol{x}_{1:T}, d_{T_1:T_N}|\Theta)\}, \end{aligned} \tag{6.4}$$

where $\Theta^k$ is the estimation of $\Theta$ after $k$ iteration.

In order to evaluate the above $Q$ function, the posteriori $p(x_{1:T}, d_{T_1:T_N}|u_{1:T}, y_{T_1:T_N}, \Theta^k)$ is required to calculate; by the Bayesian theory, we have

$$p(\boldsymbol{x}_{1:T}, d_{T_1:T_N}|u_{1:T}, y_{T_1:T_N}, \Theta^k) = C_1 p(y_{T_1:T_N}|u_{1:T}, \boldsymbol{x}_{1:T} d_{T_1:T_N}, \Theta^k) p(\boldsymbol{x}_{1:T}|u_{1:T}, \Theta^k), \tag{6.5}$$

where

$$C_1 = \frac{p(d_{T_1:T_N}|u_{1:T}, \Theta^k)}{p(y_{T_1:T_N}|u_{1:T}, \Theta^k)}.$$

In (6.4), the term $p(y_{T_1:T_N}, x_{1:T}, d_{T_1:T_N}|\Theta)$ which is the joint density function of outputs, states, and time delays can be decomposed using the Bayesian property as

$$\begin{aligned} p(y_{T_1:T_N}, \boldsymbol{x}_{1:T}, d_{T_1:T_N}|\Theta) &= p(y_{T_1:T_N}|\boldsymbol{x}_{1:T}, d_{T_1:T_N}, \Theta) p(\boldsymbol{x}_{1:T}, d_{T_1:T_N}|\Theta) \\ &= p(y_{T_1:T_N}|\boldsymbol{x}_{1:T}, d_{T_1:T_N}, \Theta) p(\boldsymbol{x}_{1:T}|d_{T_1:T_N}, \Theta) p(d_{T_1:T_N}|\Theta), \end{aligned} \tag{6.6}$$

where the first term of the joint density function in Eq. (6.6) can be

$$\begin{aligned} p(y_{T_1:T_N}|\boldsymbol{x}_{1:T}, d_{T_1:T_N}, \Theta) &= p(y_{T_N}|y_{T_1:T_{N-1}}, \boldsymbol{x}_{1:T}, d_{T_1:T_N}, \Theta) p(y_{T_1:T_{N-1}}|\boldsymbol{x}_{1:T}, d_{T_1:T_N}, \Theta) \\ &= p(y_{T_N}|\boldsymbol{x}_{T_N-d_{T_N}}, d_{T_N}, \Theta) p(y_{T_{N-1}}|\boldsymbol{x}_{T_{N-1}-d_{T_{N-1}}}, d_{T_{N-1}}, \Theta)\ldots \\ &\quad \times p(y_{T_1}|\boldsymbol{x}_{T_1-d_{T_1}}, d_{T1}, \Theta) \\ &= \prod_{i=1}^{N} p(y_{T_i}|\boldsymbol{x}_{T_i-d_{T_i}}, \Theta). \end{aligned} \tag{6.7}$$

We use the Markov property in the equation above, and the condition distribution of $y$ is only related to state and delay according to the above expression of $y$. Similarly, the second term of Eq. (6.6) can be

$$\begin{aligned} p\left(\boldsymbol{x}_{1:T}, d_{T_1:T_N}, \Theta\right) &= p\left(\boldsymbol{x}_T|\boldsymbol{x}_{1:T-1}, d_{T_1:T_N}, \Theta\right)p\left(\boldsymbol{x}_{1:T-1}|d_{T_1:T_N}, \Theta\right) \\ &= p\left(\boldsymbol{x}_T|\boldsymbol{x}_{1:T-1}, d_{T_1:T_N}, \Theta\right)p\left(\boldsymbol{x}_{T-1}|\boldsymbol{x}_{1:T-2}, d_{T_1:T_N}, \Theta\right)\ldots \\ &\quad \times p\left(\boldsymbol{x}_2|\boldsymbol{x}_1, d_{T_1:T_N}, \Theta\right)p\left(\boldsymbol{x}_1|d_{T_1:T_N}, \Theta\right) \\ &= p(\boldsymbol{x}_1|\Theta)\prod_{i=2}^{T}p(\boldsymbol{x}_t|\boldsymbol{x}_{t-1}, \Theta). \end{aligned} \tag{6.8}$$

The Markov property can also be used in state decomposition, the conditional distribution of $x$ is independent of delay $d$, and the present state is only related to the previous state according to the state equation. Substituting (6.7) and (6.8) into (6.6), the joint density of the likelihood of the full data set can be rewritten as

$$p\left(y_{T_1:T_N}, \boldsymbol{x}_{1:T}, d_{T_1:T_N}|\Theta\right) = \prod_{i=1}^{N} p(y_{T_i}|\boldsymbol{x}_{T_i-d_{T_i}}, d_{T_i}, \Theta).p(\boldsymbol{x}_1|\Theta)\prod_{i=2}^{T} p(\boldsymbol{x}_t|\boldsymbol{x}_{t-1}, \Theta).C_2, \tag{6.9}$$

where $C_2 = p(d_{T_1:T_N}|\Theta)$ is considered as a constant for delay is the uniform distribution. Furthermore, combining (6.4), (6.5), and (6.9), the Q function can be rearranged as

$$\begin{aligned} Q(\Theta|\Theta^k) &= E_{\boldsymbol{x}_{1:T}, d_{T_1:T_N}|C_{obs}, \Theta^k}\left\{\log\left[p\left(y_{T_1:T_N}, \boldsymbol{x}_{1:T}, d_{T_1:T_N}|\Theta\right)\right]\right\} \\ &= C_1\sum_{i=1}^{N}\int_{\boldsymbol{x}_{T_i-d_{T_i}}, d_{T_i}} p(y_{T_i}|\boldsymbol{x}_{T_i-d_{T_i}}, d_{T_i}, \Theta^k)p(\boldsymbol{x}_{T_i-d_{T_i}}|\Theta^k) \\ &\quad \times \log p(y_{T_i}|\boldsymbol{x}_{T_i-d_{T_i}}, d_{T_i}, \Theta)\mathrm{d}\boldsymbol{x}_{T_i-d_{T_i}}, d_{T_i} \\ &\quad + C_1\int_{\boldsymbol{x}_{1:T}, d_{T_1:T_N}}\prod_{i=1}^{N}p(y_{T_i}|\boldsymbol{x}_{T_i-d_{T_i}}, d_{T_i}, \Theta^k)\prod_{i=1}^{T}p(\boldsymbol{x}_t|\boldsymbol{x}_{t-1}, \Theta^k) \\ &\quad \times \sum_{i=1}^{T}\log p(\boldsymbol{x}_t|\boldsymbol{x}_{t-1}, \Theta)\mathrm{d}\boldsymbol{x}_{1:T}, d_{T_1:T_N} + C_3. \end{aligned}$$

The above expression contains two integral items, and for the first integral term, the posteriori can be extracted as

$$\begin{aligned} & p\left(y_{T_i}|\boldsymbol{x}_{T_i-d_{T_i}}, d_{Ti}, \Theta^k\right)p\left(\boldsymbol{x}_{T_i-d_{T_i}}|\Theta^k\right) \\ &= \frac{1}{\sqrt{2\pi\sigma_v}}\exp\left(-\frac{1}{2}\left(y_{T_1} - \boldsymbol{c}\boldsymbol{x}_{T_i-d_{T_i}}\right)^{\mathrm{T}}\sigma_v^{-2}\left(y_{T_1} - \boldsymbol{c}\boldsymbol{x}_{T_i-d_{T_i}}\right)\right) \\ &\quad \times \frac{1}{\sqrt{2\pi|\boldsymbol{P}_{T_i-d_{T_i}}|}}\exp\left(-\frac{1}{2}\left(\boldsymbol{x}_{T_i-d_{T_i}} - \boldsymbol{\mu}_{T_i-d_{T_i}}\right)^{\mathrm{T}}\boldsymbol{P}_{T_i-d_{T_i}}^{-1}\left(\boldsymbol{x}_{T_i-d_{T_i}} - \boldsymbol{\mu}_{T_i-d_{T_i}}\right)\right). \end{aligned}$$

Adding the index parts, we can obtain

$$\begin{aligned}
&\left(y_{T_1}-\boldsymbol{c}\boldsymbol{x}_{T_i-d_{T_i}}\right)^{\mathrm{T}}\sigma_v^{-2}\left(y_{T_1}-\boldsymbol{c}\boldsymbol{x}_{T_i-d_{T_i}}\right)\\
&+\left(\boldsymbol{x}_{T_i-d_{T_i}}-\boldsymbol{\mu}_{T_i-d_{T_i}}\right)^{\mathrm{T}}\boldsymbol{P}_{T_i-d_{T_i}}^{-1}\left(\boldsymbol{x}_{T_i-d_{T_i}}-\boldsymbol{\mu}_{T_i-d_{T_i}}\right)\\
&=y_{T_i}^2\sigma_v^{-2}-y_{T_i}\sigma_v^{-2}\boldsymbol{c}\boldsymbol{x}_{T_i-d_{T_i}}-y_{T_1}\sigma_v^{-2}\boldsymbol{x}_{T_i-d_{T_i}}^{\mathrm{T}}\boldsymbol{c}^{\mathrm{T}}+\boldsymbol{x}_{T_i-d_{T_i}}^{\mathrm{T}}\sigma_v^{-2}\boldsymbol{c}^{\mathrm{T}}\boldsymbol{c}\boldsymbol{x}_{T_i-d_{T_i}}\\
&+\boldsymbol{x}_{T_i-d_{T_i}}^{\mathrm{T}}\boldsymbol{P}_{T_i-d_{T_i}}^{-1}\boldsymbol{x}_{T_i-d_{T_i}}-\boldsymbol{x}_{T_i-d_{T_i}}^{\mathrm{T}}\boldsymbol{P}_{T_i-d_{T_i}}^{-1},\boldsymbol{\mu}_{T_i-d_{T_i}}-\boldsymbol{\mu}_{T_i-d_{T_i}}^{\mathrm{T}}\boldsymbol{P}_{T_i-d_{T_i}}^{-1}\boldsymbol{x}_{T_i-d_{T_i}}\\
&+\boldsymbol{\mu}_{T_i-d_{T_i}}^{\mathrm{T}}\boldsymbol{P}_{T_i-d_{T_i}}^{-1}\boldsymbol{\mu}_{T_i-d_{T_i}}.
\end{aligned}$$

Packing similar terms gives the following two items:

$$\begin{aligned}
\boldsymbol{x}_{T_i-d_{T_i}}^{\mathrm{T}}\sigma_v^{-2}\boldsymbol{c}^{\mathrm{T}}\boldsymbol{c}\boldsymbol{x}_{T_i-d_{T_i}}+\boldsymbol{x}_{T_i-d_{T_i}}^{\mathrm{T}}\boldsymbol{P}_{T_i-d_{T_i}}^{-1}\boldsymbol{x}_{T_i-d_{T_i}}&=\boldsymbol{x}_{T_i-d_{T_i}}^{\mathrm{T}}\left(\sigma_v^{-2}\boldsymbol{c}^{\mathrm{T}}\boldsymbol{c}+\boldsymbol{P}_{T_i-d_{T_i}}^{-1}\right)\boldsymbol{x}_{T_i-d_{T_i}}\\
&=:\boldsymbol{x}_{T_i-d_{T_i}}^{\mathrm{T}}\boldsymbol{P}_{new}^{-1}\boldsymbol{x}_{T_i-d_{T_i}}
\end{aligned}$$

and

$$\begin{aligned}
\boldsymbol{x}_{T_i-d_{T_i}}^{\mathrm{T}}\boldsymbol{c}^{\mathrm{T}}y_{T_i}\sigma_v^{-2}+\boldsymbol{x}_{T_i-d_{T_i}}^{\mathrm{T}}\boldsymbol{P}_{T_i-d_{T_i}}^{-1}\boldsymbol{\mu}_{T_i-d_{T_i}}&=\boldsymbol{x}_{T_i-d_{T_i}}^{\mathrm{T}}\left(\boldsymbol{c}^{\mathrm{T}}y_{T_i}\sigma_v^{-2}+\boldsymbol{P}_{T_i-d_{T_i}}^{-1}\boldsymbol{\mu}_{T_i-d_{T_i}}\right)\\
&=:2\boldsymbol{x}_{T_i-d_{T_i}}^{\mathrm{T}}\boldsymbol{P}_{new}^{-1}\boldsymbol{\mu}_{new}.
\end{aligned}$$

Thus, we can obtain the new covariance matrix $P_{new}$ and mean value $\mu_{new}$ of the first integral item:

$$\begin{aligned}
\boldsymbol{P}_{new}&=\left(\sigma_v^{-2}\boldsymbol{c}^{\mathrm{T}}\boldsymbol{c}+\boldsymbol{P}_{T_i-d_{T_i}}^{-1}\right)^{-1},\\
\boldsymbol{\mu}_{new}&=\frac{1}{2}\left(\sigma_v^{-2}\boldsymbol{c}^{\mathrm{T}}\boldsymbol{c}+\boldsymbol{P}_{T_i-d_{T_i}}^{-1}\right)^{-1}\left(y_{T_i}\sigma_v^{-2}\boldsymbol{c}^{\mathrm{T}}+\boldsymbol{P}_{T_i-d_{T_i}}^{-1}\boldsymbol{\mu}_{T_i-d_{T_i}}\right).
\end{aligned}$$

The log function is about the fast-rate function in the second integral term. The posterior probability contains not only the fast-rate function but also the slow-rate function, so we cannot calculate the Q function directly, and we need to discuss three different conditions.

**Case 1**: When $t=T_i-d_{T_i}$, the second integral term becomes

$$\sum_{i=1}^{N}\sum_{k=0}^{q}\int_{\boldsymbol{x}_t,\boldsymbol{x}_{t-1}}p\left(y_{t+d_{T_i}}|\boldsymbol{x}_t,d_{T_i}=k,\Theta^k\right)p\left(\boldsymbol{x}_t|\boldsymbol{x}_{t-1},\Theta^k\right)p\left(\boldsymbol{x}_{t-1}|\Theta^k\right)\log p(\boldsymbol{x}_t|\boldsymbol{x}_{t-1},\Theta)d\boldsymbol{x}_t,\boldsymbol{x}_{t-1},$$

where the posteriori equations are, respectively,

$$p\left(y_{t+d_{T_i}}|\boldsymbol{x}_t, d_{T_i}=k, \Theta^k\right) = \frac{1}{\sqrt{2\pi\sigma_v}}\exp\left(-\tfrac{1}{2}\left(y_{t+d_{T_i}}-\boldsymbol{c}\boldsymbol{x}_t\right)^{\mathrm{T}}\sigma_v^{-2}\left(y_{t+d_{T_i}}-\boldsymbol{c}\boldsymbol{x}_t\right)\right),$$

$$p(\boldsymbol{x}_t|\boldsymbol{x}_{t-1}, \Theta^k) = \frac{1}{\sqrt{2\pi|\boldsymbol{P}_0|}}\exp\left(-\tfrac{1}{2}\boldsymbol{x}_t-\boldsymbol{A}\boldsymbol{x}_{t-1}-\boldsymbol{b}u_{t-1}\right)^{\mathrm{T}}\boldsymbol{P}_0^{-1}(\boldsymbol{x}_t-\boldsymbol{A}\boldsymbol{x}_{t-1}-\boldsymbol{b}u_{t-1}),$$

$$p(\boldsymbol{x}_{t-1}|\Theta^k) = \frac{1}{\sqrt{2\pi|\boldsymbol{P}_{t-1}|}}\exp\left(-\tfrac{1}{2}\boldsymbol{x}_{t-1}-\boldsymbol{\mu}_{t-1}\right)^{\mathrm{T}}\boldsymbol{P}_{t-1}^{-1}(\boldsymbol{x}_{t-1}-\boldsymbol{\mu}_{t-1}).$$

Combining the indexes of the above three expressions gives

$$\begin{aligned}
&\left(y_{t+d_{T_i}}-\boldsymbol{c}\boldsymbol{x}_t\right)^{\mathrm{T}}\sigma_v^{-2}\left(y_{t+d_{T_i}}-\boldsymbol{c}\boldsymbol{x}_t\right)+(\boldsymbol{x}_t-\boldsymbol{A}\boldsymbol{x}_{t-1}-\boldsymbol{b}u_{t-1})^{\mathrm{T}}\boldsymbol{P}_0^{-1}(\boldsymbol{x}_t-\boldsymbol{A}\boldsymbol{x}_{t-1}-\boldsymbol{b}u_{t-1})\\
&+\left(\boldsymbol{x}_{t-1}-\boldsymbol{\mu}_{t-1}\right)^{\mathrm{T}}\boldsymbol{P}_{t-1}^{-1}\left(\boldsymbol{x}_{t-1}-\boldsymbol{\mu}_{t-1}\right)\\
&=y_{t+d_{T_i}}^2\sigma_v^{-2}-2y_{t+d_{T_i}}\sigma_v^{-2}\boldsymbol{c}\boldsymbol{x}_t+\boldsymbol{x}_t^{\mathrm{T}}\sigma_v^{-2}\boldsymbol{c}^{\mathrm{T}}\boldsymbol{c}\boldsymbol{x}_t\\
&+\boldsymbol{x}_t^{\mathrm{T}}\boldsymbol{P}_0^{-1}\boldsymbol{x}_t-\boldsymbol{x}_t^{\mathrm{T}}\boldsymbol{P}_0^{-1}\boldsymbol{A}\boldsymbol{x}_{t-1}-\boldsymbol{x}_{t-1}^{\mathrm{T}}\boldsymbol{A}^{\mathrm{T}}\boldsymbol{P}_0^{-1}\boldsymbol{x}_t+\boldsymbol{x}_{t-1}^{\mathrm{T}}\boldsymbol{A}^{\mathrm{T}}\boldsymbol{P}_0^{-1}\boldsymbol{A}\boldsymbol{x}_{t-1}\\
&-2\boldsymbol{b}^{\mathrm{T}}\boldsymbol{P}_0^{-1}\boldsymbol{x}_t u_{t-1}+2\boldsymbol{b}^{\mathrm{T}}\boldsymbol{P}_0^{-1}\boldsymbol{A}\boldsymbol{x}_{t-1}u_{t-1}+\boldsymbol{b}^{\mathrm{T}}\boldsymbol{P}_0^{-1}\boldsymbol{b}u_{t-1}^2\\
&+\boldsymbol{x}_{t-1}^{\mathrm{T}}\boldsymbol{P}_{t-1}^{-1}\boldsymbol{x}_{t-1}-2\boldsymbol{\mu}_{t-1}^{\mathrm{T}}\boldsymbol{P}_{t-1}^{-1}\boldsymbol{x}_{t-1}+\boldsymbol{\mu}_{t-1}^{\mathrm{T}}\boldsymbol{P}_{t-1}^{-1}u_{t-1}.
\end{aligned}$$

The object here is to categorize similar index items of the above function to one item, and then, we can get new covariance matrix and mean value.

Assume $\begin{bmatrix}\boldsymbol{x}_t\\ \boldsymbol{x}_{t-1}\end{bmatrix}\sim N(\boldsymbol{\mu}_1, \boldsymbol{\Sigma}_1), \boldsymbol{\Sigma}_1=\begin{bmatrix}\boldsymbol{P}_1^1 & \boldsymbol{P}_2^1\\ \boldsymbol{P}_3^1 & \boldsymbol{P}_4^1\end{bmatrix}^{-1}$, we have

$$\begin{bmatrix}\boldsymbol{x}_t\\ \boldsymbol{x}_{t-1}\end{bmatrix}^{\mathrm{T}}\boldsymbol{\Sigma}_1^{-1}\begin{bmatrix}\boldsymbol{x}_t\\ \boldsymbol{x}_{t-1}\end{bmatrix}-2\boldsymbol{\mu}_1^{\mathrm{T}}\boldsymbol{\Sigma}_1^{-1}\begin{bmatrix}\boldsymbol{x}_t\\ \boldsymbol{x}_{t-1}\end{bmatrix}+\boldsymbol{\mu}_1^{\mathrm{T}}\boldsymbol{\Sigma}_1^{-1}\boldsymbol{\mu}_1, \tag{6.10}$$

where

$$\begin{aligned}
&\begin{bmatrix}\boldsymbol{x}_t\\ \boldsymbol{x}_{t-1}\end{bmatrix}^{\mathrm{T}}\boldsymbol{\Sigma}_1^{-1}\begin{bmatrix}\boldsymbol{x}_t\\ \boldsymbol{x}_{t-1}\end{bmatrix}=\left[\boldsymbol{x}_t^{\mathrm{T}},\boldsymbol{x}_{t-1}^{\mathrm{T}}\right]\begin{bmatrix}\boldsymbol{P}_1^1 & \boldsymbol{P}_2^1\\ \boldsymbol{P}_3^1 & \boldsymbol{P}_4^1\end{bmatrix}\begin{bmatrix}\boldsymbol{x}_t\\ \boldsymbol{x}_{t-1}\end{bmatrix}\\
&=\boldsymbol{x}_t^{\mathrm{T}}\boldsymbol{P}_1^1\boldsymbol{x}_t+\boldsymbol{x}_t^{\mathrm{T}}\boldsymbol{P}_2^1\boldsymbol{x}_{t-1}+\boldsymbol{x}_{t-1}^{\mathrm{T}}\boldsymbol{P}_3^1\boldsymbol{x}_t+\boldsymbol{x}_{t-1}^{\mathrm{T}}\boldsymbol{P}_4^1\boldsymbol{x}_{t-1}\\
&=\boldsymbol{x}_t^{\mathrm{T}}\boldsymbol{P}_0^1\boldsymbol{x}_t-\boldsymbol{x}_t^{\mathrm{T}}\boldsymbol{P}_0^{-1}\boldsymbol{A}\boldsymbol{x}_{t-1}-\boldsymbol{x}_{t-1}^{\mathrm{T}}\boldsymbol{A}^{\mathrm{T}}\boldsymbol{P}_0^{-1}\boldsymbol{x}_t+\boldsymbol{x}_{t-1}^{\mathrm{T}}\boldsymbol{A}^{\mathrm{T}}\boldsymbol{P}_0^{-1}\boldsymbol{A}\boldsymbol{x}_{t-1}-\boldsymbol{x}_{t-1}^{\mathrm{T}}\boldsymbol{P}_{t-1}^{-1}\boldsymbol{x}_{t-1}+\boldsymbol{x}_t^{\mathrm{T}}\sigma_v^{-2}\boldsymbol{c}^{\mathrm{T}}\boldsymbol{c}\boldsymbol{x}_t\\
&=\boldsymbol{x}_t^{\mathrm{T}}\left(\boldsymbol{P}_0^{-1}+\sigma_v^{-2}\boldsymbol{c}^{\mathrm{T}}\boldsymbol{c}\right)\boldsymbol{x}_t-\boldsymbol{x}_t^{\mathrm{T}}\boldsymbol{P}_0^{-1}\boldsymbol{A}\boldsymbol{x}_{t-1}-\boldsymbol{x}_{t-1}^{\mathrm{T}}\boldsymbol{A}^{\mathrm{T}}\boldsymbol{P}_0^{-1}\boldsymbol{x}_t+\boldsymbol{x}_{t-1}^{\mathrm{T}}\left(\boldsymbol{A}^{\mathrm{T}}\boldsymbol{P}_0^{-1}\boldsymbol{A}+\boldsymbol{P}_{t-1}^{-1}\right)\boldsymbol{x}_{t-1}.
\end{aligned}$$

Thus, mapping the same item, we have new covariance matrix

$$\begin{aligned}
\boldsymbol{P}_1^1&=\boldsymbol{P}_0^{-1}+\frac{\boldsymbol{c}^{\mathrm{T}}\boldsymbol{c}}{\sigma_v^2},\\
\boldsymbol{P}_2^1&=-\boldsymbol{P}_0^{-1}\boldsymbol{A},\\
\boldsymbol{P}_3^1&=-\boldsymbol{A}^{\mathrm{T}}\boldsymbol{P}_0^{-1},\\
\boldsymbol{P}_4^1&=\boldsymbol{A}^{\mathrm{T}}\boldsymbol{P}_0^{-1}\boldsymbol{A}+\boldsymbol{P}_{t-1}^{-1}.
\end{aligned}$$

Substituting the above four expressions into $\sum_1$, the covariance matrix will be

$$\boldsymbol{\Sigma}_1=\begin{bmatrix}\boldsymbol{P}_0^{-1}+\sigma_v^{-2}\boldsymbol{c}^{\mathrm{T}}\boldsymbol{c} & -\boldsymbol{P}_0^{-1}\boldsymbol{A}\\ -\boldsymbol{A}^{\mathrm{T}}\boldsymbol{P}_0^{-1} & \boldsymbol{A}^{\mathrm{T}}\boldsymbol{P}_0^{-1}\boldsymbol{A}+\boldsymbol{P}_{t-1}^{-1}\end{bmatrix}^{-1}.$$

The second item of Eq. (6.10) can be expressed as

$$\begin{aligned}-2\boldsymbol{\mu}_1^{\mathrm{T}}\boldsymbol{\Sigma}_1^{-1}\begin{bmatrix}\boldsymbol{x}_t\\ \boldsymbol{x}_{t-1}\end{bmatrix}&=-2u_{t-1}\boldsymbol{b}^{\mathrm{T}}\boldsymbol{P}_0^{-1}\boldsymbol{x}_t-2y_{t+d_{T_i}}\sigma_v^{-2}\boldsymbol{c}\boldsymbol{x}_t\\&\quad+2u_{t-1}\boldsymbol{b}^{\mathrm{T}}\boldsymbol{P}_0^{-1}\boldsymbol{A}\boldsymbol{x}_{t-1}-2\boldsymbol{\mu}_{t-1}^{\mathrm{T}}\boldsymbol{P}_{t-1}^{-1}\boldsymbol{x}_{t-1}\\&=-2\left[u_{t-1}\boldsymbol{b}^{\mathrm{T}}\boldsymbol{P}_0^{-1}+y_{t+d_{T_i}}\sigma_v^{-2}\boldsymbol{c},\ \boldsymbol{\mu}_{t-1}^{\mathrm{T}}\boldsymbol{P}_{t-1}^{-1}-u_{t-1}\boldsymbol{b}^{\mathrm{T}}\boldsymbol{P}_0^{-1}\boldsymbol{A}\right]\begin{bmatrix}\boldsymbol{x}_t\\ \boldsymbol{x}_{t-1}\end{bmatrix}.\end{aligned}$$

Using the linear transformation, we can get the mean value

$$\boldsymbol{\mu}_1^{\mathrm{T}}=\left[u_{t-1}\boldsymbol{b}^{\mathrm{T}}\boldsymbol{P}_0^{-1}+y_{t+d_{T_i}}\sigma_v^{-2}\boldsymbol{c},\ \boldsymbol{\mu}_{t-1}^{\mathrm{T}}\boldsymbol{P}_{t-1}^{-1}-u_{t-1}\boldsymbol{b}^{\mathrm{T}}\boldsymbol{P}_0^{-1}\boldsymbol{A}\right]\cdot\Sigma_1.$$

**Case 2**: When $t-1=T_i-d_{T_i}$, the second integral term of Q function becomes

$$\sum_{i=1}^{N}\sum_{k=0}^{q}\int_{\boldsymbol{x}_t,\boldsymbol{x}_{t-1}}p\left(y_{t+d_{T_i}}-1|\boldsymbol{x}_{t-1},\lambda_i=k,\Theta^k\right)p(\boldsymbol{x}_t|\boldsymbol{x}_{t-1},\Theta^k)p(\boldsymbol{x}_{t-1}|\Theta^k)\log p(\boldsymbol{x}_t|\boldsymbol{x}_{t-1},\Theta)d\boldsymbol{x}_t,\boldsymbol{x}_{t-1}.$$

Using the similar way of the above condition (6.1) to get the following new covariance matrix and mean value of condition (6.2)

$$\boldsymbol{\Sigma}_2=\begin{bmatrix}\boldsymbol{P}_1^2 & \boldsymbol{P}_2^2\\ \boldsymbol{P}_3^2 & \boldsymbol{P}_4^2\end{bmatrix}^{-1}=\begin{bmatrix}\boldsymbol{P}_0^{-1} & -\boldsymbol{P}_0^{-1}\boldsymbol{A}\\ -\boldsymbol{A}^{\mathrm{T}}\boldsymbol{P}_0^{-1} & \sigma_v^{-2}\boldsymbol{c}^{\mathrm{T}}\boldsymbol{c}+\boldsymbol{A}^{\mathrm{T}}\boldsymbol{P}_0^{-1}\boldsymbol{A}+\boldsymbol{P}_{t-1}^{-1}\end{bmatrix}^{-1},$$

$$\boldsymbol{\mu}_2^{\mathrm{T}}=\left[u_{t-1}\boldsymbol{b}^{\mathrm{T}}\boldsymbol{P}_0^{-1},\ y_{t+d_{T_i}-1}\sigma_v^{-2}\boldsymbol{c}+\boldsymbol{\mu}_{t-1}^{\mathrm{T}}\boldsymbol{P}_{t-1}^{-1}-u_{t-1}\boldsymbol{b}^{\mathrm{T}}\boldsymbol{P}_0^{-1}\boldsymbol{A}\right]\cdot\Sigma_2.$$

**Case 3**: When $t\neq T_i-d_{T_i},\ T_i-d_{T_i}+1$ we have the second integral term of Q function

$$p\left(y_{T_i}|\boldsymbol{x}_{t_i-d_{T_i}},d_{T_i}=k,\Theta^k\right)\sum_{t\neq T_i-d_{T_i},T_i-d_{T_i}+1}\int_{\boldsymbol{x}_t,\boldsymbol{x}_{t-1}}p(\boldsymbol{x}_t|\boldsymbol{x}_{t-1},\Theta^k)p(\boldsymbol{x}_{t-1}|\Theta^k)\log p(\boldsymbol{x}_t|\boldsymbol{x}_{t-1},\Theta)d\boldsymbol{x}_t,\boldsymbol{x}_{t-1}$$

and the corresponding covariance matrix and mean value will be

$$\Sigma_3 = \begin{bmatrix} P_1^3 & P_2^3 \\ P_3^3 & P_4^3 \end{bmatrix}^{-1} = \begin{bmatrix} P_0^{-1} & -P_0^{-1}A \\ -A^{\mathrm{T}}P_0^{-1} & A^{\mathrm{T}}P_0^{-1}A + P_{t-1}^{-1} \end{bmatrix}^{-1},$$

$$\mu_3^{\mathrm{T}} = \left[u_{t-1}b^{\mathrm{T}}P_0^{-1}, \mu_{t-1}^{\mathrm{T}}P_{t-1}^{-1} - u_{t-1}b^{\mathrm{T}}P_0^{-1}A\right] \cdot \Sigma_3.$$

Finally, we can get the Q function

$$\begin{aligned} Q(\Theta|\Theta^k) = {} & C_1 \sum_{i=1}^{N}\sum_{k=0}^{q} \mathrm{E}_{x_{T_i-d_{T_i}}}\left[\log p\left(y_{T_i}\middle|x_{T_i-d_{T_i}}, d_{T_i}=k,\Theta\right)\right] \\ & + C_1 \sum_{i=1}^{N}\sum_{k=0}^{q} \mathrm{E}_{x_{T_i-d_{T_i}}, x_{T_i-d_{T_i}}-1}\left[\log p\left(x_{T_i-d_{T_i}}|x_{T_i-d_{T_i}-1}, d_{T_i}=k,\Theta\right)\right] \\ & + C_1 \sum_{i=1}^{N}\sum_{k=0}^{q} \mathrm{E}_{x_{T_i-d_{T_i}+1}, x_{T_i-d_{T_i}}}\left[\log p\left(x_{T_{i-d_{T_i}+1}}|x_{T_{i-d_{T_i}}}, d_{T_i}=k,\Theta\right)\right] \\ & + C_1 \textstyle\sum_{t\neq T_{i-d_{T_i}}, T_{i-d_{T_i}+1}} \mathrm{E}_{x_t,x_{t-1}}\left[\log p(x_t|x_{t-1},\Theta)\right] + C_3. \end{aligned}$$

In order to calculate the Q-function, the unknown terms should be computed first.

$$\log p\left(y_{T_i}|x_{T_{i-d_{T_i}}}, d_{T_i}=k,\Theta\right) = -\log\sqrt{2\pi\sigma_v} - \frac{1}{2\sigma_v^2}\left[y_{T_i} - c^{\mathrm{T}}x_{T_{i-k}}\right]^2,$$

$$\log p(x_t|x_{t-1},\Theta) = -\log\sqrt{2\pi|P_0|} - \frac{1}{2}\frac{(x_t - Ax_{t-1} - bu_{t-1})^{\mathrm{T}}(x_t - Ax_{t-1} - bu_{t-1})}{|P_0|}.$$

Thus, for the first expectation expression of the Q-function, we have

$$\begin{aligned} & \mathrm{E}_{x_{T_i-d_{T_i}}}\left[\log p\left(y_{T_i}|x_{T_i-d_{T_i}}, d_{T_i}=k,\Theta\right)\right] = \mathrm{E}_{x_{T_i-d_{T_i}}}\left\{-\log\sqrt{2\pi\sigma_v} - \frac{1}{2\sigma_v^2}\left[y_{T_i} - cx_{T_{i-k}}\right]^2\right\} \\ & = -\log\sqrt{2\pi\sigma_v} - \frac{1}{2\sigma_v^2}\left\{y_{T_i}^2 - 2y_{T_i}c\mu_{new} + \mathrm{trace}\left[c^{\mathrm{T}}cP_{new}\right] + \mu_{new}^{\mathrm{T}}c^{\mathrm{T}}c\mu_{new}\right\}. \end{aligned}$$

The second expectation expression can be expanded as follows

$$\begin{aligned} & \mathrm{E}_{x_{T_i-d_{T_i}}, x_{T_{i-d_{T_i}-1}}}\left[\log p\left(x_{T_i-d_{T_i}}|x_{T_i-d_{T_i}-1}, d_{T_i}=k,\Theta\right)\right] \\ & = -\log\sqrt{2\pi|P_0|} - \frac{1}{2}\mathrm{E}_x\left[\left(x_{T_{i-k}} - Ax_{T_{i-k-1}}bu_{T_{i-k-1}}\right)^{\mathrm{T}}P_0^{-1}\left(x_{T_{i-k}} - Ax_{T_{i-k-1}} - bu_{T_{i-k-1}}\right)\right] \\ & = -\log\sqrt{2\pi|P_0|} - \frac{1}{2}[E_x(x_{T_{i-k}}^{\mathrm{T}}P_0^{-1}x_{T_{i-k}}) - \mathrm{E}_x(x_{T_{i-k}}^{\mathrm{T}}P_0^{-1}Ax_{T_{i-k-1}}) - \mu_{1,T_{i-k}}^{\mathrm{T}}P_0^{-1}bu_{T_{i-k-1}} \\ & \quad - \mathrm{E}_x(x_{T_{i-k-1}}^{\mathrm{T}}A^{\mathrm{T}}P_0^{-1}x_{T_{i-k}}) + \mathrm{E}_x(x_{T_{i-k-1}}^{\mathrm{T}}A^{\mathrm{T}}P_0^{-1}Ax_{T_{i-k-1}}) + \mu_{1,T_{i-k-1}}^{\mathrm{T}}A^{\mathrm{T}}P_0^{-1}bu_{T_{i-k-1}} \\ & \quad - b^{\mathrm{T}}P_0^{-1}\mu_{1,T_{i-k}u_{T_{i-k-1}}} + b^{\mathrm{T}}P_0^{-1}A\mu_{1,T_{i-k-1}u_{T_{i-k-1}}} + b^{\mathrm{T}}P_0^{-1}bu_{T_{i-k-1}}^2], \end{aligned}$$

where

$$\begin{aligned}
\mathrm{E}_{\boldsymbol{x}}\left(\boldsymbol{x}_{T_i-k}^{\mathrm{T}}\boldsymbol{P}_0^{-1}\boldsymbol{x}_{T_i-k}\right) &= \mathrm{trace}\left(\boldsymbol{P}_0^{-1}\boldsymbol{P}_1^1\right) + \boldsymbol{\mu}_{1,T_i-k}^{\mathrm{T}}\boldsymbol{P}_0^{-1}\boldsymbol{\mu}_{1,T_i-k},\\
\mathrm{E}_{\boldsymbol{x}}\left(\boldsymbol{x}_{T_i-k}^{\mathrm{T}}\boldsymbol{P}_0^{-1}\boldsymbol{A}\boldsymbol{x}_{T_i-k-1}\right) &= \mathrm{trace}\left(\boldsymbol{P}_0^{-1}\boldsymbol{A}\mathrm{cov}\left(\boldsymbol{x}_{T_i-k},\boldsymbol{x}_{T_i-k-1}\right)\right) + \boldsymbol{\mu}_{1,T_i-k}^{\mathrm{T}}\boldsymbol{P}_0^{-1}\boldsymbol{A}\boldsymbol{\mu}_{1,T_i-k-1}\\
&= \mathrm{trace}\left(\boldsymbol{P}_0^{-1}\boldsymbol{A}\boldsymbol{P}_2^1\right) + \boldsymbol{\mu}_{1,T_i-k}^{\mathrm{T}}\boldsymbol{P}_0^{-1}\boldsymbol{A}\boldsymbol{\mu}_{1,T_i-k-1},\\
\mathrm{E}_{\boldsymbol{x}}\left(\boldsymbol{x}_{T_i-k-1}^{\mathrm{T}}\boldsymbol{A}^{\mathrm{T}}\boldsymbol{P}_0^{-1}\boldsymbol{x}_{T_i-k}\right) &= \mathrm{trace}\left(\boldsymbol{A}^{\mathrm{T}}\boldsymbol{P}_0^{-1}\boldsymbol{P}_0^{-1}\boldsymbol{P}_3^1\right) + \boldsymbol{\mu}_{1,T_i-k-1}^{\mathrm{T}}\boldsymbol{A}^{\mathrm{T}}\boldsymbol{P}_0^{-1}\boldsymbol{\mu}_{1,T_i-k},\\
\mathrm{E}_{\boldsymbol{x}}\left(\boldsymbol{x}_{T_i-k-1}^{\mathrm{T}}\boldsymbol{A}^{\mathrm{T}}\boldsymbol{P}_0^{-1}\boldsymbol{A}\boldsymbol{x}_{T_i-k-1}\right) &= \mathrm{cov}\left(\boldsymbol{x}_{T_i-k-1}^{\mathrm{T}}\boldsymbol{x}_{T_i-k-1}^{\mathrm{T}}\boldsymbol{A}^{\mathrm{T}}\boldsymbol{P}_0^{-1\mathrm{T}}\boldsymbol{A}\right) + \boldsymbol{\mu}_{1,T_i-k-1}^{\mathrm{T}}\boldsymbol{A}^{\mathrm{T}}\boldsymbol{P}_0^{-1}\boldsymbol{A}\boldsymbol{\mu}_{1,T_i-k-1}\\
&= \mathrm{trace}\left(\boldsymbol{A}^{\mathrm{T}}\boldsymbol{P}_0^{-1}\boldsymbol{A}\boldsymbol{P}_4^1\right) + \boldsymbol{\mu}_{1,T_i-k-1}^{\mathrm{T}}\boldsymbol{A}^{\mathrm{T}}\boldsymbol{P}_0^{-1}\boldsymbol{A}\boldsymbol{\mu}_{1,T_i-k-1}.
\end{aligned}$$

Thus, substituting the above expectations to the second expectation equation, we can get the final expression

$$\begin{aligned}
&\mathrm{E}_{\boldsymbol{x}_{T_i-d_{T_i}},\boldsymbol{x}_{T_i-d_{T_i}-1}}\left[\log p\left(\boldsymbol{x}_{T_i-d_{T_i}}|\boldsymbol{x}_{T_i-d_{T_i}-1},d_{T_i}=k,\Theta\right)\right]\\
&= -\log\sqrt{2\pi|\boldsymbol{P}_0|}-\tfrac{1}{2}[\mathrm{trace}(\boldsymbol{P}_0^{-1}\boldsymbol{P}_1^1) + \boldsymbol{\mu}_{1,T_i-k}^{\mathrm{T}}\boldsymbol{P}_0^{-1}\boldsymbol{\mu}_{1,T_i-k}\\
&- \mathrm{trace}(\boldsymbol{P}_0^{-1}\boldsymbol{A}\boldsymbol{P}_2^1) - \boldsymbol{\mu}_{1,T_i-k}^{\mathrm{T}}\boldsymbol{P}_0^{-1}\mathrm{A}\boldsymbol{\mu}_{1,T_i-k-1} - \boldsymbol{\mu}_{1,T_i-k}^{\mathrm{T}}\boldsymbol{P}_0^{-1}\boldsymbol{b}u_{T_i-k-1}\\
&- \mathrm{trace}(\boldsymbol{A}^{\mathrm{T}}\boldsymbol{P}_0^{-1}\boldsymbol{P}_3^1) - \boldsymbol{\mu}_{1,T_i-k-1}^{\mathrm{T}}\boldsymbol{A}^{\mathrm{T}}\boldsymbol{P}_0^{-1}\boldsymbol{\mu}_{1,T_i-k} + \mathrm{trace}(\boldsymbol{A}^{\mathrm{T}}\boldsymbol{P}_0^{-1}\boldsymbol{A}\boldsymbol{P}_4^1) + \boldsymbol{\mu}_{1,T_i-k-1}^{\mathrm{T}}\boldsymbol{A}^{\mathrm{T}}\boldsymbol{P}_0^{-1}\mathrm{A}\boldsymbol{\mu}_{1,T_i-k-1}\\
&+ \boldsymbol{\mu}_{1,T_i-k-1}^{\mathrm{T}}\boldsymbol{A}^{\mathrm{T}}\boldsymbol{P}_0^{-1}\boldsymbol{b}u_{T_i-k-1} - \boldsymbol{b}^{\mathrm{T}}\boldsymbol{P}_0^{-1}\boldsymbol{\mu}_{1,T_i-k}u_{T_i-k-1} + \boldsymbol{b}^{\mathrm{T}}\boldsymbol{P}_0^{-1}\mathrm{A}\boldsymbol{\mu}_{1,T_i-k-1}u_{T_i-k-1} + \boldsymbol{b}^{\mathrm{T}}\boldsymbol{P}_0^{-1}\boldsymbol{b}u_{T_i-k-1}^2].
\end{aligned}$$

Similarly, the third and the fourth expectation expressions will be

$$\begin{aligned}
&\mathrm{E}_{\boldsymbol{x}_{T_i-d_{T_i}+1},\boldsymbol{x}_{T_i-d_{T_i}}}[\log p\left(\boldsymbol{x}_{T_i-d_{T_i}+1}\middle|\boldsymbol{x}_{T_i-d_{T_i}},d_{T_i}=k,\Theta\right)]\\
&= -\log\sqrt{2\pi|\boldsymbol{P}_0|}-\tfrac{1}{2}[\mathrm{trace}(\boldsymbol{P}_0^{-1}\boldsymbol{P}_1^2) + \boldsymbol{\mu}_{2,T_i-k+1}^{\mathrm{T}}\boldsymbol{P}_0^{-1}\boldsymbol{\mu}_{2,T_i-k+1}\\
&- \mathrm{trace}(\boldsymbol{P}_0^{-1}\boldsymbol{A}\boldsymbol{P}_2^2) - \boldsymbol{\mu}_{2,T_i-k+1}^{\mathrm{T}}\boldsymbol{P}_0^{-1}\mathrm{A}\boldsymbol{\mu}_{2,T_i-k} - \boldsymbol{\mu}_{2,T_i-k+1}^{\mathrm{T}}\boldsymbol{P}_0^{-1}\boldsymbol{b}u_{T_i-k}\\
&- \mathrm{trace}(\boldsymbol{A}^{\mathrm{T}}\boldsymbol{P}_0^{-1}\boldsymbol{P}_3^2) - \boldsymbol{\mu}_{2,T_i-k}^{\mathrm{T}}\boldsymbol{A}^{\mathrm{T}}\boldsymbol{P}_0^{-1}\boldsymbol{\mu}_{2,T_i-k+1} + \mathrm{trace}(\boldsymbol{A}^{\mathrm{T}}\boldsymbol{P}_0^{-1}\boldsymbol{A}\boldsymbol{P}_4^2) + \boldsymbol{\mu}_{2,T_i-k}^{\mathrm{T}}\boldsymbol{A}^{\mathrm{T}}\boldsymbol{P}_0^{-1}\mathrm{A}\boldsymbol{\mu}_{2,T_i-k}\\
&+ \boldsymbol{\mu}_{1,T_i-k}^{\mathrm{T}}\boldsymbol{A}^{\mathrm{T}}\boldsymbol{P}_0^{-1}\boldsymbol{b}u_{T_i-k} - \boldsymbol{b}^{\mathrm{T}}\boldsymbol{P}_0^{-1}\boldsymbol{\mu}_{2,T_i-k+1}u_{T_i-k} + \boldsymbol{b}^{\mathrm{T}}\boldsymbol{P}_0^{-1}\mathrm{A}\boldsymbol{\mu}_{2,T_i-k}u_{T_i-k} + \boldsymbol{b}^{\mathrm{T}}\boldsymbol{P}_0^{-1}\boldsymbol{b}u_{T_i-k}^2],
\end{aligned}$$

and

$$\begin{aligned}
&\mathrm{E}_{\boldsymbol{x}}[\log p(\boldsymbol{x}_t|\boldsymbol{x}_{t-1},\Theta)] = -\log\sqrt{2\pi|\boldsymbol{P}_0|}-\tfrac{1}{2}[\mathrm{trace}(\boldsymbol{P}_0^{-1}\boldsymbol{P}_1^3) + \boldsymbol{\mu}_{3,t}^{\mathrm{T}}\boldsymbol{P}_0^{-1}\boldsymbol{\mu}_{3,t}\\
&- \mathrm{trace}(\boldsymbol{P}_0^{-1}\boldsymbol{A}\boldsymbol{P}_2^3) - \boldsymbol{\mu}_{3,t}^{\mathrm{T}}\boldsymbol{P}_0^{-1}\mathrm{A}\boldsymbol{\mu}_{3,t-1} - \boldsymbol{\mu}_{3,t}^{\mathrm{T}}\boldsymbol{P}_0^{-1}\boldsymbol{b}u_{t-1}\\
&- \mathrm{trace}(\boldsymbol{A}^{\mathrm{T}}\boldsymbol{P}_0^{-1}\boldsymbol{P}_3^1) - \boldsymbol{\mu}_{3,t-1}^{\mathrm{T}}\boldsymbol{A}^{\mathrm{T}}\boldsymbol{P}_0^{-1}\boldsymbol{\mu}_{3,t} + \mathrm{trace}(\boldsymbol{A}^{\mathrm{T}}\boldsymbol{P}_0^{-1}\boldsymbol{A}\boldsymbol{P}_4^1) + \boldsymbol{\mu}_{3,t-1}^{\mathrm{T}}\boldsymbol{A}^{\mathrm{T}}\boldsymbol{P}_0^{-1}\mathrm{A}\boldsymbol{\mu}_{3,t-1}\\
&+ \boldsymbol{\mu}_{3,t-1}^{\mathrm{T}}\boldsymbol{A}^{\mathrm{T}}\boldsymbol{P}_0^{-1}\boldsymbol{b}u_{t-1} - \boldsymbol{b}^{\mathrm{T}}\boldsymbol{P}_0^{-1}\boldsymbol{\mu}_{3,t}u_{t-1} + \boldsymbol{b}^{\mathrm{T}}\boldsymbol{P}_0^{-1}\mathrm{A}\boldsymbol{\mu}_{3,t-1}u_{t-1} + \boldsymbol{b}^{\mathrm{T}}\boldsymbol{P}_0^{-1}\boldsymbol{b}u_{t-1}^2].
\end{aligned}$$

In order to calculate the parameter estimates, the gradients should be taken over the Q-function with respect to unknown parameters. Taking the gradient of $Q(\Theta|\Theta^k)$ with respect to $A$, $b$, and $c$ and equating them to zero, the estimates of the parameters are separately derived as

$$\begin{aligned}
\boldsymbol{A}^{k+1} = \frac{1}{12}\boldsymbol{P}_0\sum_{i=1}^{N}\sum_{k=0}^{2}\sum_{t=1}^{T}\boldsymbol{W}_{ikt}\big(&\boldsymbol{P}_4^1 + \boldsymbol{P}_4^2 + \boldsymbol{P}_4^3 + \boldsymbol{P}_4^{1\mathrm{T}} + \boldsymbol{P}_4^{2\mathrm{T}} + \boldsymbol{P}_4^{3\mathrm{T}}\\
&+ 2\boldsymbol{\mu}_{1,T_i-k-1}\boldsymbol{\mu}_{1,T_i-k-1}^{\mathrm{T}} + 2\boldsymbol{\mu}_{2,T_i-k}\boldsymbol{\mu}_{2,T_i-k}^{\mathrm{T}} + 2\boldsymbol{\mu}_{3,t-1}\boldsymbol{\mu}_{3,t-1}^{\mathrm{T}}\big)^{-1},
\end{aligned}$$

where

$$\begin{aligned}
W_{ikt} = & P_0^{-1}P_2^{1\mathrm{T}} + P_0^{-1}P_3^1 + 2P_0^{-1}\mu_{1,T_{i-k}}\mu_{1,T_{i-k}}^{\mathrm{T}} \\
& - 2P_0^{-1}b^{k+1}\mu_{1,T_{i-k}}^{\mathrm{T}}u_{T_{i-k-1}} + P_0^{-1}P_2^{2\mathrm{T}} + P_0^{-1}P_3^2 \\
& + 2P_0^{-1}\mu_{2,T_{i-k+1}}\mu_{2,T_{i-k}}^{\mathrm{T}} - 2P_0^{-1}b^{k+1}\mu_{2,T_{i-k}}^{\mathrm{T}}u_{T_{i-k}} \\
& + P_0^{-1}P_2^{3\mathrm{T}} + P_0^{-1}P_3^3 + 2P_0^{-1}\mu_{3,t}\mu_{3,t-1}^{\mathrm{T}} - 2P_0^{-1}b^{k+1}\mu_{3,t-1}^{\mathrm{T}}u_{T_{i-k}},
\end{aligned}$$

$$\begin{aligned}
b^{k+1} = & \sum_{i=1}^{N}\sum_{k=0}^{2}\sum_{t=1}^{T}\left(u_{T_{i-k-1}}^2 + u_{T_{i-k}}^2 + u_{t-1}^2\right)^{-1} \\
& \left(\mu_{1,T_{i-k}}u_{T_{i-k-1}} - A^{k+1}\mu_{1,T_{i-k-1}}u_{T_{i-k-1}} + \mu_{2,T_{i-k+1}}u_{T_{i-k}}\right. \\
& \left. - A^{k+1}\mu_{2,T_{i-k}}u_{T_{i-k}} + \mu_{3,t}u_{t-1} - A^{k+1}\mu_{3,t-1}u_{t-1}\right),
\end{aligned}$$

$$c^{k+1} = -2\sum_{i=1}^{N}\sum_{k=0}^{2}y_{T_i}\left(P_{new}^{\mathrm{T}} + P_{new} + 2\mu_{new}\mu_{new}^{\mathrm{T}}\right)^{-1}\mu_{new}.$$

The probability of delay can be derived using the Bayesian rule as

$$\begin{aligned}
& p(d_{T_i} = k \mid x_{T_{i-k}}, y_{T_i}, u_{1:T}, \Theta^k) \\
& = \frac{p(y_{T_i} \mid x_{T_{i-k}}, d_{T_i} = k, u_{1:T}, \Theta^k)\, p(x_{T_{i-k}}, d_{T_i} = k \mid u_{1:T}, \Theta^k)}{\sum_{k=0}^{q} p(y_{T_i} \mid x_{T_{i-k}}, d_{T_i} = k, u_{1:T}, \Theta^k)\, p(x_{T_{i-k}}, d_{T_i} = k \mid u_{1:T}, \Theta^k)} \\
& = \frac{p(y_{T_i} \mid x_{T_{i-k}}, d_{T_i} = k, u_{1:T}, \Theta^k)\, p(x_{T_{i-k}} \mid u_{1:T}, \Theta^k)\, p(d_{T_i} = k \mid u_{1:T}, \Theta^k)}{\sum_{k=0}^{q} p(y_{T_i} \mid x_{T_{i-k}}, d_{T_i} = k, u_{1:T}, \Theta^k)\, p(x_{T_{i-k}} \mid u_{1:T}, \Theta^k)\, p(d_{T_i} = k \mid u_{1:T}, \Theta^k)} \\
& = \frac{\frac{1}{\sqrt{2\pi\sigma_v}}\exp\left(-\frac{1}{2\sigma_v^2}\left[y_{T_i} - c\mu_{T_{i-k}}\right]^2\right)p\left(d_{T_i} = k\right)}{\sum_{k=0}^{q}\frac{1}{\sqrt{2\pi\sigma_v}}\exp\left(-\frac{1}{2\sigma_v^2}\left[y_{T_i} - c\mu_{T_{i-k}}\right]^2\right)p\left(d_{T_i} = k\right)}.
\end{aligned}$$

The above probability expression shows that the sampling delay of the ith measurement output $y_{T_i}$ is equal to the conditional probability of $k$. A diagram of explaining the proposed algorithm is shown in Figs. 6.1 and 6.2.

### 6.1.3 Simulation study

Consider the following multirate state space system with time-varying time delay:

$$\begin{aligned}
x_t &= 1.8x_t - u_t + w_t, \\
y_{T_i} &= 0.4x_{T_{i-d_{T_i}}} + v_{T_i},
\end{aligned}$$

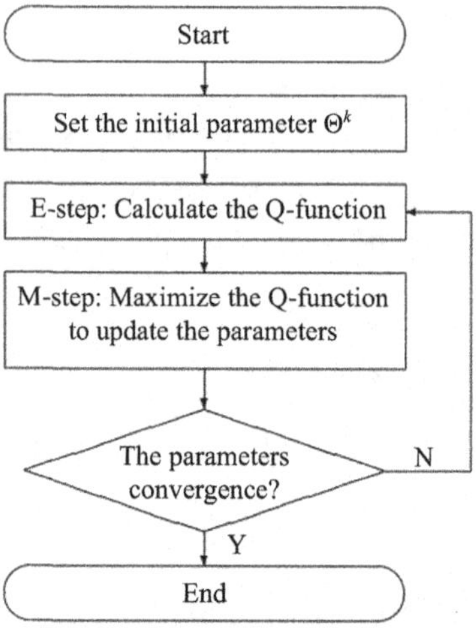


**Figure 6.1** A diagram explaining the proposed algorithm.

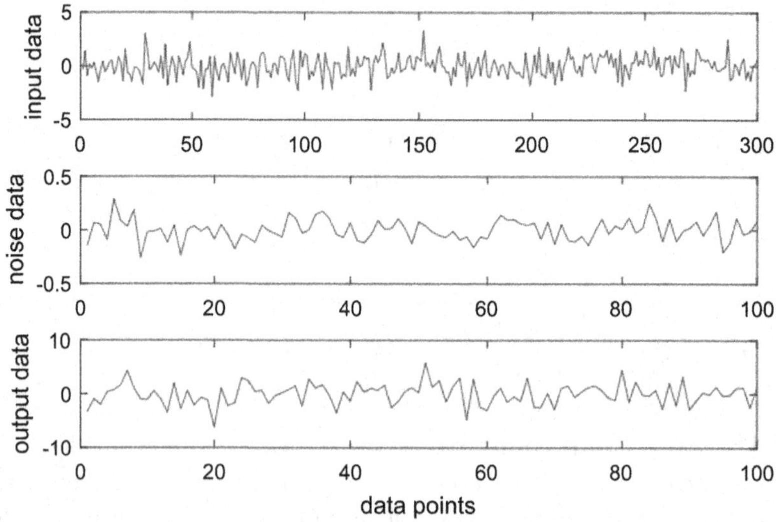


**Figure 6.2** The multirate sampled inputs and outputs.

where the fast-rate input sequence $\{u_t\}$ is generated from Gaussian distribution $N(0, \sigma_u^2)$ with $\sigma_u^2 = 1$; the slow-rate output $\{y_{T_i}\}$ is available at time instant $T_i \cdot \Delta t(T_i = 5i)$ with random time delay $d_{T_i} \cdot \Delta t(d_{T_i} \in [0, 2])$; the variance of the process noise $\{\omega_t\}$ and measurement noise $\{\nu_{T_i}\}$ are

$\sigma_v^2 = 0.01$ and $\sigma_w^2 = 0.01$, respectively; thus, the noise-to-signal ratio is $\delta_{ns} = \sqrt{\sigma_v^2/\sigma_u^2} = 10\%$. In simulation, $L = 300$ fast-rate inputs and $N = 100$ slow-rate outputs are collected for system identification.

Applying the proposed EM algorithm with a randomly generated initial guess to identify the unknown parameters, the estimated state space model with time-varying time delay parameters, the estimated time delay, and the Q-function are shown in Figs. 6.3–6.6, respectively. From these figures, it can be observed that the proposed EM algorithm has good identification performance since the estimated parameters approach the real ones after a few iterations. In order to further verify the effectiveness of the proposed algorithm, using the Monte Carlo simulations with 15 sets of noise realizations, the parameter estimates and their estimation biases of the EM algorithm are shown in Table 6.1, and the mean and standard deviation of parameter estimates from Monte Carlo simulations at SNR = 26.89dB and SNR = 46.48dB are calculated in Table 6.2.

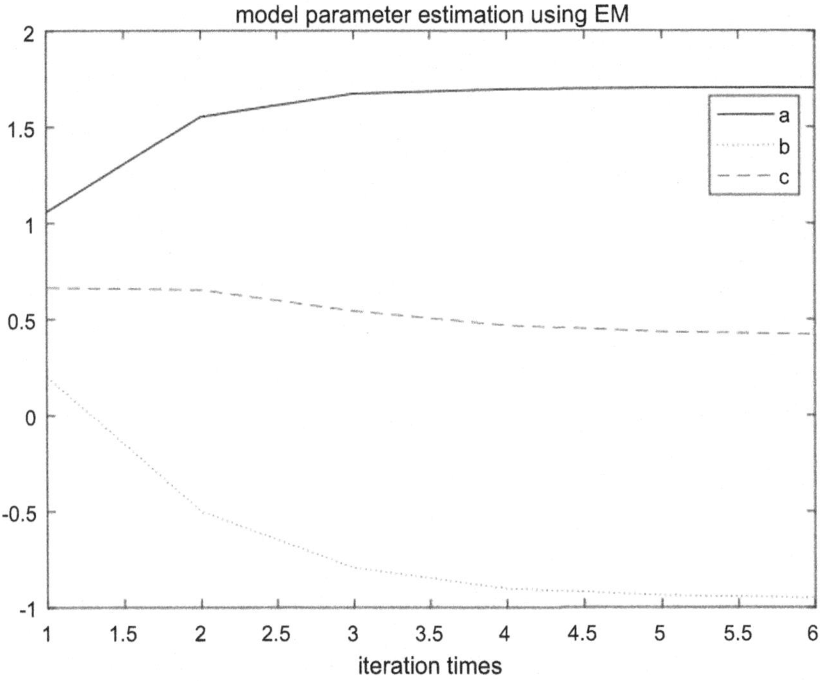


**Figure 6.3** The parameter estimates versus iteration k.

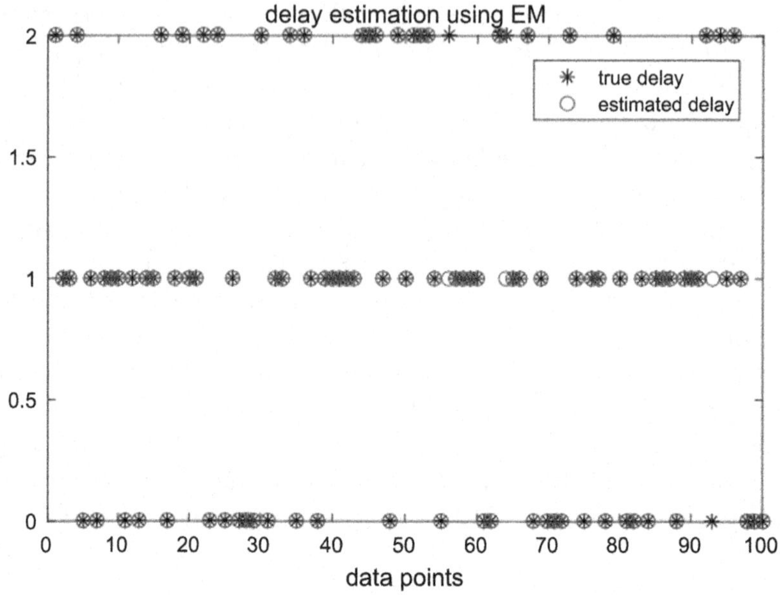


**Figure 6.4** The EM estimates of the uncertain delays.

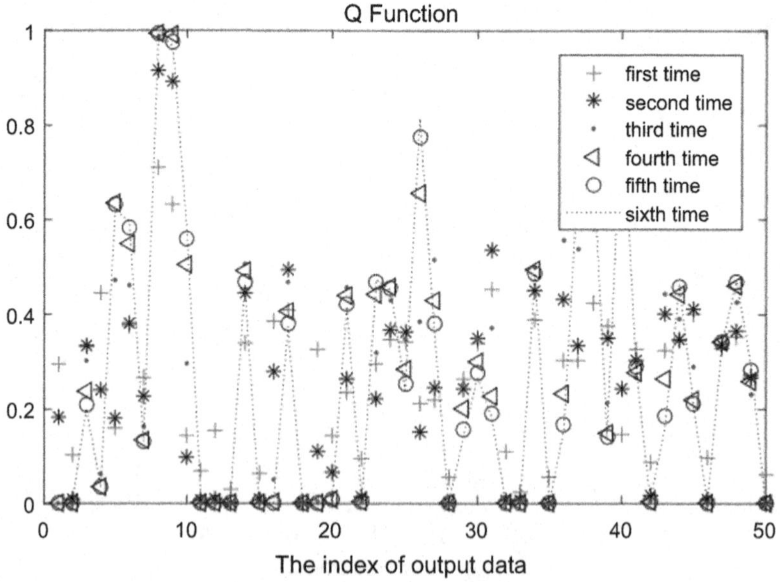


**Figure 6.5** The Q-function.

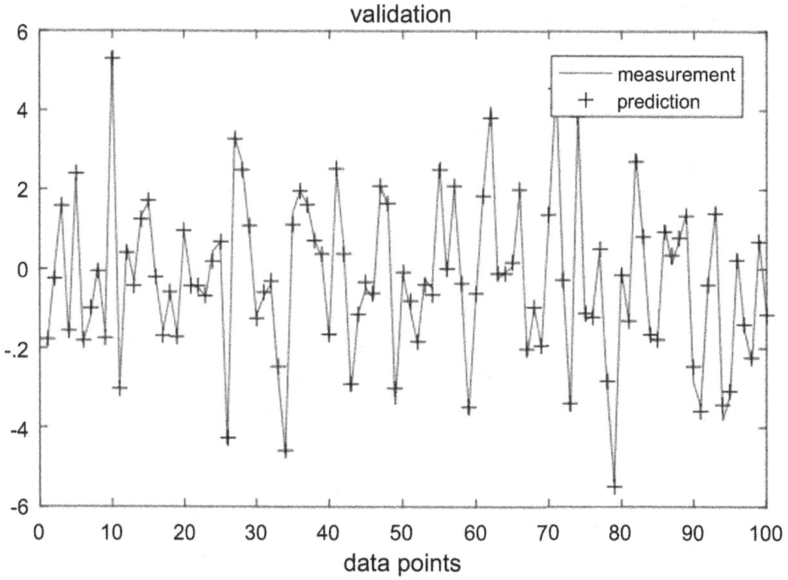


**Figure 6.6** Cross-validation results of the estimated model.

**Table 6.1** The EM estimates and variances based on 15 Monte Carlo simulations.

| *t* | *a* | *b* | *c* |
|---|---|---|---|
| 1 | 1.6945 ± 0.2346 | −0.8389 ± 0.1531 | 0.4489 ± 0.3565 |
| 3 | 1.8008 ± 0.4323 | −0.9621 ± 0.5177 | 0.4425 ± 0.4747 |
| 5 | 1.7935 ± 0.4732 | −1.0176 ± 0.5354 | 0.4161 ± 0.4672 |
| 7 | 1.8142 ± 0.4021 | −0.9976 ± 0.5345 | 0.4484 ± 0.4856 |
| 9 | 1.7877 ± 0.4670 | −1.0080 ± 0.5047 | 0.4096 ± 0.4798 |
| 11 | 1.7970 ± 0.4655 | −1.0198 ± 0.4850 | 0.4307 ± 0.4834 |
| 13 | 1.8109 ± 0.3907 | −1.0220 ± 0.4967 | 0.4117 ± 0.5032 |
| 15 | 1.7898 ± 0.4281 | −0.9980 ± 0.5178 | 0.4188 ± 0.4821 |
| True values | 1.80000 | −1.00000 | 0.40000 |

**Table 6.2** The mean and standard deviation of parameter estimates from Monte Carlo simulations.

| **True values** | **SNR = 26.89 dB** | | **SNR = 46.48 dB** | |
|---|---|---|---|---|
| | **Mean** | **Std** | **Mean** | **Std** |
| $a = 1.8$ | 1.7280 | 0.1683 | 1.8035 | 0.1511 |
| $b = -1.0$ | −0.9023 | 0.1938 | −0.9773 | 0.0967 |
| $c = 0.4$ | 0.4503 | 0.1965 | 0.4015 | 0.1132 |

### 6.1.4 Conclusions

This chapter considers the identification of MR systems with unknown random delays and continuous states, which includes conventional nonuniformly sampled-data systems and multirate systems as special cases. Due to the impact of random delay, the irregularly sampled output cannot appropriately indicate current noise-free output. Thus, traditional identification methods like least squares algorithm fail to identify such MR systems if the uncertain delay problem is overlooked. To address this challenge, an EM algorithm is applied in this article to identify the multirate state space model with time-varying time delay, in which unknown delays and states are treated as hidden variables and are estimated along with the model parameters. The proposed EM algorithm has been evaluated in a simulation example, and the obtained results confirm that the algorithm is effective with high estimation accuracy and fast convergence rate. The proposed approaches in the chapter can combine other mathematical tools and statistical strategies to study the performances of some parameter estimation algorithms and can be applied to other multivariable systems with different structures and disturbance noises and other literature.

## 6.2 Moving horizon estimation for multirate system with time-varying time delays

State estimation is one of the most important elements in development and implementation of advanced process control techniques. The performance of state estimation has a significant effect on the performance of control. The Kalman filter (KF), extended Kalman filter (EKF), and unscented Kalman filter (UKF) are among the most widely used techniques for state estimation.

In practice, many processes may exhibit both continuous and discrete-valued dynamic behaviors alone (known as hybrid system), and in such a process, state estimation is more challenging than continuous systems. In this paper, we consider processes that are described by state space models with time-varying time delays. In other words, time delays can change at every sampling instant. Thus, to estimate system states, the integer-valued time delays must be estimated simultaneously. Generally, there exist two approaches to estimate the state of a hybrid system, respectively: (1) sequential approach that uses two separate algorithms to estimate the continuous and noncontinuous states. In this method, two objective functions are used to estimate the continuous states and noncontinuous states sequentially. Although this approach is simpler and more intuitive, one can hardly obtain the optimal solution for two objective functions simultaneously. (2) Simultaneous approach uses a single objective function to estimate the continuous and noncontinuous states simultaneously. Such an approach can provide a systematic way to estimate both types of states.

State estimation with unknown time-varying time delay is a hybrid state estimation problem. In order to estimate the continuous states and noncontinuous states simultaneously, the moving horizon estimation (MHE) algorithm has been introduced. In general, the moving horizon estimation approach can address a hybrid state estimation problem in the presence of bounded system and measurement noises. It is an efficient optimization-based strategy to produce estimates of unknown variables or parameters. In addition, the increased interest in MHE has resulted from its proven superiority over other approaches such as the Kalman filter.

Although MHE has been well studied in simultaneous estimation of continuous and noncontinuous states, few results have been reported for the simultaneous estimation of state and time-varying time delays for multirate systems. Compared with single-sensor system, the multiple-sensor system enhances the performance of estimation. However, the sampling rates of process measurements may vary in different sensors which constitute a multirate system. Time delay inevitably exists in many practical processes, and it has a significant effect on control performance. Normally, in traditional state estimation, the time delay of the system is assumed to be known and constant. However, in reality, it is often unknown and can be time-variant. Based on these considerations, this study aims at developing MHE schemes to simultaneously estimate the continuous states and discrete time delay sequence of multirate state space systems. First, the arrival cost for a MHE will be derived. Second, the optimal and the approximate MHE schemes are formulated for simultaneous estimation of continuous state and discrete time delay sequences. Then, the effects of different measurement noise levels on the estimation of the state and time delay are conducted. Finally, a numerical example is given to verify the derived estimation algorithm.

## 6.2.1 Problem statement

Consider a time delay system described by the following multirate state space model,

$$\boldsymbol{x}_t = \boldsymbol{A}\boldsymbol{x}_{k-1} + \boldsymbol{B}\boldsymbol{u}_{k-1} + \omega_{k-1}, \tag{6.11}$$

$$y_{T_i} = \boldsymbol{c}\boldsymbol{x}_{T_i - d_{T_i}} + \upsilon_T, \tag{6.12}$$

where $x_k \in R^n$ is the state vector, $u_k \in R^n$ is the fast-rate input, $y_{T_i} \in R$ is the slow-rate output, and $d_{T_i}$ is the discrete time delay which can vary at every sample. $\omega_k \in \mathcal{N}(0, Q_k)$ and $\nu_{T_i} \in \mathcal{N}(0, \sigma^2_{\nu_{T_i}})$ are process noise and measurement noise, respectively, and follow the Gaussian distribution. $A \in R^{n \times n}$, $B \in R^{n \times n}$, and $c \in R^{1 \times n}$ are the system parameter matrices/vector of appropriate dimensions. For estimation, the previous multirate measurement input–output data $\{u_1, u_2, \ldots, u_K, y_{T_1}, y_{T_2}, \ldots, y_{T_L}\}$ are available. The term "multirate" here refers to the multiple sampling rates. The input data are measured at a fast rate; that is, at each unit sampling instant, the inputs are available. The output data are available at varying and slow rates; that is, the sampling interval of output data is also time-varying and larger than the unit sampling interval.

Therefore, this chapter focuses on state and time-varying time delay estimation of a multirate system given known input and output data. The fast-rate variables, continuous states, and inputs are denoted as $x_{1:K} = \{x_1, \ldots, x_K\}$ and $u_{1:K} = \{u_1, \ldots, u_K\}$, respectively. For convenience, the slow-rate variables, time delay sequences, and outputs are denoted as $d_{1:K} = \{d_{T_1}, \ldots, d_{T_L}\}$ and $y_{1:K} = \{y_{T_1}, \ldots, y_{T_L}\}$, where $d_{A:B}$ and $y_{A:B}$ represent slow-rate delays and outputs available during time period $A{:}B$. To maximize the conditional joint probability density function (PDF) $P(d_{1:T}, x_{1:T} | y_{1:T})$, an optimal batch estimation of the continuous states and discrete time delay values will be obtained. The optimization problem is formulated by minimizing the following negative logarithm function of joint distribution,

$$\overset{\min}{\hat{d}_{1:T}, \hat{\boldsymbol{x}}_{1:T}} J = -\ln P(\boldsymbol{x}_{1:T}, d_{1:T}, y_{1:T}).$$

Commonly used approaches, such as the Kalman filter, take advantages of all the available measurements $y_{1:T}$, $u_{1:T}$ to estimate the sequences $\widehat{d}_{1:T}$ and $\widehat{x}_{1:T}$ through a recursive estimation.

## 6.2.2 Moving horizon estimation

### *6.2.2.1 MHE formulation*

In this chapter, we will take a probability approach to derive MHE and its arrival cost for the multirate time-varying time delay system described by the state space models. By taking into account all the available measurements, the MHE approach optimizes the negative logarithm of the joint distribution of the state $x_{T-N:T}$ and time delay $d_{T-N:T}$ given all measurement data up to time $T$, that is,

$$\min_{\hat{d}_{T-N:T}, \hat{\boldsymbol{x}}_{T-N:T}} J = -\ln P(\boldsymbol{x}_{T-N:T}, d_{T-N:T}, y_{1:T}). \tag{6.13}$$

Then by applying the chain rule, $J$ can also be further derived as

$$\begin{aligned} J = & -\ln P(\boldsymbol{x}_{T-N}, d_{T-N}, y_{1:T-N}) \\ & -\ln P(\boldsymbol{x}_{T-N+1:T}, d_{T-N+1:T}, y_{T-N+1:T} | \boldsymbol{x}_{T-N}, d_{T-N}, y_{1:T-N}), \end{aligned}$$

where $d_{T-N}$ and $d_{T-N+1:T}$ can be nonexistent if there is no slow-rate output at time $T-N$ or in the period of $T-N+1:T$, since it is the multirate system. The same applies to the output $y$. The various cases are shown in Fig. 6.7. For the first case, $d_{T-N}$ exists and $d_{T-N+1:T}$ does not exist. The second case is opposite (nonexistent $d_{T-N}$ and existent $d_{T-N+1:T}$). For the third one, there are $d_{T-N}$ and $d_{T-N+1:T}$ in the moving window. $d_{T-N}$ and $d_{T-N+1:T}$ do not exist in the last case.

Since $x_{T-N+1:T}$, $d_{T-N+1:T}$, and $y_{T-N+1:T}$ are conditionally independent of $y_{1:T-N}$, given $x_{T-N}$ and $d_{T-N}$, the above equation can be simplified as

$$\begin{aligned} J = & -\ln P(\boldsymbol{x}_{T-N}, d_{T-N}, y_{1:T-N}) \\ & -\ln P(\boldsymbol{x}_{T-N+1:T}, d_{T-N+1:T}, y_{T-N+1:T} | \boldsymbol{x}_{T-N}, d_{T-N}). \end{aligned} \tag{6.14}$$

In MHE, the first term of Eq. (6.14) is called arrival cost as shown below. This term represents the cost function up to time $T-N$, and it will be further discussed shortly.

$$\phi_{T-N} = -\ln P(\boldsymbol{x}_{T-N}, d_{T-N}, y_{1:T-N}). \tag{6.15}$$

The second term of Eq. (6.14) is the cost function in the selected horizon $T-N+1:T$

$$J_{T-N+1:T} = -\ln P(\boldsymbol{x}_{T-N+1:T}, d_{T-N+1:T}, y_{T-N+1:T} | \boldsymbol{x}_{T-N}, d_{T-N}). \tag{6.16}$$

Thus, Eq. (6.13) can be simplified as

$$\min_{\hat{d}_{T-N:T}, \hat{\boldsymbol{x}}_{T-N:T}} J = \phi_{T-N} + J_{T-N+1:T}. \tag{6.17}$$

The joint probability distribution of the discrete time delay sequence $d_{T-N+1:T}$ and the continuous state $x_{T-N+1:T}$ with the available measurements

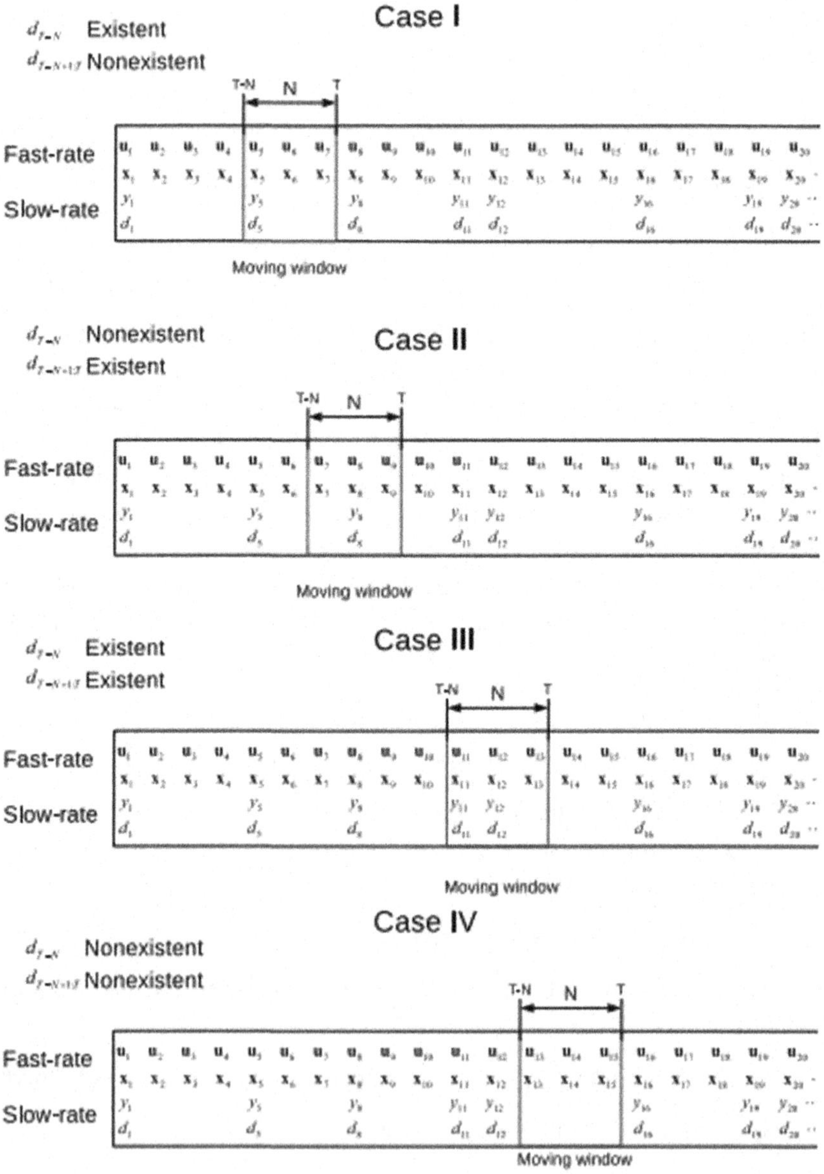


**Figure 6.7** Moving horizon of multirate system.

$y_{T-N+1:T}$ can be expressed as

$$\begin{aligned}&P(\boldsymbol{x}_{T-N+1:T}, d_{T-N+1:T}, y_{T-N+1:T}|\boldsymbol{x}_{T-N}, d_{T-N})\\&\quad= P(y_{T-N+1:T}|\boldsymbol{x}_{T-N:T}, d_{T-N:T})P(\boldsymbol{x}_{T-N+1:T}, d_{T-N+1:T}|\boldsymbol{x}_{T-N}, d_{T-N})\\&\quad= \prod_{i=g}^{h} P\left(d_{T_i}|d_{T_{i-1}}\right) \prod_{k=T-N+1}^{T} P(\boldsymbol{x}_k|\boldsymbol{x}_{k-1}), \prod_{i=g}^{h} P(y_{T_i}|\boldsymbol{x}_{T_i-d_{T_i}}, d_{T_i}),\end{aligned} \tag{6.18}$$

where $g$ and $h$ denote the starting and ending time of available slow-rate data in the period of $T-N+1{:}T$. Based on Eqs. (6.16), (6.17), and (6.18), we can formulate the following optimization problem,

$$\begin{aligned}&\min_{\hat{d}_{T-N:T}, \hat{\boldsymbol{x}}_{T-N:T}} J = \phi_{T-N} - \sum_{i=g}^{h} \ln P\left(d_{T_1}|d_{T_{i-1}}\right)\\&- \sum_{k=T-N+1}^{T} \ln P(\boldsymbol{x}_k|\boldsymbol{x}_{k-1}) - \sum_{i=g}^{h} \ln P(y_{T_i}|\boldsymbol{x}_{T_i-d_{T_i}}, d_{T_i}).\end{aligned} \tag{6.19}$$

The discrete time delay $d_{T_i}$ is considered to follow a Markov chain model; hence, the distribution of time delay is given as:

$$P\left(d_{T_i} = m|d_{T_{i-1}} = n\right) = \mathbf{P}^{(m,n)}_{T_{i-1},T_i},$$

where $P^{(m,n)}_{T_{i-1},T_i}$ is the transition probability.

The likelihoods of the continuous state $x_k$ and output measurement $y_{T_i}$ depend on the distribution of noise. The additive noises are assumed to follow Gaussian distribution. So we can derive the explicit expressions of the following terms:

- Given the state $x_{k-1}$, the conditional probability distribution of the state $x_k$ is:

$$\begin{aligned}&\boldsymbol{x}_k \sim \mathcal{N}(\boldsymbol{A}x_{k-1} + \boldsymbol{B}\boldsymbol{u}_{k-1}, \mathbf{Q}_k),\\&P(\boldsymbol{x}_k|\boldsymbol{x}_{k-1}) = \frac{1}{\sqrt{2\pi|\mathbf{Q}_k}} \exp\left[-\frac{1}{2}(\boldsymbol{x}_k - \boldsymbol{A}x_{k-1} - \boldsymbol{B}\boldsymbol{u}_{k-1})^T \mathbf{Q}_k^{-1}(\boldsymbol{x}_k - \boldsymbol{A}x_{k-1} - \boldsymbol{B}\boldsymbol{u}_{k-1})\right].\end{aligned} \tag{6.20}$$

- Conditioned on the time delay $d_{T_i}$ and $x_{T_i-d_{T_i}}$, the probability distribution of the observation $y_{T_i}$ is

$$\begin{aligned}&y_{T_i} \sim \mathcal{N}\left(\boldsymbol{c}\boldsymbol{x}_{T_i-d_{T_i}}, \sigma^2_{\upsilon_{T_i}}\right),\\&P\left(y_{T_i}\middle|\boldsymbol{x}_{T_i-d_{T_i}}, d_{T_i}\right) = \frac{1}{\sqrt{2\pi\sigma_{\upsilon_{T_i}}}} \exp\left[-\frac{1}{2\sigma^2_{\upsilon_{T_i}}}\left(y_{T_i} - \boldsymbol{c}\boldsymbol{x}_{T_i-d_{T_i}}\right)^2\right].\end{aligned} \tag{6.21}$$

Then, we introduce the following terms:

$$\hat{\omega}_k = \hat{x}_k - A\hat{x}_{k-1} - Bu_{k-1},$$
$$\hat{\upsilon}_{T_i} = y_{T_i} - c\hat{x}_{T_i - \hat{d}_{T_i}}$$

and by taking the negative logarithm of (6.20) and (6.21), we have

$$-\ln P(x_k|x_{k-1}) = B_k + \tfrac{1}{2}\hat{\omega}_k^T Q_k^{-1}\hat{\omega}_k,$$
$$-\ln P\left(y_{T_i}|x_{T_i - d_{T_i}}, d_{T_i}\right) = C_{T_i} + \tfrac{1}{2\sigma^2_{\upsilon_{T_i}}}\hat{\upsilon}^2_{T_i},$$

where $B_k = -\ln\frac{1}{\sqrt{2\pi|Q_k|}}$ and $C_{T_i} = -\ln\frac{1}{\sqrt{2\pi}\sigma_{\upsilon_{T_i}}}$. If we define $\gamma^{(m)}_{T_{i-1}}$, $\alpha^{(n)}_{T_i}$, and $\mathbb{P}^{(m,n)}_{T_{i-1},T_i}$ as

$$\gamma^{(m)}_{T_{i-1}} = \begin{cases} 1 & \text{if } m = \hat{d}_{T_{i=1}} \\ 0 & \text{if } m \neq \hat{d}_{T_{i=1}} \end{cases},$$

$$\alpha^{(n)}_{T_i} = \begin{cases} 1 & \text{if } n = \hat{d}_{T_i} \\ 0 & \text{if } n \neq \hat{d}_{T_i} \end{cases},$$

$$\mathbb{P}^{(m,n)}_{T_{i-1},T_i} = P\left(d_{T_i} = n \middle| d_{T_{i-1}} = m, \mathit{for}, m, n \in D\right),$$

we can obtain

$$\begin{aligned}\min_{\hat{d}_{T-N:T},\hat{\omega}_{T-N:T}} J_T = \phi_{T-N} - \sum_{i=g}^{h}\sum_{m=1}^{D}\sum_{n=1}^{D}\gamma^{(m)}_{T_{i-1}}\alpha^{(n)}_{T_i}\ln\mathbb{P}^{(m,n)}_{T_{i-1},T_i} \\ + \sum_{k=T-N}^{T-1}\hat{\omega}_k^T Q_k^{-1}\hat{\omega}_k + \sum_{i=g}^{h}\sum_{n=1}^{D}\alpha^{(n)}_{T_i}\hat{\upsilon}^2_{T_i}\sigma_\upsilon^{-2}.\end{aligned} \tag{6.22}$$

### 6.2.3 Arrival cost

Since this is a multirate system, the arrival cost must be derived with respect to two different cases, respectively:

**1.** If $d_{T-N}$ and $y_{T-N}$ exist

The arrival cost in Eq. (6.15) can be expanded into two parts as follows:

$$\begin{aligned}\phi_{T-N} &= -\ln\left[P\left(x_{T-N}\middle|d_{T-N}, y_{0:\,T-N}\right)P(d_{T-N}, y_{0:\,T-N})\right] \\ &= \phi^x_{T-N} + \phi^d_{T-N}.\end{aligned}$$

Arrival cost: $\varphi^d_{T-N}$

$$\phi^d_{T-N}(n) = -\ln P(d_{T-N}, y_{0:T-N}), \tag{6.23}$$

where $\phi^d_{T-N}(n)$ is the cost of arriving at state $d_{T-N} = n$ at time $T-N$ given all of the measurements up to time $T-N$. A recursive solution of (6.23) can be derived using a forward procedure technique,

$$\phi^d_{T-N}(n)\sum_{m=1}^{D}\left[\phi_{T-N-1}(m) - \ln \mathbb{P}^{(m,n)}_{T-N-1,T-N}\right] - \ln P(y_{T-N}|d_{T-N} = n). \tag{6.24}$$

Under the assumption of Gaussian noise distribution, the conditional distribution density function $P(y_{T-N}|d_{T-N} = n)$ from (6.14) can be expressed as

$$P(y_{T-N}|d_{T-N}) = \frac{1}{\sqrt{2\pi\sigma_{v_{T-N}}}}\exp\left[-\frac{1}{2\sigma^2_{v_{T-N}}}\left(y_{T-N} - \boldsymbol{c}\boldsymbol{x}_{T-N-d_{T-N}}\right)^2\right].$$

Therefore, the last term in (6.24) becomes

$$-\ln P(y_{T-N}|d_{T-N} = n) = C_{T-N} + \frac{1}{2\sigma^2_{v_{T-N}}}\hat{v}^2_{T-N}, \tag{6.25}$$

where $C_{T-N} = -\ln\frac{1}{\sqrt{2\pi\sigma_{v_{T-N}}}}$. By substituting (6.25) into (6.24), we have the following cost function:

$$\phi^d_{T-N}(n) = \sum_{m=1}^{D}\left[\phi_{T-N-1}(m) - \ln \mathbb{P}^{(m,n)}_{T-N-1:T-N}\right] + \frac{1}{2\sigma^2_{v_{T-N}}}\hat{v}^2_{T-N} + C_{T-N}.$$

Arrival cost: $\phi^x_{T-N}$

$$\phi^x_{T-N} = -\ln P(\boldsymbol{x}_{T-N}|d_{T-N}, y_{0:T-N}).$$

Using Bayesian rule, $P(x_{T-N}\big|d_{T-N}, y_{0:T-N})$ can be written as

$$P(\boldsymbol{x}_{T-N}|y_{0:T-N}, d_{T-N}) = \frac{P(\boldsymbol{x}_{T-N}|d_{T-N})P(y_{0:T-N}|\boldsymbol{x}_{T-N}, d_{T-N})}{P(y_{0:T-N}|d_{T-N})}$$

or

$$P(\boldsymbol{x}_{T-N}|y_{0:T-N}, d_{T-N}) = D_{T-N}P(\boldsymbol{x}_{T-N}|d_{T-N}),$$

where $D_{T-N} = \frac{P(y_{0:T-N}|x_{T-N}, d_{T-N})}{P(y_{0:T-N}|d_{T-N})}$; then, the above equation becomes

$$\phi^{x}_{T-N} = -\ln D_{T-N} - \ln P(\boldsymbol{x}_{T-N}|d_{T-N})$$
$$= \tfrac{1}{2}\hat{\boldsymbol{\omega}}^{T}_{T-N-1}\boldsymbol{P}^{-1}_{T-N}\hat{\boldsymbol{\omega}}_{T-N-1} + E_{T-N},$$

where

$$\begin{aligned}\boldsymbol{P}_{T-N} &= \boldsymbol{A}\boldsymbol{P}_{T-N-1}\boldsymbol{A}^{\mathrm{T}} + \boldsymbol{Q}_{T-N-1}\\ &\quad - \boldsymbol{A}\boldsymbol{P}_{T-N-1}\boldsymbol{c}\left(\boldsymbol{c}\boldsymbol{P}_{T-N-1}\boldsymbol{c}^{\mathrm{T}} + \boldsymbol{R}_{T-N}\right)^{-1}\boldsymbol{c}\boldsymbol{P}_{T-N-1}\boldsymbol{A}^{\mathrm{T}},\\ E_{T-N} &= B_{T-N} + C_{T-N} - \ln D_{T-N}.\end{aligned}$$

**2.** If $d_{T-N}$ and $y_{T-N}$ do not exist
The arrival cost in (6.15) can be expanded as follows:

$$\begin{aligned}\phi_{T-N} &= -\ln\left[P\left(\boldsymbol{x}_{T-N}\middle|y_{0:T-N}\right)P(y_{0:T-N})\right]\\ &= -\ln P\left(\boldsymbol{x}_{T-N}\middle|y_{0:T-N}\right) - \ln P(y_{0:T-N}),\end{aligned} \tag{6.26}$$

where $F_{T-N} = P(y_{0:T-N})$. Then, Eq. (6.26) becomes

$$\phi_{T-N} = \tfrac{1}{2}\hat{\boldsymbol{\omega}}^{T}_{T-N-1}\boldsymbol{P}^{-1}_{T-N}\hat{\boldsymbol{\omega}}_{T-N-1} + G_{T-N},$$

where $G_{T-N} = B_{T-N} + C_{T-N} - \ln F_{T-N}$.

### 6.2.4 Objective function

The objective function can now be formulated as

$$\begin{aligned}\min_{\hat{d}_{T_g:T_h},\,\hat{\boldsymbol{\omega}}_{T-N-1:T-1}} J_T &= \sum_{m=1}^{D}\sum_{n=1}^{D}\gamma^{(m)}_{T-N}\alpha^{(n)}_{T-N-1}\left[\phi_{T-N-1}(m) - \ln \mathbb{P}^{(m,n)}_{T-N-1:T-N}\right]\\ &\quad + \frac{1}{2\sigma^2_{\upsilon_{T-N}}}\hat{\upsilon}^2_{T-N} + \frac{1}{2}\hat{\boldsymbol{\omega}}^{T}_{T-N-1}\boldsymbol{P}^{-1}_{T-N}\hat{\boldsymbol{\omega}}_{T-N-1}\\ &\quad - \sum_{i=g}^{h}\sum_{m=1}^{D}\sum_{n=1}^{D}\gamma^{(m)}_{T_{i-1}}\alpha^{(n)}_{T_i}\ln \mathbb{P}^{(m,n)}_{T_{i-1},T_i} + \frac{1}{2}\sum_{k=T-n}^{T-1}\hat{\boldsymbol{\omega}}^{T}_{k}Q^{-1}_{k}\hat{\boldsymbol{\omega}}_{k}\\ &\quad + \frac{1}{2}\sum_{i=g}^{h}\sum_{n=1}^{D}\alpha^{(n)}_{T_i}\hat{\upsilon}^2_{T_i}\sigma^{-2}_{\upsilon_{T_i}}.\end{aligned} \tag{6.27}$$

Since this is a multirate system, the slow-rate output may not be available at time $T-N$. Thus, the corresponding time delay is also empty.

It leads to another case: if $d_{T-N}$ and $y_{T-N}$ do not exist, $J_T$ reduces to

$$\min_{\hat{d}_{T_g:T_h},\hat{\boldsymbol{\omega}}_{T-N-1:T-1}} J_T = \tfrac{1}{2}\hat{\boldsymbol{\omega}}_{T-N-1}^T \boldsymbol{P}_{T-N}^{-1}\hat{\boldsymbol{\omega}}_{T-N-1} - \sum_{i=g}^{h}\sum_{m=1}^{D}\sum_{n=1}^{D}\gamma_{T_{i-1}}^{(m)}\alpha_{T_i}^{(n)}\ln \mathbb{P}_{T_{i-1},T_i}^{(m,n)}$$
$$+\tfrac{1}{2}\sum_{k=T-n}^{T-1}\hat{\omega}_k^T Q_k^{-1}\hat{\omega}_k + \tfrac{1}{2}\sum_{i=g}^{h}\sum_{n=1}^{D}\alpha_{T_i}^{(n)}\hat{\upsilon}_{T_i}^2\sigma_{\upsilon_{T_i}}^{-2} \tag{6.28}$$

subject to the following constraints:

$$\hat{\boldsymbol{x}}_k = \boldsymbol{A}\hat{\boldsymbol{x}}_{k-1} + \boldsymbol{B}\boldsymbol{u}_{k-1} + \hat{\omega}_{k-1}, k = T-N:T-1, \tag{6.29}$$

$$\hat{\upsilon}_{T_i} = y_{T_i} - \boldsymbol{c}\hat{\boldsymbol{x}}_{T_i-\hat{d}_{T_i}}, T_i = T-N:T, \tag{6.30}$$

$$\gamma_{T_{i-1}}^{(m)} = \begin{cases} 1 & if\ m = \hat{d}_{T_{i-1}} \\ 0 & if\ m \neq \hat{d}_{T_{i-1}} \end{cases}, \tag{6.31}$$

$$\alpha_{T_i}^{(n)} = \begin{cases} 1 & if\ n = \hat{d}_{T_i} \\ 0 & if\ n \neq \hat{d}_{T_i} \end{cases}, \tag{6.32}$$

$$\mathbf{Q}_k = \begin{bmatrix} \sigma_{\omega,1}^2 & 0 \\ 0 & \sigma_{\omega,2}^2 \end{bmatrix}. \tag{6.33}$$

#### 6.2.4.1 Optimization procedure

Let the continuous state space model be $A$, $B$, and $c$ and the discrete time delay Markov chain transition model be P, and the optimization procedure can be summarized as in the following steps:

- Select a window size $N$.
- Given the initial guess of the time delay $d_{T_0}$.
- For $T = N+1:K$
  1. If $d_{T-N}$ and $y_{T-N}$ exist, solve the optimization problem in Eq. (6.27). If $d_{T-N}$ and $y_{T-N}$ do not exist, solve the optimization problem in Eq. (6.28). Obtain the estimation of time delay $\widehat{d}_{T_g:T_h}$ and additive noise $\hat{w}_{T-N-1:T-1}$.
  2. Calculate the state estimation $\widehat{x}_k (k = T-N:T-1)$ using Eq. (6.29).
- Obtain the estimates for the states $\widehat{x}_{1:K}$ and time delays $\widehat{d}_{T_1:T_L}$.

### 6.2.5 Simulation study

In this section, a numerical example is considered to illustrate the proposed algorithm. For this purpose, let us consider the following multirate state space system with time-varying time delay:

$$\begin{aligned} \boldsymbol{x}_k &= \begin{bmatrix} 1.00 & 1.00 \\ -1.20 & -0.60 \end{bmatrix} \boldsymbol{x}_{k-1} + \begin{bmatrix} 1.00 & 0.20 \\ -1.20 & -0.60 \end{bmatrix} \boldsymbol{u}_{k-1} + \omega_k, \\ y_{T_i} &= \begin{bmatrix} 1.20, & 1.75 \end{bmatrix} \boldsymbol{x}_{T_i - d_{T_i}} + v_{T_i}. \end{aligned}$$

To evaluate the performance of the MHE in presence of constraints, we consider the following constraints:

$$\begin{aligned} 0 \leqslant d_{T_i} \leqslant 2, \\ -H\sigma_\omega \leqslant \omega_{k_i} \leqslant H\sigma_\omega, i = 1, 2, \end{aligned}$$

where $H \geq 1$ may be considered as a tuning parameter to constrain the continue states; the fast-rate input sequence $\{u_k\}$ follows the Gaussian distribution $N(0, \sigma_u^2)$; the slow-rate output $\{y_{T_i}\}$ can be measured at time instant $T_i \times \Delta t$ with time delay $d_{T_i} \times \Delta t$ ($d_{T_i}$ ranging from 0 to 2) which follows hidden Markov model; and the variances of the noise (we assume $\sigma_\omega^2 = \sigma_v^2 = \sigma^2$) are $\sigma^2 = 0.1$ and $\sigma^2 = 0.3$, respectively. The transition probability matrix of time delay is given as:

$$\mathbb{P} = \begin{bmatrix} 0.3673 & 0.3011 & 0.3241 \\ 0.2857 & 0.2903 & 0.3426 \\ 0.3469 & 0.4086 & 0.3333 \end{bmatrix}$$

In the simulation, fast-rate inputs of size $K = 1000$ and slow-rate outputs of size $L = 100$ are selected, which are shown in Fig. 6.8 with $\sigma^2 = 0.1$ and Fig. 6.9 with $\sigma^2 = 0.3$, respectively.

First, a proper window size and initial time delay are specified. Next, the MHE algorithm is applied to estimate the unknown continuous state and discrete time delay sequences. The state and time delay estimates and their estimation errors with different variances of the noise are shown in Figs. 6.10–6.14 and Table 6.3.

The state estimation results are shown in Figs. 6.10 and 6.11, where the dot and the solid lines indicate the estimated states and the real states, respectively. From these two figures, we can see that the proposed MHE algorithm has achieved good performance and the state estimates get closer to their true values as $t$ increases.

The delay estimations are shown in Figs. 6.12 and 6.13, and the graphic and tabular results of accuracy of delay estimation with different

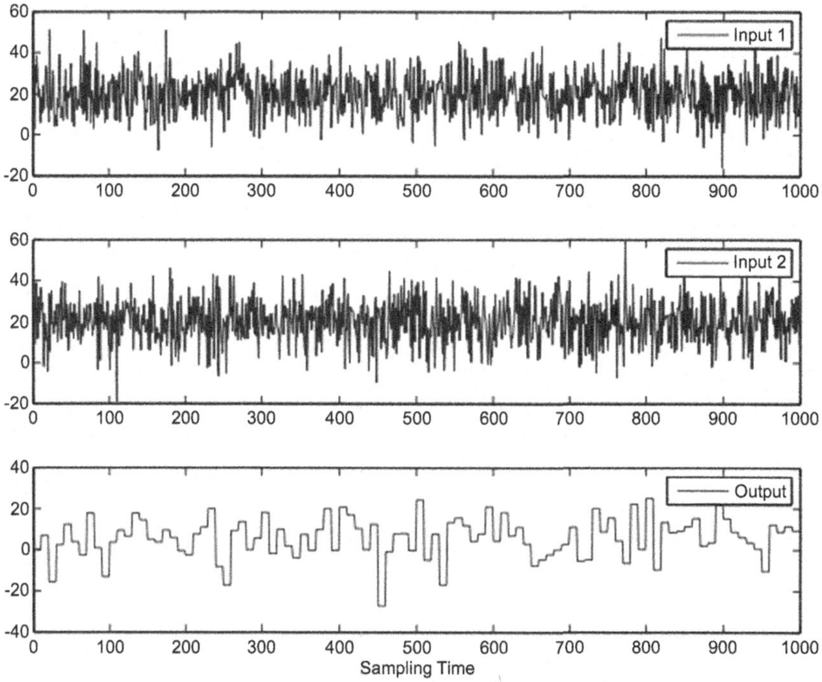

Figure 6.8 The multirate sampled inputs and outputs with $\sigma^2 = 0.1$.

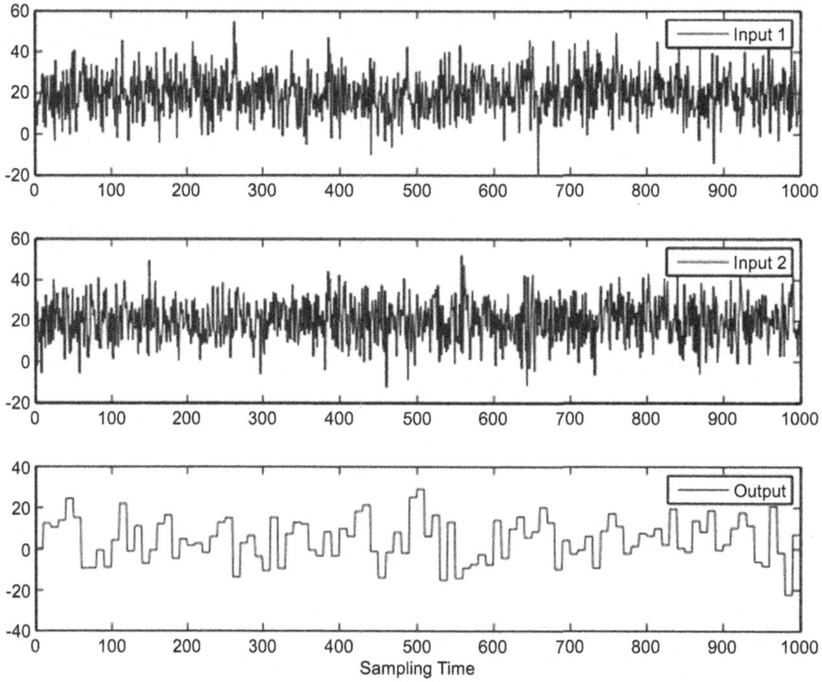

Figure 6.9 The multirate sampled inputs and outputs with $\sigma^2 = 0.3$.

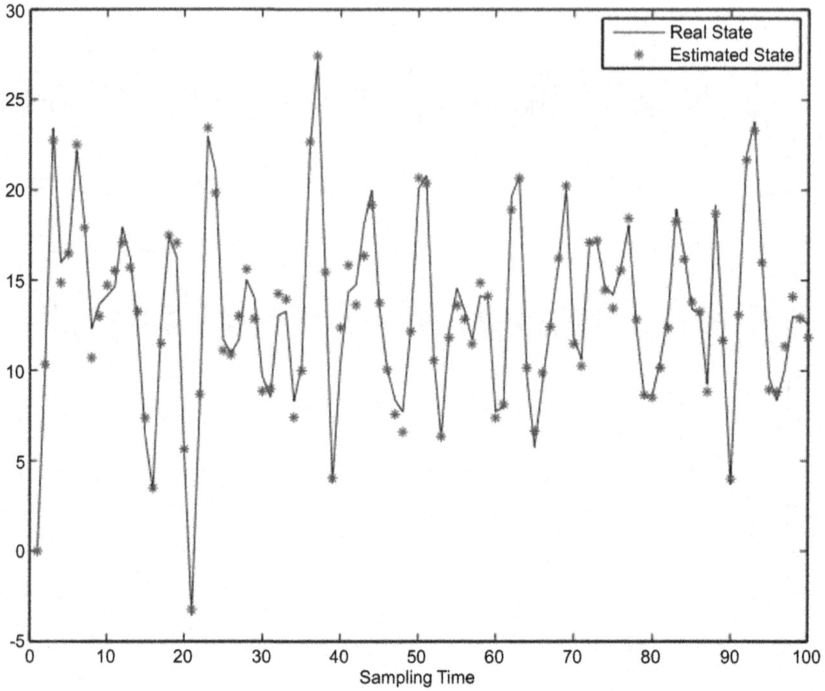


**Figure 6.10** The state estimate with $\sigma^2 = 0.1$.

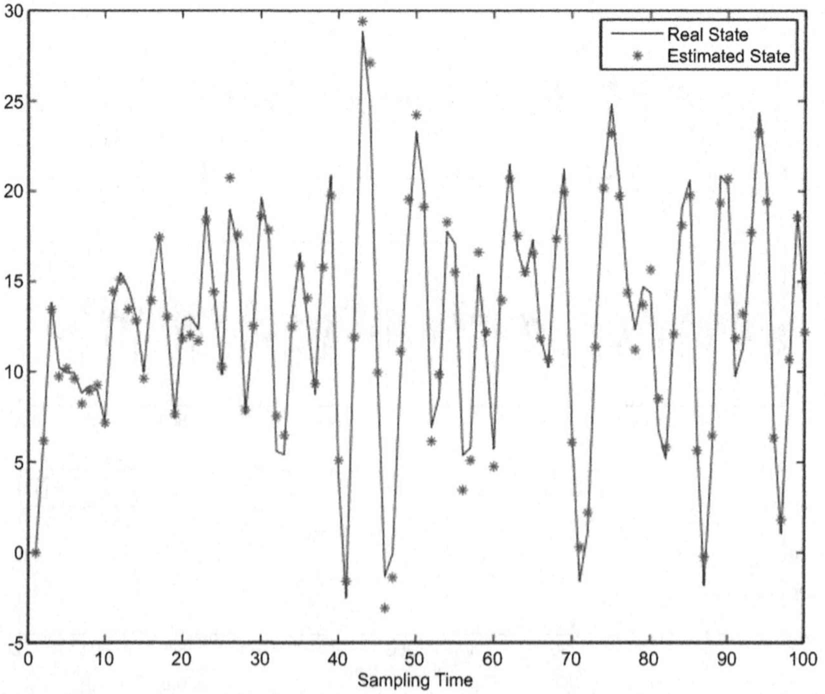


**Figure 6.11** The state estimate with $\sigma^2 = 0.3$.

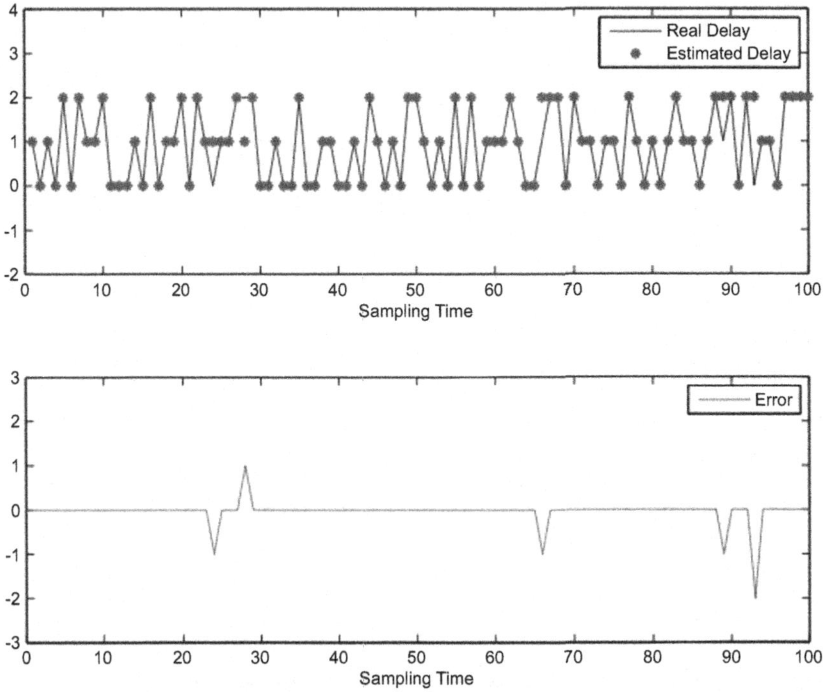


**Figure 6.12** The delay estimation with $\sigma^2 = 0.1$.

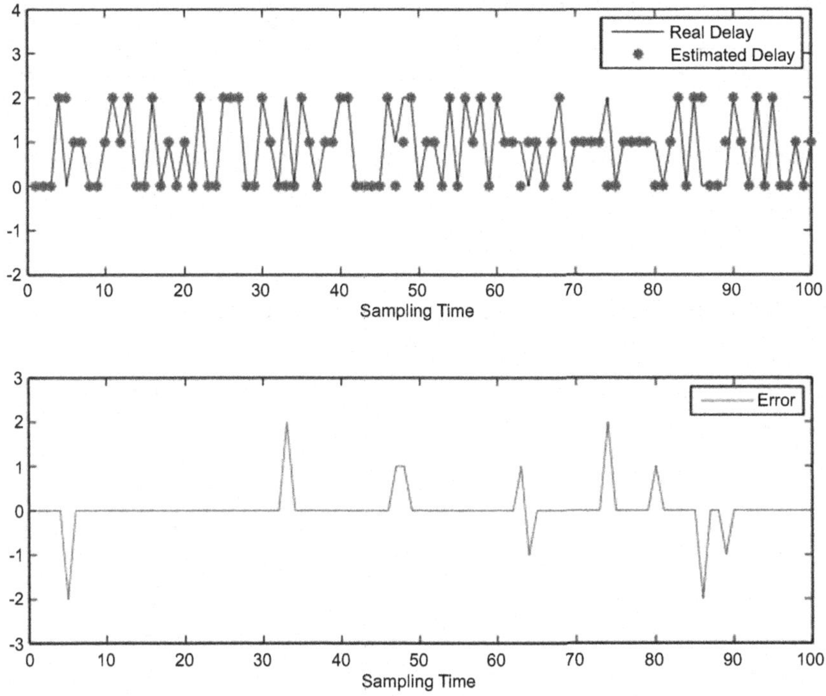


**Figure 6.13** The delay estimation with $\sigma^2 = 0.3$.

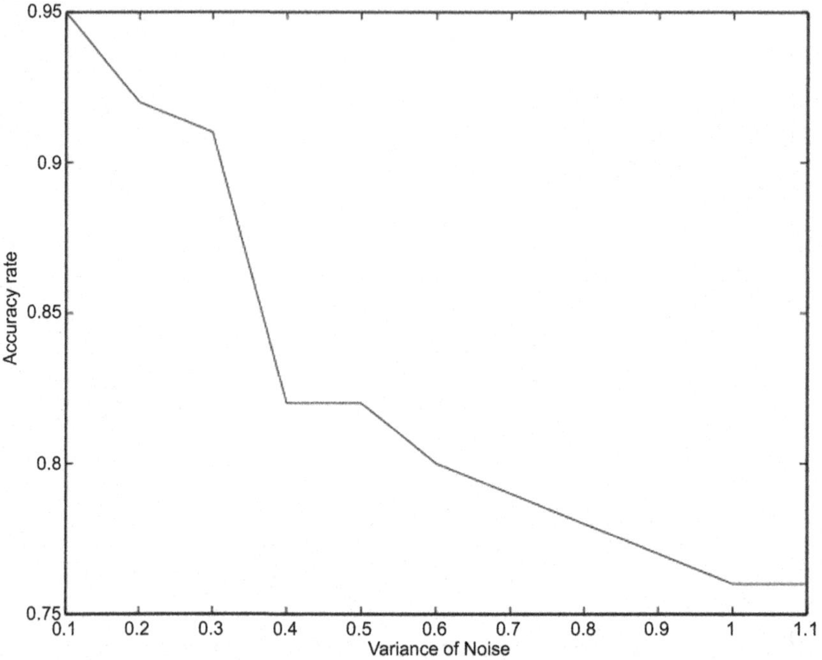


**Figure 6.14** The accuracy of delay estimation for different variances of noise.

**Table 6.3** The accuracy of delay estimation for different variances of noise.

| $\sigma^2$ | 0.1 | 0.2 | 0.3 | 0.4 | 0.5 | 0.6 |
|---|---|---|---|---|---|---|
| Accuracy | 0.95 | 0.92 | 0.91 | 0.82 | 0.82 | 0.80 |

noise levels are shown in Fig. 6.14 and Table 6.3, respectively, which demonstrate that the uncertainty of delay estimation mainly depends on the level of the measurement noise.

### 6.2.6 Wood-Berry distillation column simulation

In this section, the Wood-Berry distillation column process is used to validate the effectiveness of the proposed algorithm. The transfer function of the process is expressed as:

$$\begin{bmatrix} D(s) \\ B(s) \end{bmatrix} = \begin{bmatrix} \frac{12.8e^{-s}}{16.7s+1} & \frac{-18.9e^{-3s}}{21.0s+1} \\ \frac{1.6e^{-7s}}{10.9s+1} & \frac{-19.4e^{-3s}}{14.4s+1} \end{bmatrix} \begin{bmatrix} R(s) \\ S(S) \end{bmatrix} + \begin{bmatrix} d_1(s) \\ d_2(S) \end{bmatrix}, \tag{6.34}$$

where $D$ is the overhead product composition, $B$ is the bottom product composition, $R$ is the reflux flow rate, and $S$ is the steam flow rate. The overhead product composition and bottom composition are the output variables, and the reflux flow rate and steam flow rate are the input variables. Thus, the model of the product composition can be expressed as:

$$y_1(s) = \frac{12.8e^{-s}}{16.7s+1}u_1(s) - \frac{18.9e^{-3s}}{21.0s+1}u_2(s) + d_1(s),$$

$$y_2(s) = \frac{1.6e^{-7s}}{10.9s+1}u_1(s) - \frac{19.4e^{-3s}}{14.4s+1}u_2(s) + d_2(s).$$

The four state variables are defined at the influence of the reflux flow rate $u_1(s)$ and steam flow rate $u_2(s)$, that is,

$$x_1(s) = \frac{12.8e^{-s}}{16.7s+1}u_1(s), x_2(s) - \frac{18.9e^{-3s}}{21.0s+1}u_2(s),$$

$$x_3(s) = \frac{1.6e^{-7s}}{10.9s+1}u_1(s), x_4(s) - \frac{19.4e^{-3s}}{14.4s+1}u_2(s).$$

As a result, we have

$$y_1(s) = [1 - 1]\begin{bmatrix} x_1(s) \\ x_2(s) \end{bmatrix} + d_1(s),$$

$$y_2(s) = [1 - 1]\begin{bmatrix} x_1(s) \\ x_2(s) \end{bmatrix} + d_2(s).$$

Assuming the sampling interval is 1 min, then the continuous time system can be transferred into a discrete time system shown below:

$$\begin{cases} x_{1t} = \dfrac{12.8}{16.7(z-1)+1}u_{1(t-1)} \\ x_{2t} = \dfrac{18.9}{21.0(z-1)+1}u_{2(t-3)} \\ y_{1t} = [1 - 1]\begin{bmatrix} x_{1t} \\ x_{2t} \end{bmatrix} + d_{1t} \end{cases} \Rightarrow \begin{cases} x_{1(t+1)} = 0.940x_{1t} + 0.766u_{1(t-1)} \\ x_{2(t+1)} = 0.952x_{1t} + 0.900u_{1(t-3)} \\ y_{1t} = [1 - 1]\begin{bmatrix} x_{1t} \\ x_{2t} \end{bmatrix} + d_{1t} \end{cases}, \tag{6.35}$$

$$\begin{cases} x_{3t} = \dfrac{1.6}{10.9(z-1)+1}u_{1(t-7)} \\ x_{4t} = \dfrac{19.4}{14.4(z-1)+1}u_{2(t-3)} \\ y_{2t} = [1 - 1]\begin{bmatrix} x_{3t} \\ x_{4t} \end{bmatrix} + d_{2t} \end{cases} \Rightarrow \begin{cases} x_{3(t+1)} = 0.908x_{3t} + 0.147u_{1(t-7)} \\ x_{4(t+1)} = 0.931x_{4t} + 1.347u_{2(t-3)} \\ y_{2t} = [1 - 1]\begin{bmatrix} x_{3t} \\ x_{4t} \end{bmatrix} + d_{2t} \end{cases}. \tag{6.36}$$

We define the state vectors and input as:

$$\boldsymbol{x}_{1t} = \begin{bmatrix} x_{1t} \\ x_{2t} \end{bmatrix}, \boldsymbol{x}_{2t} = \begin{bmatrix} x_{3t} \\ x_{4t} \end{bmatrix}, \boldsymbol{u}_t = \begin{bmatrix} u_{1t} \\ u_{2t} \end{bmatrix}. \tag{6.37}$$

The additive noise is added to the states, and then, the state space equations are derived as:

$$\begin{cases} \boldsymbol{x}_{1(t+1)} = \boldsymbol{A}_1\boldsymbol{x}_{1t} + \boldsymbol{B}_1\boldsymbol{u}_t + \omega_{1t} \\ y_{1t} = \boldsymbol{c}^{\mathrm{T}}\boldsymbol{x}_{1t} + v_{1t} \end{cases},$$

$$\boldsymbol{A}_1 = \begin{bmatrix} 0.940 & 0 \\ 0 & 0.952 \end{bmatrix}, \boldsymbol{B}_1 = \begin{bmatrix} 0.766 & 0 \\ 0 & 0.900 \end{bmatrix}, \boldsymbol{c}^{\mathrm{T}} = [1 - 1], \tag{6.38}$$

$$\begin{cases} \boldsymbol{x}_{2(t+1)} = \boldsymbol{A}_2\boldsymbol{x}_{2t} + \boldsymbol{B}_2\boldsymbol{u}_t + \omega_{2t} \\ y_{2t} = \boldsymbol{c}^{\mathrm{T}}\boldsymbol{x}_{2t} + v_{2t} \end{cases},$$

$$\boldsymbol{A}_2 = \begin{bmatrix} 0.908 & 0 \\ 0 & 0.931 \end{bmatrix}, \boldsymbol{B}_2 = \begin{bmatrix} 0.147 & 0 \\ 0 & 1.347 \end{bmatrix}, \boldsymbol{c}^{\mathrm{T}} = [1 - 1]. \tag{6.39}$$

Since both overhead product and bottom product compositions are carried out by laboratory analysis, the output is available only at a slow rate. The unknown time delay exists in the measurements. This leads to a multirate system with unknown time delay:

$$\begin{cases} \boldsymbol{x}_{1(t+1)} = \boldsymbol{A}_1\boldsymbol{x}_{1t} + \boldsymbol{B}_1\boldsymbol{u}_t + \omega_{1t} \\ y_{1T_i} = \boldsymbol{c}^{\mathrm{T}}\boldsymbol{x}_{1T_i - d_{1T_i}} + v_{1t} \end{cases}, \tag{6.40}$$

and

$$\begin{cases} \boldsymbol{x}_{2(t+1)} = \boldsymbol{A}_2\boldsymbol{x}_{2t} + \boldsymbol{B}_2\boldsymbol{u}_t + \omega_{2t} \\ y_{2T_i} = \boldsymbol{c}^{\mathrm{T}}\boldsymbol{x}_{2\left(T_i - d_{2T_i}\right)} + v_{2t} \end{cases}. \tag{6.41}$$

The multirate inputs and outputs are shown in Figs. 6.15 and 6.16.

The state estimation results are shown in Figs. 6.17 and 6.18, where the dot and the solid lines indicate the estimated states and the real states, respectively. From these two figures, we can see that the proposed MHE algorithm has achieved good performance.

The delay estimations are shown in Figs. 6.19 and 6.20, which demonstrate the proposed MHE estimator can also achieve fairly accurate time delay estimation.

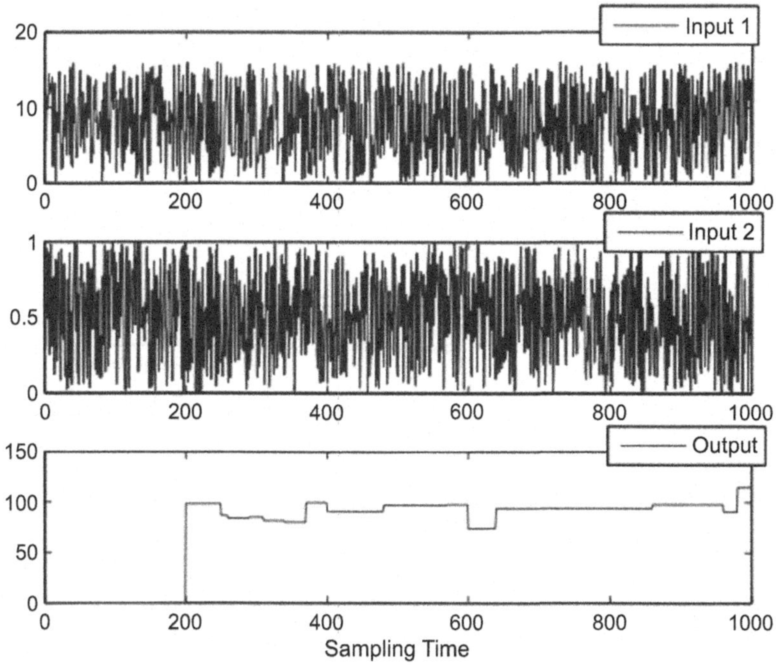


**Figure 6.15** The multirate sampled inputs and output $y_1$ with $\sigma^2 = 0.3$, distillation column simulation.

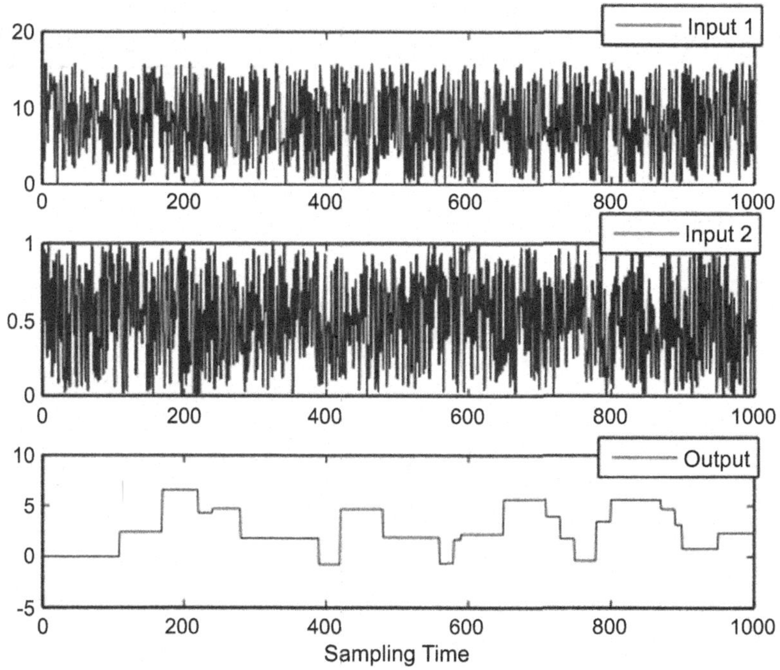


**Figure 6.16** The multirate sampled inputs and output $y_2$ with $\sigma^2 = 0.3$, distillation column simulation.

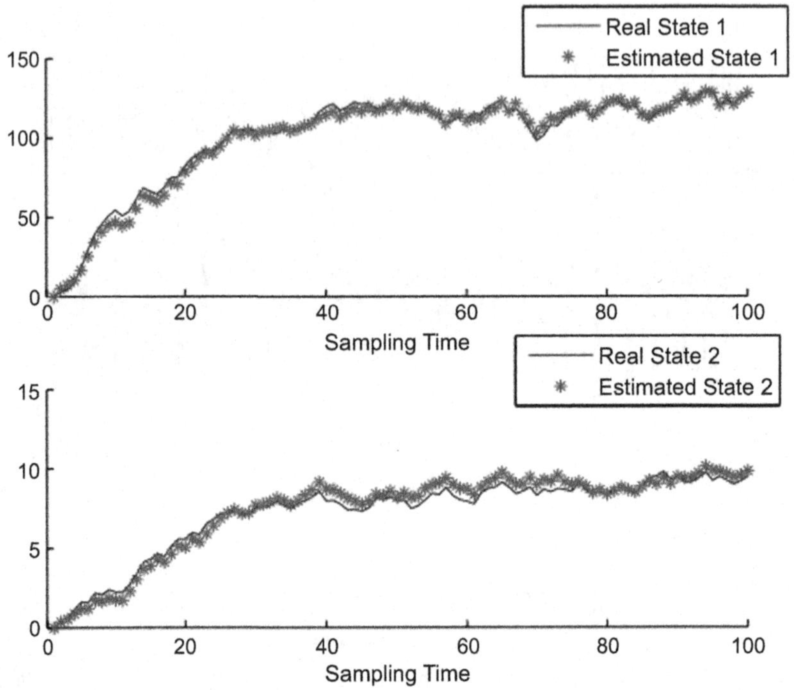


**Figure 6.17** The state estimation $(x_{1t}, x_{2t})$ with $\sigma^2 = 0.3$, distillation column simulation.

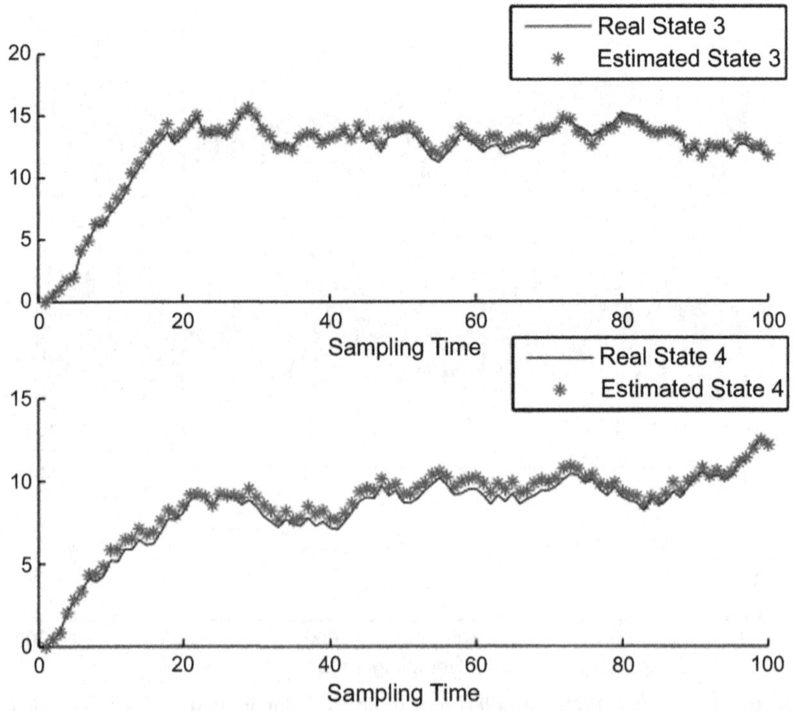


**Figure 6.18** The state estimation $(x_{3t}, x_{4t})$ with $\sigma^2 = 0.3$, distillation column simulation.

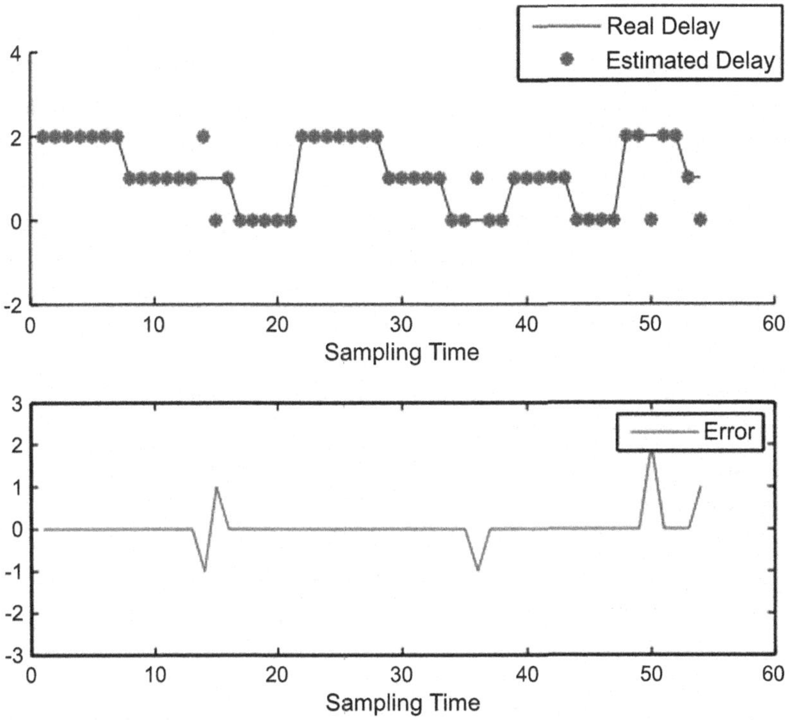


**Figure 6.19** The delay estimation ($d_1$) with $\sigma^2 = 0.3$, distillation column simulation.

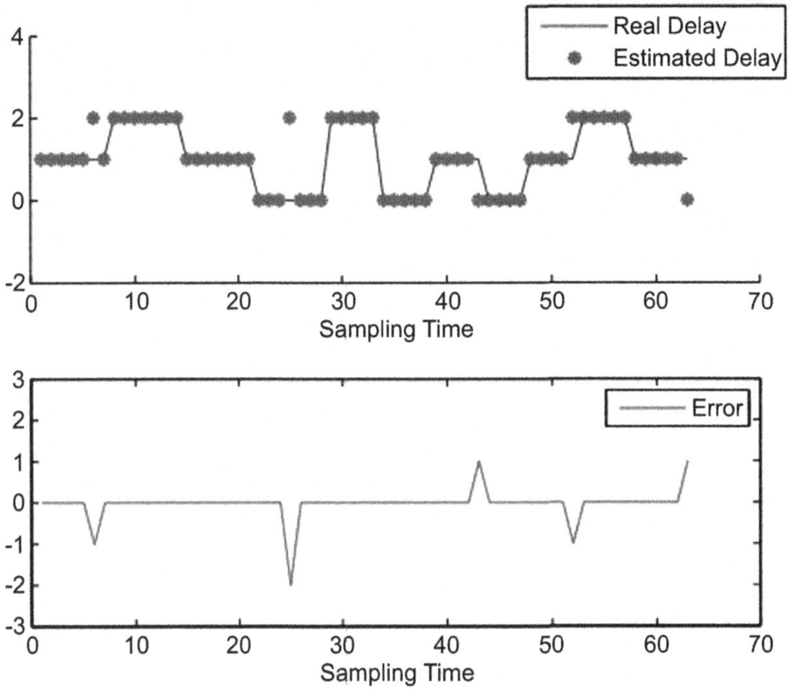


**Figure 6.20** The delay estimation ($d_1$) with $\sigma^2 = 0.3$, distillation column simulation.

### 6.2.7 Conclusions

In this chapter, we investigated moving horizon estimation for multirate state space models with time-varying time delay. The challenge of this problem lies in the simultaneous discrete time delay sequence estimation and continuous state estimation. This issue is addressed by proper formulation of the estimation problem under the MHE framework and by deriving an appropriate arrival cost which summarizes the effect of past and a prior information on the current states. The main contributions of this paper are:

- It derives moving horizon estimation for the multirate system with time-varying time delay.
- It derives the arrival cost for the multirate system with time-varying time delay.
- Although the MHE algorithm is presented for the linear state space model with time delay, the basic idea can also be extended to estimation of nonlinear model.

## References

Bretas, N.G., Bretas, A.S., 2015. A two steps procedure in state estimation gross error detection, identification, and correction. International Journal of Electrical Power & Energy Systems 73, 484–490.

Rafal, S., Marek, R., Krzysztof, L., 2017. Modeling of discrete-time fractional-order state space systems using the balanced truncation method. Journal of the Franklin Institute 354, 3008–3020.

Zamani, M., Bottegal, G., Anderson, B.D.O., 2016. On the zero-freeness of tall multirate linear systems. IEEE Transactions on Automatic Control 61, 3606–3611.

# Index

*Note*: Page numbers followed by "*f*" and "*t*" refer to figures and tables, respectively.

## T

Printed in the United States
by Baker & Taylor Publisher Services